Military Sociology

This textbook introduces the reader to the field of military sociology through narrative reviews of selected key studies in the discipline.

The book provides *a guided introduction*. In each chapter, the authors set the stage and then immerse the reader in Spotlights – that is, descriptions of essential studies that inform the discipline of military sociology. The goal is to afford readers a ready pathway into how sociologists and social scientists have thought about topics in the study of the military and war.

Topics covered in the book include:

- What is *military sociology*? What does it have to offer for understanding armed forces, wars, and societies?
- What basic tools are needed to ply *sociological*, or more broadly, *social science* perspectives for studying war and the military?
- What are the bio-social bases of war? What does the spectrum of such societally organized violence look like?
- How do societies raise and maintain formal militaries? What are variations in their social composition and in the profiles of civil–military relations?
- How and why is military organization and war changing so dramatically in the 21st-century? What does the future hold?

This book will be of great interest to students of military sociology, the armed forces and society, peace studies, and international relations.

Wilbur J. Scott is Professor Emeritus, Department of Behavioral Sciences & Leadership, U.S. Air Force Academy, and Professor Emeritus, Department of Sociology, University of Oklahoma, U.S.A.

Karin Modesto De Angelis is Professor, Department of Behavioral Sciences & Leadership, U.S. Air Force Academy, U.S.A.

David R. Segal is Professor Emeritus, Department of Sociology, University of Maryland, U.S.A.

"This book by renowned scholars Wilbur Scott, Karin De Angelis, and David Segal is a more than welcome contribution to the field of military sociology. In times when international affairs have put the armed forces center stage again, this is a book with a wealth of empirical studies and relevant insights. It will turn out to be a must-read, not only for (student-)officers throughout the world, but for everyone interested in international relations and the role of the military therein."

Joseph Soeters, *Netherlands Defense Academy and Tilburg University*

"A timely and long overdue textbook focused on applying social scientific thinking to military problem sets. This book will prove useful for military professionals who want to understand the utility that social science brings to the profession of arms and will provide the motivated student with a foundational set of tools to better account for the socio-cultural complexities of 21st-century competition and conflict."

Lt Col. Matthew Linford, *USAF, Director, USAF Information Operations Training*

"*Military Sociology* is a path-breaking textbook examining the proposition that a soldier is first and foremost a social being. The authors explore various dimensions of that assertion, examining military cultures and service cultures, and their intersection with factors such as race and sex, with additional examination of military families and veterans. Their spotlighting of seminal texts provides a terrific pedagogical springboard. Highly recommended."

Valerie M. Hudson, *Bush School of Government, Texas A&M University, U.S.A.*

"This volume provides a tsunami of intellectual waves for scholars and students to surf their way across the shore that is military sociology and the sociology of war. Not since 1965 has such a gnarly pipeline of smart studies been compiled, synthesized, and presented into such efficient brilliance. This compendium of knowledge reveals the oft invisible institution that is the armed forces in the United States and other countries around the world."

Morten G. Ender, *United States Military Academy, West Point*

"Now more than ever, militaries are coming into focus as the great unknown in social and political thought. Due to long-standing disciplinary neglect across the social sciences as well as the growing remoteness of military lifeworlds and practices from mainstream social life, the agency of militaries in shaping their social and political environments has been obscured. This excellent collection of important works demonstrates that, below the surface, scholars have already laid the groundwork for a resurgence in military sociology. With this book as a guide, the next generation with have a big advantage in resituating the military as among the most defining institutions of our time."

Thomas Crosbie, *Royal Danish Defence College*

Military Sociology

A Guided Introduction

Wilbur J. Scott, Karin Modesto De Angelis, and David R. Segal

Routledge
Taylor & Francis Group

LONDON AND NEW YORK

Cover image: Photo, taken in 2006 by Wilbur Scott on a railway footpath near the U.S.'s 3rd Infantry Division's base in Wurzburg, Germany, reminds of the social dimensions of all things military.

First published 2023
by Routledge
4 Park Square, Milton Park, Abingdon, Oxon OX14 4RN

and by Routledge
605 Third Avenue, New York, NY 10158

Routledge is an imprint of the Taylor & Francis Group, an informa business

British Library Cataloguing-in-Publication Data
A catalogue record for this book is available from the British Library

Library of Congress Cataloging-in-Publication Data
Names: Scott, Wilbur J., 1946- author. | De Angelis, Karina Modesto, author. | Segal, David R., author.
Title: Military sociology: a guided introduction / Wilbur J. Scott, Karina Modesto De Angelis, and David R. Segal.
Description: Abingdon, Oxon; New York, NY: Routledge, [2023] | Includes bibliographical references and index. | Identifiers: LCCN 2022031806 (print) | LCCN 2022031807 (ebook) | ISBN 9781032252926 (hardback) | ISBN 9781032252919 (paperback) | ISBN 9781003282549 (ebook)
Subjects: LCSH: Sociology, Military. | Armed Forces.
Classification: LCC U21.5 .S36 2023 (print) | LCC U21.5 (ebook) | DDC 306.2/7--dc23/eng/20220921
LC record available at https://lccn.loc.gov/2022031806
LC ebook record available at https://lccn.loc.gov/2022031807

ISBN: 978-1-032-25292-6 (hbk)
ISBN: 978-1-032-25291-9 (pbk)
ISBN: 978-1-003-28254-9 (ebk)

DOI: 10.4324/9781003282549

Typeset in Times New Roman
by SPi Technologies India Pvt Ltd (Straive)

My loving family,
Carol,
Melanie/Mark, Matt/Christie,
and, of course,
Sammy, Katie, and Alex.
And my military family,
my brothers in Alpha Co., 1/14th Inf, 4th Inf Div
Vietnam, 1968–1969
(W. J. Scott)

My family: Tom, Luca, and Ana
whose love, support, and overall goodness make me whole.
And Colonel Edward Modesto, U.S. Army,
who will always be my hero
(K. M. De Angelis)

My loving family,
Mady and Eden,
Rhea, Roger, and Brad.
And the generations of colleagues and graduate students,
civilian and military,
who have done their best to keep me
young and relevant
(D. R. Segal)

Contents

Spotlighted Studies[1]

Note

1 *Spotlighted* studies are summarized and synthesized into each chapter's discussion by the authors of this book.

Figures

Tables

Acknowledgments

This project began in 2012 when David Segal floated the idea to Wilbur Scott of writing an *Introduction to Sociology* textbook in which the material primarily would be drawn from military sociology. The intended audience was the cadets, midshipmen, and officer candidates enrolled in introductory sociology courses at U.S. and overseas service academies, taught in these institutions at the junior level or above. In response, Scott spent a semester on sabbatical leave at the University of Maryland to sit in on Segal's military sociology class. From there, the writing of draft chapters slowly lifted off the runway.

One problem quickly became apparent when using some of the early drafts in the intended course at the Air Force Academy. Cadets wanted in their first sociology course to immerse themselves in the field's novel and exotic studies, but ones dealing with anything but the military. This did not mean they were uninterested in a sociology of the military. They were perfectly willing to go there in more specialized courses – such as "Military and Society" and "Sociology of Violence and War" – so long as they could see the usefulness of social science thinking in understanding war and the military.

Another important change occurred in 2015. Looking over what Scott and Segal had written up to that point, Mady Wechsler Segal suggested we add a younger, perhaps female, coauthor to "bring in some fresh air." Scott's colleague at the Air Force Academy, Karin De Angelis, who also is a former student of both the Segals at the University of Maryland, was the logical choice. This collaboration has been critical in producing a more diverse range of selections spotlighted in our work, particularly in the chapters on race/ethnicity, gender/sexual orientation, and military families. More importantly, the text is not the work of only one or two of us, but a blend of our inspirations, strengths, and academic biases.

The writing of the "Spotlights," as described in the Preface, thus moved to topics and research most directly useful in courses beyond "Introduction to Sociology." The complete reorientation of the project into a textbook on "Military Sociology" occurred in 2021 at the suggestion of Andrew Humphrys, our editor at Routledge. The book you see is the product of this evolution, and its intended audiences extend well beyond service academy students in upper division courses.

We wish to acknowledge two special sources of indebtedness. First, all three of us have been associated with active-duty militaries and military service academies, Scott and De Angelis with the U.S. Air Force Academy, and Segal with the U.S. Military Academy, West Point, and the U.S. Naval Academy. Segal also is a founder of the Center for Research on Military Organization at the University of Maryland, has spent time as a researcher at the Walter Reed Army Institute of Research and the U.S. Army Research Institute, and has served as a special assistant for peace operations to the Army Chief of

Staff. Scott and De Angelis both served in the U.S. military, he as an infantry platoon leader in Vietnam, and she as a force support officer, including a deployment to Al-Udeid, Qatar, in 2004 with the Air Force's 510th Expeditionary Fighter Squadron.

These platforms have provided us with an insiders' access to the issues and concerns of all things military. It is difficult to envision how we could have acquired many of the insights expressed in the text without these advantages. Of course, nothing we say in the book should be taken as official positions of the U.S. service academies, the Departments of the Army, Navy, or Air Force, or the U.S. Department of Defense. For better or for worse, the book reflects the considered opinions solely of the authors.

Our second source of indebtedness is that the three of us share a common intellectual and professional home: military sociology's major professional organization, the Inter-University Seminar on Armed Forces & Society (IUS). This is our base, the space within which it has been possible for decades to share ideas, be exposed to new ones, and enjoy the collegiality and support of others, both here and abroad, who share our passion for military sociology. We could list literally legions of fellow IUS members who have encouraged and sustained us.

Segal has been a member of IUS since its inception and has served as a member of its Executive Council (1974–1982) and Board of Directors (1987–1994 and 2004 to the present). He has a good excuse for his absence from the IUS's Board from 1995 to 2003: in those years, he was its President. Scott has been a member since 1987 and served on the Executive Council from 1993 to 1997, and De Angelis has been a member of the Executive Council since 2012. Segal's list of honors and awards is lengthy, but foremost on it is the Morris Janowitz Career Achievement Award, bestowed on him in 2007 for a career of excellence in the study of armed forces and society, and service to the discipline of military sociology. Scott also received this award in 2019.

Finally, we need to disclose that the authors have not received any financial support or remuneration, monetarily or in kind, for the research, authorship, or publication of this book, other than what we were paid for doing our duties at our respective institutions. We have arranged for all proceeds that may be generated from the sale and distribution of this book to go directly to IUS's general fund.

Preface

The military and war have been objects of occasional fascination since sociology's 19th-century origins. The horrors of World War I rekindled these interests and inspired systematic social science analyses of war in the 1930s.[1] However, the dust had barely settled when a follow-on world war dwarfed that first "war that will end war," turning an idealistic turn of phrase by the father of science fiction, British writer H. G. Wells, into an ironic one.[2]

At the onset of World War II, the U.S. Department of War – renamed the Department of Defense (DoD) in 1947 – tapped civilian psychologists, sociologists, and other social scientists, pressing them into service as researchers working in multidisciplinary teams. Foremost among them was one headed by sociologist Samuel Stouffer.[3] Many more were drafted to fill the usual military ranks. World War II thus would supply a generation of American social scientists with insights born out of professional and personal experience for a new specialty, *military sociology*, under the leadership of Morris Janowitz and, later, "second generation" military sociologist Charles Moskos.

Over the years since then, the discipline has become more broadly international and comparative in scope. The "American example," as Italian general officer and sociologist Giuseppe Caforio put it, inspired a social science reawakening in Europe of all things military.[4] It is this enterprise we are referring to in this book as *military sociology*: an approach to the study of war and the military that draws on the work of social science disciplines – sociology, political science, history, anthropology, psychology – and is carried out both in academic and civilian settings in many countries. Though some, including Caforio, use the term "sociology of the military" instead (because of the awkward way "military sociology" translates into some languages), we reserve that label for the body of work that is strictly sociological.

Despite its impressive growth and the pressing relevance of its subject matter, military sociology has hovered near the margins of sociology itself. Consider for instance a review by West Point sociology professor Morten Ender of textbooks designed for use in "Introductory Sociology" courses.[5] Ender and his associate did a content analysis of the 31 top-selling "Introductory Sociology" textbooks from the late 1990s and early 2000s with an eye toward their coverage of the military, war, and peace. They considered the extent to which these topics were introduced and developed as concepts, received treatment as important issues in their own right, were featured in photographs or other graphics, or appeared in listings of primary and secondary references.

They found that only *one* of the 31 textbooks devoted an entire chapter to the military as a social institution or to war and peace studies. If these topics received any coverage at all, it usually was in a chapter devoted to "political sociology" or as a mention in another of the substantive chapters (on "socialization" or "social groups," for example). One-third of the textbooks did contain at least one reference to the works of Janowitz, Moskos, or Stouffer. The *New York Times* was the most cited source of material on war, peace, and the military.

Ender did stumble onto an unexpected feature. Military and war-related scenes appeared regularly in the textbooks' photographs. Two-thirds of these images were of people such as servicemembers (e.g., male recruits with clean-shaven heads), victims of war or other political conflicts, and anti-war protestors. The remaining one-third of military photos consisted of famous places (the Pentagon, the Cold War-era Berlin Wall), war scenes (aerial views of planes and bombers doing their thing), or conflict-laden locations (apartheid South Africa or Belfast, Northern Ireland). The single photograph used most often was that of the lone student staring down four tanks in China's Tiananmen Square during the 1989 Democracy Movement protests, brutally put down by the Chinese Army. Descriptions accompanying the images typically lacked contextual or conceptual explanation, this despite ample sociological work on war and militaries.

This neglect may be attributable to military sociology's multidisciplinary nature and to the changing roles of the social sciences during the U.S.A.'s 20th-century wars. One might argue that the World War II studies helped establish the legitimacy of social science more broadly on the American scene. In contrast, the war in Vietnam and other politically imbued disputes of the day embroiled both the American Anthropological Association and the American Sociological Association (ASA) in controversy. At its annual meeting in 1965, well-known anthropologist Marshall Sahlins "delivered a scathing attack" on the discipline's involvement in Project Camelot, a Latin American field study of revolutionary movements funded by the Central Intelligence Agency.[6] And, in 1968, ASA's membership voted on a resolution calling for an end to U.S. bombing in Vietnam and an immediate withdrawal of all American troops.[7] (It did not pass.) However, these events marked an end to any easy relationship with DoD, and to a waning of military studies within these disciplines.

This is unfortunate, for its subject matter is rich, interesting, and critically important. It can inform virtually every one of social sciences' "big questions," and often points to respected correctives of DoD policies. Simply put, it is not possible to adequately address the nuances of war and the military without being well-grounded in the social sciences. The military and war, after all, are highly organized forms of human activity with interacting macro- and micro-dimensions. Who better to tackle all this than social scientists?

The first and only textbook devoted to a sociology of the military was published more than fifty years ago. Written by Charles Coates and Roland Pellegrin, both veterans of World War II and sociology professors at the University of Maryland and the University of Oregon, respectively, it was a thoughtful and comprehensive treatment of the World War II American military.[8] Unfortunately, the work did not receive the exposure it deserved. Published in 1965, the year the U.S.A.'s first ground troops splashed ashore in Vietnam, the book's impact on the field was extremely limited. Thus, present-day newcomers to military sociology are left with a series of excellent but daunting edited texts in which definitive thought-pieces and studies have been compiled, several by Caforio.[9]

The timing for a fresh text in military sociology seems ripe. Sociology programs at military academies are growing. West Point now offers a major in sociology and a minor in diversity and inclusion studies, and the Air Force Academy (AFA) has added a sociology emphasis in its Behavioral Sciences major. All three military academies have had military officers with Ph.D.s in Sociology as Department Heads of Leadership (Annapolis) and Behavioral Sciences & Leadership (West Point and AFA). And, a 2001 article in

Military Review, the professional journal of the U.S. Army, argued that every military officer should study sociology.[10] The growth of sociology in American military academies has diffused to academies in other countries and to non-academy officer training programs, all of which tend to share curriculum materials.

Likewise, on the civilian front, interest in war and peace studies and the military is growing. The most recent generation of students did not directly experience World War II and the many regional conflicts that followed in its wake. Rather, they have grown up in a volatile and violent post-9/11 world[11] and its spate of nasty "forever wars."[12] For many, their knowledge of this world and its dynamics is limited and their exposure to it often comes through social media and the entertainment industry. Many have been left wondering why attempts to reduce the incidence of 21st-century wars seem fraught with even more conflict.

We offer perspectives and tools for thinking about all these issues smartly. Our book is meant as a textbook for introducing the military and war by walking an intelligent but uninitiated reader through selected, "spotlighted" studies. In this sense, it is a repository of what we consider essential sociological and social science thinking and research on the military and war. This introduction to the field was written with the following readers in mind:

- **upper-division undergraduate students** at military academies and civilian universities where Military Sociology, Sociology of War, War Studies, Security Studies, Peace Studies, and such are taught in the third or fourth year;
- **instructors and students** who wish to expand or sharpen their thinking on war and the military;
- **military officers** seeking social science frameworks for thinking about military and war;
- **any and all curious others** with an interest in the whys and wherefores of organized intergroup violence and military affairs.

Two stylistic notes. The book is written so that individual chapters largely stand on their own. Of course, we feel much is to be gained by reading them all in their entirety. However, the judicious selection of individual chapters also can be a viable option for instructors, students, and others seeking perspective on some specific topic.

Two, a review of textbooks reveals strategies authors typically employ when referencing the social and behavioral science literatures. Some texts cite numerous studies and articles, giving each a rather cursory, few-sentences treatment. A less frequent approach, exemplified by Rodney Stark's "Over the Shoulder of a Professional Sociologist" technique, focuses instead on a much smaller number of selections, each presented in detail.[13] Stark's "Over the Shoulder" approach starts with the events and questions that motivate a study in the first place. It then follows the study's twists and turns, moving on to its ensuing impacts and controversies.

Hence, this text, as the subtitle indicates, provides *a guided introduction*. In each chapter, we set the stage and then immerse the reader in *Spotlights*, that is, our descriptions of definitive studies that inform the discipline of military sociology. The goal is to afford readers a ready pathway into how sociologists and kindred social scientists have thought about and studied key topics in the study of the military and war. Among these are all-time classics most military sociologists would agree must be on the list. Others reflect our first choices and, yes, biases, in covering relevant issues for each topic. We admit there are many other highly meritorious studies which *could*, maybe even *should*, have been included. We feel we have given the matter due diligence and are confident the reader will find the spotlighted material highly interesting, relevant, and informative.

Table 0.1 presents some relevant characteristics of the authors in our spotlighted studies. Altogether, we spotlight 48 studies. For certain topics, two complementary studies are featured together. As the table indicates, just under half of the spotlighted studies were done by a sociologist or a team of sociologists (44%). Reflective of military sociology's broad disciplinary composition, the remaining studies were conducted by political scientists (25%), an interdisciplinary team of social scientists (15%), historians (8%), or psychologists (4%). Two studies are by scholars of law.

At its inception, military sociology was a near-male-exclusive field. No longer, as can be seen in our numbers. Well over a third of the research we spotlight, 42%, was carried out by a woman or a team that included at least one female co-investigator. Finally, though military sociology began as an American venture, it is these days international and comparative. Our selection of studies, admittedly, reflects a U.S.A.-centric partiality. Still, scholars from seven nations other than the U.S. – the U.K. (8), Israel (3), France (2), the Netherlands (2), and one each from Bosnia-Hercegovina, Portugal, and Vietnam – make up 38% of the contributors.

Turning now to the individual chapters, Chapter 1 describes the advent of military sociology. Spotlighted are Samuel Stouffer's Chapter 10 of *The American Soldier*, "The Negro Soldier", and its study for the War Department on race relations in World War II's segregated American military;[14] Dorothy Swaine Thomas and Richard Nishimoto's study of the Japanese American internment camps, and Tamotsu Shibutani's *The Derelicts of Company K*, a sociological account of Japanese American army draftees jaded by their experience in forced internment camps;[15] and Edward Shils and Morris Janowitz's classic study, "Cohesion and Disintegration in the Wehrmacht in World War II." The latter is based on their wartime job of analyzing German war propaganda and interviewing German prisoners-of-war.[16] The chapter concludes with an account of how Janowitz and others in the decades after the war molded the study of war and the military into a coherent field of study.

Chapter 2 attunes the reader to levels of social scientific theoretical thinking applied to the military and war. In the Spotlight are: Émile Durkheim's timeless classic, *Suicide: A Study in Sociology*, with applications to *altruistic* and *heroic suicide* in the military;[17] Bosnia-Herzegovinian-born Siniša Malešević's *The Sociology of War and Violence*, which provides a macro-level theory to account for large-scale inter-group violence in modern wars;[18] and the recent and pathbreaking work by Mary Caprioli, "Gendered Conflict" and Valerie Hudson and her associates, *Sex and World Peace*.[19] Their studies show how gender issues are linked to the likelihood of war.

Table 0.1 Authors' Characteristics for Spotlighted Studies

	Male	*Female*	*U.S.*	*International*	*Total*	*Percentage*
Sociology	12	9	18	3	21	**44**
Political Science	9	3	8	4	12	**25**
Interdisciplinary Team	2	5	1	6	7	**15**
History	3	1	1	3	4	**8**
Psychology	0	2	1	1	2	**4**
Law	2	0	1	1	2	**4**
Total	**28**	**20**	**30**	**18**	**48**	
Percentage	**58**	**42**	**62**	**38**		**100**

Chapter 3 turns to the biological and cultural bases of warfare. Socio-biology makes many sociologists nervous, but we offer some suggestions about how to join the conversation. Our first Spotlight is on Joshua Goldstein's *War and Gender: How Gender Shapes the War System and Vice Versa*.[20] He addresses this puzzle: societies vary considerably in how they have done gender and how they have done war, so, why is it that war, almost without exception, has been the exclusive domain of males? For our Spotlights on culture, we review "The Small Warship" by George Homans,[21] who introduces social exchange theory via his experiences as a junior officer on a World War II ship at sea, and an insightful treatment of "Military Culture," or more correctly, *military cultures*, by Joseph Soeters, Donna Winslow, and Alise Weibull.[22]

The military as a bureaucracy and as a profession are the topics of Chapter 4. We introduce Max Weber's classic statement on *bureaucracy* as a form of social organization, and *mass-production plants* and *mass armies* as prime examples. In the 1960s, military sociology shifted its focus from the study of ordinary soldiers to the professional officer corps. Spotlighted are the studies that led the way: political scientist Samuel Huntington's *The Soldier and the State* and Morris Janowitz's *The Professional Soldier*.[23] The 1980s saw another shift with questions about the wisdom of an All-Volunteer Force, headed by Charles Moskos's "From Institution to Occupation".[24] The collapse of the Soviet Union produced yet another shift as new strategic realities generated studies of "post-modern" militaries. The third Spotlight features a critic of that concept. In "The Post-Fordist Military," sociologist Anthony King offers an alternative label and set of considerations.[25]

Chapter 5 focuses on two enduring questions in the study of war: *who* fights, and *why* do soldiers, sailors, airmen, and Marines, and for that matter, terrorists, fight? Our first Spotlight is on David Segal's *Recruiting for Uncle Sam*, a classic analysis of the who and of the how the U.S.A. has filled its military ranks over the last 150 years.[26] Second is Nora Kinzer Stewart's, *Unit Cohesion in the Falklands/Malvinas War*, based upon her interviews with opposing British and Argentine officers and enlisted men who fought in the conflict.[27] Finally, we focus on Marc Sageman's *Understanding Terror Networks*.[28] A political sociologist and former Central Intelligence Agency analyst, Sageman constructed his own sampling frame to assess who is attracted to join al-Qaeda and its affiliates, and why.

Race and ethnicity are the topics of Chapter 6. We explain why race and ethnicity are important, enduring considerations in society and the military, and how these have played out on the American scene. Our first Spotlight focuses on a unique dimension of the American Indian World War II experience with the story of Samuel Holiday, *Under the Eagle: Samuel Holiday, Code Talker*, with annotations by historian Robert McPherson.[29] We continue with historian Ruth Ginio's *The French Army and Its African Soldiers*, a look at racist policies under European colonialism in West Africa (*racisme de type coloniale*).[30] We conclude with a study of the experiences of Muslims in the U.S. military after 9/11 by sociologist Michelle Sandhoff, *Service in a Time of Suspicion*.[31]

Chapter 7 explores gender and sexual orientation in society and the military. We begin with Brenda Moore's studies of African American and Japanese American women in the Women's Army Corps in World War II. Our first Spotlight features political scientist Judith Stiehm's classic on mandated gender integration at the U.S. military service academies, *Bring Me Men and Women*.[32] We continue with a Spotlight on two sociological studies of women in the military, "Not Just Weapons of the Weak" by Laura Miller, and "From Loyalty to Dissent: How Military Women Respond to Integration Dilemmas" in Portugal and the Netherlands, by Helena Carreiras.[33] We close with a Spotlight on Aaron Belkin's, "Don't Ask, Don't Tell: Is the Gay Ban Based on Military Necessity?" and

Jonathan Lee's, "The Comprehensive Review Working Group and Don't Ask, Don't Tell Repeal at the Department of Defense."[34]

Chapter 8 is the first of three chapters on war. When most people think of war, they have in mind "big" ones, that is, conventional wars pitting the militaries of two or more states, or maybe a nuclear one. We spotlight three studies. The first is renowned British historian Richard Overy's *Why the Allies Won*.[35] Overy notes that the crushing Allied victory in World War II was by no means a given. Rather, fortunate but unlikely outcomes combined to turn initial advantages of the Axis powers to dust. We follow with Helen Fein's analysis of variations in death rates for Jews by country during the Holocaust in her classic sociological treatise, *Accounting for Genocide*.[36] Lastly, we spotlight political scientist Scott Sagan's "Why Do States Build Nuclear Weapons?"[37] One obvious reason is "self-defense," but the realities and rationales are more complex than that.

Chapter 9 considers "small" wars, that is, armed conflicts between very unequal militaries or between state-based militaries and non-state paramilitary groups. The first Spotlight is on historian Alf Heggoy's *Insurgency and Counterinsurgency in Algeria*, and on officer-turned-scholar David Galula's *Pacification in Algeria, 1956–1958*.[38] Heggoy documents the problematic use of torture by the French army, while Galula's work reads as a textbook example of how the French military could have, probably should have, proceeded. The second Spotlight features two contrasting books on the Vietnam war: political scientist and activist Trúóng Nhú Táng's *A Viet Cong Memoir*, and Lt Colonel John Nagl's *Learning to Eat Soup with a Knife*.[39] We conclude with two studies of the euphemistically termed "Troubles" in Northern Ireland by political scientist Richard English, *Armed Struggle: The History of the IRA*, and Sandhurst security studies professor Aaron Edwards, *UVF: Behind the Mask*.[40]

We conclude our three-part look at war in Chapter 10 with a look at 21st-century "new" wars. We begin with two studies of recent violence in Israel and the Palestinian territories: Richard Davis's innovative study, *Hamas, Popular Support, and War in the Middle East*, and *Rethinking Contemporary Warfare* by Israeli anthropologist Eyal Ben-Ari and his associates.[41] Davis links cycles of violence by Hamas with public opinion polling data from the Palestinian West Bank. Ben-Ari et al. offer new theoretical concepts for analyzing the violence they observed during the *Al-Aqsa Intifada* and adaptations to it by the Israeli Defense Forces. We then spotlight work by British security and legal analyst, Jack McDonald. *Enemies Known and Unknown: Targeted Killings in America's Transnational War* assesses the *individuated warfare* associated with drones as remote killing platforms.[42] We conclude with a discussion of cyber and information warfare, highlighted by Russian security analyst Keir Giles's *Handbook of Russian Information Warfare* doctrine.[43]

Chapter 11 turns to military families, a special group intimately connected to and affected by all this. We begin with a Spotlight on one of the all-time classics in military sociology, Mady Wechsler Segal's "The Military and the Family as Greedy Institutions," and an update of it by Karin De Angelis and Mady Segal, "Transitions in the Military and the Family as Greedy Institutions."[44] The second Spotlight presents Meytal Eran-Jona's survey of the Israeli Defense Force's spouses, "Married to the Military," who finds the heavy demands on military personnel create family arrangements that deviate from the norm in Israeli-Jewish life.[45] We conclude with two studies, one by Manon Andres and René Moelker, "There and Back," of children in Dutch military families during and after deployments by one of their parents to Bosnia or Afghanistan, the other by Edna Hunter-King, "Children of Military Personnel Missing in Action in Southeast Asia."[46]

Our Chapter 12 examines the issues surrounding another special group: veterans. Author Wilbur Scott provides a personal and research account of the experiences of French veterans of the war in Algeria in contrast with those of American veterans of Vietnam. The first Spotlight turns to the work of cultural historian Timothy Ashplant and his associates, *The Politics of War Memory and Commemoration*.[47] We follow with analyses of perceptions of veterans in the U.K. and then the U.S.A., the first by Rita Phillips and her associates, "Exploring the Victimization of British Veterans: Comparing British Beliefs about Veterans with Beliefs about Soldiers,"[48] and the second by Meredith Kleykamp and her associates, "Who Supports U.S. Veterans and Who Exaggerates Their Support?"[49] Finally, we address the issue of war trauma with a Spotlight on a study by Natalie Purcell and her associates at the San Francisco Veterans Affairs Health Care System, "Healing from Moral Injury."[50]

We conclude with several "threat pictures" and their likely impact on future topics in military sociology. Of course, Russia's recent invasion of Ukraine has cast a large shadow on these projections.

Notes

1 See, Quincy Wright, 1942, *The Study of War*, Chicago: University of Chicago Press.
2 H.G. Wells, 1914, *The War That Will End War*, London: Frank and Cecil Palmer.
3 Samuel A. Stouffer, Edward A. Suchman, Leland C. DeVinney, Shirley A. Star, and Robin M. Williams, Jr., 1949, *The American Soldier: Adjustment during Army Life*, Princeton, N.J.: Princeton University Press.
4 Giuseppe Caforio (ed.), 1998, *The Sociology of the Military*, Cheltenham, U.K. and Northampton, Mass.: Edward Elgar.
5 Morten G. Ender and Ariel A. Gibson, 2005, "Invisible Institution: The Military, War, and Peace in Pre-9/11 Introductory Sociology Textbooks," *Journal of Political and Military Sociology*, 33 (Winter): 249–266.
6 Herbert S. Lewis, 2012, "The Radical Transformation of Anthropology – Herb Lewis' Review of Dramatic Changes in Anthropology," Association of Senior Anthropologists, posted on https://asa.americananthro.org, 28 January 2012.
7 ASA Press Release, 2003, "Sociological Association Takes Position on Iraq," Washington, D.C.: ASA, 1 August. The ASA membership passed a resolution in 2003 calling for "an immediate end to the war against Iraq." The resolution clearly stated that favoring this policy in no way indicated support for "Saddam Hussien's [prior] dictatorship." The release went on to contrast this 2003 resolution with the earlier one during the Vietnam war which did not pass. A subsequent poll of members revealed that most favored the policy stated in the resolution but most also did not view its passage as "consistent with the role of a professional, scientific organization."
8 Charles H. Coates and Roland J. Pellegrin, 1965, *Military Sociology: A Study of American Military Institutions and Military Life*, University Park, Md.: The Social Science Press.
9 Caforio, op. cit., and Guiseppi Caforio, 2009, *Advances in Military Sociology: Essays in Honor of Charles C. Moskos*, Vols A and B, Bingley, England: Emerald Publishing Group.
10 Scott Efflandt and Brian Reed, 2001, "Developing the Warrior-Scholar: The Role of Sociology in Military Leadership," *Military Review*, 4 (Jul–Aug): 82–89.
11 "9/11" refers to 11 September 2001, the date on which civilian jetliners, hijacked by al-Qaeda operatives, crashed into the World Trade Center in New York City and the Pentagon in Washington, D.C. (and a final one in a field in Somerset County, Pennsylvania).
12 The term "forever war" may be traced to a 1974 military science-fiction novel by Joe Haldeman (*The Forever War*, New York: St. Martin's Press) and the more recent nonfiction memoir by reporter Dexter Filkins (2009, *The Forever War*, New York: Vintage Books), known for his coverage of the wars in Afghanistan and Iraq.
13 Rodney Stark, 2001, *Sociology*, 8th edition, Belmont, Calif.: Wadsworth/Thompson. See this website's description of Stark's approach: www.thomsonedu.com/thomsonedu/instructor. do?topicid=5E8E&sortby=copy&type=all_radio&courseid=SO18&product_ isbn=0495093440&disciplinenumber=14.

14 Stouffer, op. cit., Chapter 10.

15 Dorothy Swaine Thomas and Richard S. Nishimoto, 1946, *The Spoilage: Japanese-American Evacuation and Resettlement during World War II*, Berkeley and Los Angeles: University of California Press; and, Tamotsu Shibutani, 1978, *The Derelicts of Company K· A Sociological Study of Demoralization*, Berkeley and Los Angeles: University of California Press.

16 Edward A. Shils and Morris Janowitz, 1948, "Cohesion and Disintegration in the Wehrmacht in World War II," *Public Opinion Quarterly*, 12 (Summer): 280–315.

17 Émile Durkheim, [1897] 1951, *Suicide: A Study in Sociology*, trans. by John A. Spaulding and George Simpson. New York: The Free Press of Glencoe.

18 Siniša Malešević, 2010, *The Sociology of War and Violence*, New York: Cambridge University Press.

19 Mary Caprioli, 2000, "Gendered Conflict," *Journal of Peace Research*, 37 (January): 51–68; and Valerie M. Hudson, Bonnie Baliff-Spanvill, Mary Caprioli, and Chad F. Emmett, 2012, *Sex and World Peace*, New York: Columbia University Press.

20 Joshua Goldstein, 2001, *War and Gender: How Gender Shapes the War System and Vice Versa*, New York: Cambridge University Press.

21 George C. Homans, 1946, "The Small Warship," *American Sociological Review*, 11 (June): 294–300.

22 Joseph L. Soeters, Donna J. Winslow, and Alise Weibull, 2006, "Military Culture," pp. 237–254 in Giuseppe Caforio (ed.), *Handbook of the Sociology of the Military*, New York: Kluwer Academic/Plenum Publishers.

23 Samuel P. Huntington, [1957] 2003, *The Soldier and the State: The Theory and Politics of Civil–Military Relations*, Cambridge, Mass.: Belknap Press of Harvard University; and Morris Janowitz, [1960] 1971, *The Professional Soldier: A Social and Political Portrait*, New York: The Free Press.

24 Charles C. Moskos, Jr., 1977, "From Institution to Occupation: Trends in Military Organization," *Armed Forces & Society*, 4 (October): 41–50.

25 Anthony King, 2006, "The Post-Fordist Military," *Journal of Political and Military Sociology*, 34 (Winter): 359–374.

26 David R. Segal, 1989, *Recruiting for Uncle Sam: Citizenship and Military Manpower Policy*, Lawrence, Kan.: University of Kansas Press.

27 Nora Kinzer Stewart, 1991, *Mates and Muchachos: Unit Cohesion in the Falklands/Malvinas War*, Washington, D.C.: Brassey.

28 Marc Sageman, 2004, *Understanding Terror Networks*, Philadelphia: University of Pennsylvania Press.

29 Samuel Holiday and Robert S. McPherson, 2013, *Under the Eagle: Samuel Holiday, Code Talker*, Norman, Okla.: University of Oklahoma Press.

30 Ruth Ginio, 2017, *The French Army and Its African Soldiers*, Lincoln, Neb.: University of Nebraska Press.

31 Michelle Sandhoff, 2017, *Service in a Time of Suspicion: Experiences of Muslims in the U.S. Military 9/11*, Iowa City: University of Iowa Press.

32 Judith Hicks Stiehm, *Bring Me Men and Women: Mandated Change at the U.S. Air Force Academy*, Berkeley, Los Angeles, and London: University of California Press.

33 Laura L. Miller, 1997, "Not Just Weapons of the Weak: Gender Harassment as a Form of Protest by Army Men," *Social Psychology Quarterly*, 60 (March): 32–51; and Helena Carreiras, 2008, "From Loyalty to Dissent: How Military Women Respond to Integration Dilemmas," pp. 161–182 in Helena Carreiras and Gerhard Kümmel (eds.), *Women in the Military and Armed Conflict*, Wiesbaden, Germany: VS Verlag für Sozialwissenschaften.

34 Aaron Belkin, 2003, "Don't Ask, Don't Tell: Is the Gay Ban Based on Military Necessity?," *The U.S. Army War College Quarterly: Parameters*, 3 (Summer): 108-119; and Jonathan Lee, 2013, "The Comprehensive Review Working Group and Don't Ask, Don't Tell Repeal at the Department of Defense," *Journal of Homosexuality*, 60 (2–3): 282–311.

35 Richard Overy, 1997, *Why the Allies Won*, New York and London: W. W. Norton & Company.

36 Helen Fein, 1979, *Accounting for Genocide: National Responses and Jewish Victimization during the Holocaust*, New York: The Free Press.

37 Scott Sagan, 1996–1997, "Why Do States Build Nuclear Weapons? Three Models in Search of a Bomb," *International Security*, 21 (Winter): 54–86.

38 Alf Andrew Heggoy, 1972, *Insurgency and Counterinsurgency in Algeria*, Bloomington, Ind.: Indiana University Press; and Lt Colonel David Galula, [1963] 2006, *Pacification in Algeria, 1956–1958*, Santa Monica, Calif.: RAND Corporation.

39 Trúóng Nhú Táng, David Chanoff, and Doàn Vãn Toai, 1986, *A Viet Cong Memoir: An Inside Account of the Vietnam War and Its Aftermath*, New York: Vintage Books; and Lt Colonel John A. Nagl, 2005, *Learning to Eat Soup with a Knife: Counterinsurgency Lessons from Malaya and Vietnam*, Chicago, Ill.: University of Chicago Press.

40 Richard English, 2003, *Armed Struggle: The History of the IRA*, Oxford: Oxford University Press; and Aaron Edwards, 2017, *UVF: Behind the Mask*, Newbridge, Ireland: Merrion Press.

41 Richard Davis, 2016, *Hamas, Popular Support and War in the Middle East: Insurgency in the Holy Land*, London and New York: Routledge; and Eyal Ben-Ari, Zev Lerer, Uzi Ben-Shalom, and Ariel Vainer, 2010, *Rethinking Contemporary Warfare: A Sociological View of the Al-Aqsa Intifada*, Albany, New York: State University of New York Press.

42 Jack McDonald, 2017, *Enemies Known and Unknown: Targeted Killings in America's Transnational War*, Oxford and New York: Oxford University Press.

43 Keir Giles, 2016, *Handbook of Russian Information Warfare*, Rome, Italy: NATO Defense College Fellowship Monograph.

44 Mady Wechsler Segal, 1986, "The Military and the Family as Greedy Institutions," *Armed Forces & Society*, 13 (Fall): 9–38; and Karin De Angelis and Mady Wechsler Segal, 2015, "Transitions in the Military and the Family as Greedy Institutions: Original Concept and Current Applicability," pp. 22–44 in René Moelker, Manon Andres, Gary Bowen, and Philippe Manigart (eds.), *Military Families and War in the 21st Century: Comparative Perspectives*, London and New York: Routledge.

45 Meytal Eran-Jona, 2011, "Married to the Military: Military-family Relations in the Israeli Defense Forces," *Armed Forces & Society*, 37 (Fall): 19–41.

46 Manon D. Andres and René Moelker, 2011, "There and Back: How Parental Experiences Affect Children's Adjustments in the Course of Military Deployments," *Armed Forces & Society* 37 (Summer): 418-447; and, Edna J. Hunter-King, 1998, "Children of Military Personnel Missing in Action in Southeast Asia," pp. 243–256 in Yael Danieli (ed.), *International Handbook of Legacies of Trauma*, New York: Plenum Press.

47 T.G. Ashplant, Graham Dawson, and Michael Roper (eds.), 2000, *The Politics of War Memory and Commemoration*, London and New York: Routledge.

48 Rita Phillips, Vincent Connelly, and Michael Burgess, 2022, "Exploring the Victimization of British Veterans: Comparing British Beliefs about Veterans with Beliefs about Soldiers," *Armed Forces & Society*, 48 (April): 385–409.

49 Meredith Kleykamp, Crosby Hipes, and Alair Maclean, 2018, "Who Supports U.S. Veterans and Who Exaggerates Their Support?" *Armed Forces & Society*, 44: 92–115.

50 Natalie Purcell, Kristine Burkman, Jessica Keyser, Phillip Fucella, and Shira Maguen, 2018, "Healing From Moral Injury: A Qualitative Evaluation of the *Impact of Killing* Treatment for Combat Veterans," *Journal of Aggression, Maltreatment, & Trauma*, DOI:10.1080/10926771.2018.1463582.

About the Authors

Wilbur J. Scott is Professor Emeritus, Department of Behavioral Sciences & Leadership, U.S. Air Force Academy, and former Chair and Professor Emeritus, Department of Sociology, University of Oklahoma. His earlier work includes *The Politics of Readjustment: Vietnam Veterans Since the War* (Aldine de Gruyter, 1993), reissued as *Vietnam Veterans Since the War: The Politics of PTSD, Agent Orange, and the National Vietnam Memorial* (University of Oklahoma Press, 2004), and an edited volume with Sandra Carson Stanley, *Gays and Lesbians in the Military: Issues, Concerns, and Contrasts* (Aldine de Gruyter, 1994).

Karin Modesto De Angelis is Professor of Sociology, Department of Behavioral Sciences & Leadership, U.S. Air Force Academy. She joined the USAFA faculty in 2011, where she has taught courses on race, ethnicity, social class, and gender. Her passion for teaching is well-known. In 2016 she was named "Outstanding Academy Educator" (USAFA's highest teaching award) and, in 2022, the USAFA Class of 2022 voted her the "Heiser Outstanding Senior Faculty Award" (the Academy's only cadet-generated faculty award). Her research interests include race/ethnicity in the U.S. military, the recruitment of Hispanic servicemembers, military families, diversity in organizations, and GEN-Z and millennial attitudes toward social issues.

David R. Segal is Professor Emeritus, Distinguished Scholar-Teacher, and Founding Director of the Center for Research on Military Organization, Department of Sociology, University of Maryland, where he was also affiliated with the School of Public Policy, the Department of Government and Politics, and the Population Research Center. His earlier books include *Peacekeepers and their Wives* with Mady W. Segal (Greenwood, 1993), *Recruiting for Uncle Sam* (University Press of Kansas, 1989), *The All-Volunteer Force* with Jerald G. Bachman and John D. Blair (University of Michigan Press, 1977), *The Postmodern Military* co-edited with Charles C. Moskos and John Allen Williams (Oxford University Press, 2000), and *Military Sociology* (4 vols) co-edited with James Burk (Sage, 2012).

1 The Advent of a Field

Reader's Guide: While interest in war and the military date back to the 19th century, the systematic study of them in sociology traces its origins to World War II. Teams of social scientists conducted research for the U.S. Department of War, and other young budding sociologists served in the military as soldiers. Chapter Spotlights detail classic studies on racial segregation and semi-integration in the U.S. Army, the demeaning experience of Japanese Americans in World War II, and a classic study of unit cohesion in the German Army. Following the war, sociologist Morris Janowitz – and then a core of others – assumed the mantle of pulling these experiences together as a field. The resultant "military sociology" has become multidisciplinary, comparative, and international in scope and membership. It combines theory, research, and practical applications, informing both academic social science disciplines and decision-makers in defense agencies and the military.

War and military matters were on the minds of the first sociologists in Europe and the United States.[1] How could they not be? Bloody conflicts wracked both continents in the 19th century. Founding fathers August Comte and Herbert Spencer included militarist epochs in their models of social evolution but predicted their violent times would be supplanted by more sophisticated and peaceful stages. Similarly, Karl Marx, always the utopian, argued that conflicts born of wars maintaining capitalist hegemony would dissipate following a successful communist revolution.

War and the military were there in personal ways as well. In the late 19th century, the Prussian Army, in which he served a tortured stint as a conscript, provided Max Weber with a ready example of bureaucratic hyper-rationality he feared would eventually dominate modern life. In America, social historian Albion Small, the country's first Professor of Sociology, surveyed the state of the discipline in the U.S.A. in 1916. It is no accident his analysis commences with 1865, the date marking the end of the nation's colossal Civil War. That war, Small noted, shook American social science to the core, for it made painfully clear the country's experiment with democracy was "not altogether a success."[2] And, back in France, World War I wiped out Émile Durkheim's entire cohort of graduate students at the Sorbonne Université in Paris, including his son, André. They died in the trenches along with a million-plus other Frenchmen – a tragedy that shredded Durkheim's optimism about modern society.

Scientific communities set a precedence in World War I, where psychiatry and psychology gave input to enhance military preparedness and performance.[3] In Germany,

DOI: 10.4324/9781003282549-1

the seat of modern psychology in 1914, industrial psychologists advised on organizational issues, nutrition, and aptitude tests for channeling recruits into military specialties. Civilian psychiatrists in Great Britain staffed mental hospitals where its soldiers with "shell shock" – a stark form of psychological breakdown associated with World War I's trench warfare – were treated during the war.

The U.S.A. entered the war in 1917. Psychiatrist Thomas Salmon proposed a treatment program for shell shock and devised screenings for weeding out inductees with mental disorders and deficiencies.[4] Industrial psychologist Walter Dill Scott converted his aptitude screenings for business settings to military use, and Robert Yerkes, then president of the American Psychological Association, adapted IQ tests to assess new recruits' mental capacities (but besmirched his legacy by arguing these tests also proved the innate inferiority of Southern Europeans). These programs were small potatoes but identified a role for the social sciences in a war effort.[5]

Samuel Stouffer and *The American Soldier*

Germany's military invasion of Poland on 1 September 1939 denoted the official start of World War II. By 1941, Germany had stormed through Western Europe and, on 22 June, invaded the Soviet Union. All this before 7 December 1941, the date of U.S. entry into the war precipitated by the attack on Pearl Harbor, Hawaii, by the Imperial Japanese Navy. U.S. war planners already had pondered contingencies for ramping up the American military effort, including the reintroduction of social science research teams. Once in the war, the U.S.A. relied on social science inquiry on a broader scale than in any other country, hiring sociologists, psychologists, political scientists, historians, anthropologists, and economists. These were formed into *multidisciplinary research units*. Their tasks were *applied*, that is, designed to answer practical questions. None the less, the researchers brought with them the theoretical frames that guided their thinking as social scientists.

So it was that Samuel Stouffer, a sociologist and statistical guru at the University of Chicago, had received a call from the Department of War in the summer of 1941. Would he come to Washington, D.C. to head up the social science unit in its Research Branch? Stouffer's specialty was conducting large-scale surveys of people's attitudes, a form of research which had been prohibited by the War Department because of its anonymous character. But that was about to change.

Field work for the Research Branch's first survey commenced on 8 December 1941, even as Stouffer was still hastily assembling his research unit. Authorized initially to do only one survey in one Army infantry division, Stouffer's social science unit by the end of the war would conduct more than 250 studies, surveying and interviewing over half a million soldiers in the Army in both the European and Pacific theaters. Since the Army's strength in World War II was about 8,000,000, this amounts to interviews with about 6% of soldiers during the war. Their work culminated in a multi-volume publication in 1949, the first two volumes entitled *The American Soldier: Adjustment during Army Life* and *The American Soldier: Combat and Its Aftermath.*[6]

Sociologist Robin Williams, Jr., one of the coauthors of *The American Soldier* (TAS), assessed its impact 40 years later in a review article for *Public Opinion Quarterly*.[7] He noted it had a significant impact on military policy during and after the war, particularly in the areas of combat motivation and race relations. Further, its publication in 1949 sparked immense interest and debate within the social sciences and in the public media. We now turn to a particular chapter in TAS where Stouffer focused on the attitudes of White and Black soldiers in the segregated U.S. Army of World War II.

Spotlight on Sociological Thinking and Research

Samuel Stouffer et al., "Negro Soldiers," Ch. 10 in *The American Soldier: Adjustment during Army Life*

Samuel Stouffer was born in Sac City, Iowa, in 1900 where his father was the owner of the *Sac Sun* newspaper. As a boy, young Samuel often hung out at the *Sac Sun* office, sitting on its porch and listening to Civil War veterans tell their tales. In 1918, he went off to nearby Morningside College, then on to Harvard University. He returned in 1923 with an M.A. in English to serve as editor of his father's newspaper. Three years later, he moved on to the doctoral program in Sociology at the University of Chicago, because "people [in the newspaper business] were tossing a lot of bunk around and it seemed like a good idea to try to pin some of these things down."[8]

Stouffer's interest was in what he called scientific sociology. For him this meant collecting quantitative data in experiments or surveys followed by statistical analysis. By this time, the use of computer punch cards was available, at least at research centers like the University of Chicago. Researchers would write out questionnaires with forced-choice response formats, collect responses from a carefully selected sample, convert their responses to numerical codes for which holes could be punched in cards, run the cards through sorting machines to calculate the distribution of responses, and construct tables for displaying the results.

Stouffer was just getting the social science unit set up for the Department of War when the Japanese bombed Pearl Harbor. Survey topics ranged from the mundane – attitudes toward laundry services in Panama – to the grand, such as: What motivates soldiers to fight on when things get tough? Early in the war, the Army was experiencing a significant number of desertions by soldiers home on leave. Stouffer conducted a survey at an embarkation point in New York. The study detected the conditions under which soldiers would desert and suggested a change in policy: soldiers should be required to go home on leave in uniform rather than in civilian clothes. Going on leave in uniform meant family and friends treated the soldiers as heroes serving their country and saw them back to camp.[9] The desertion rate dropped to near zero and the military had a quick example of the research unit's value.

One of the biggest issues facing the Army was racial segregation. America, especially in the Southern states, was strictly segregated, a practice that carried into and permeated the military. Under the banner of States' Rights, individual states in the U.S.A. were allowed say over a range of issues, and historically none was bigger than racial ones. In the 19th century until 1865, the arrangement allowed slavery to exist in the South but not in the North. After 1865, individual states, mainly Southern ones, passed legislation – referred to as Jim Crow laws – regulating the use of public facilities by Whites and Blacks.[10] The laws segregated schools, public transportation, and public places including restaurants, toilet facilities, and even drinking fountains.

In World War II, the Army at first did not draft any African Americans under the assumption they were not fit to serve. However, as manpower needs became pressing in late 1942, the Army began drafting Black men with the aim of matching their numbers in the population. In keeping with Jim Crow practices, Army posts were totally segregated, and White and Black soldiers were placed in separate units.[11]

Black troops were quartered in separate barracks and had separate mess, training, and healthcare facilities, or if sharing use of these with White soldiers, had to use them at different times during the day. When Black soldiers went off-post, community norms and laws prevailed.

The scope of segregation may be seen in a controversy surrounding United Service Organization (USO) shows provided as morale-boosting events for U.S. servicemen. In the 1944 USO Christmas tour, African American singer Lena Horne requested changes in how separate shows for White-soldiers-only and Black-soldiers-only were being done. In two instances, at Camp James T. Robinson, Little Rock, Arkansas, and Ft. Riley, Kansas, White German inmates from nearby prisoner-of-war (POW) camps were allowed to see her shows with Black soldiers, and, at Ft. Riley, German POWs were seated in front of Black American soldiers. At Camp Robinson, she stopped her show and refused to continue when German POWs were brought in. At Ft. Riley, Horne is reported to have walked to the back half and presented her show facing the Black troops with her back to the German POWs.[12]

The duties of all-Black units did not parallel those of all-White units and there were almost no Black officers. Black troops often simply became janitors. As manpower shortages forced planners to reconsider, Black units were used in a wider variety of assignments. However, demoralization had by this time set in among most all-Black units. In March of 1943, Black troops comprised about 8% of the Army and Stouffer's team interviewed more than 7,000 Black soldiers stationed in the U.S.A. A larger percentage of Black soldiers than White ones expressed discontent, and they saw their dissatisfaction with the Army in racial terms. White soldiers still stateside were about twice as likely as Black soldiers to say they were anxious to go overseas and get into combat.[13]

Stouffer's team surveyed samples of Black and White soldiers in 1944 and again in 1945, this time in overseas settings. Most all-Black units by now had moved to jobs in transportation and construction. Black units also had been encouraged to volunteer for combat. Some of those who did so were assigned as smaller units within larger all-White units desperately in need of replacements. Instead of putting Black replacements individually into empty slots in a heretofore all-White company, the remaining Whites would be congregated into two or three of its platoons and an all-Black platoon would be added to bring the company up to strength – a kind of semi-integration. This was a critical shift in policy because companies moved as single entities on the battlefield and platoons in the same company were highly reliant upon each other.

The arrangement seems to have worked well. In the 1945 survey, for example, 90% of White officers and sergeants in the semi-integrated companies rated the performance of all-Black platoons compared with White ones in their company "just the same as or better than White troops."[14] The 1945 survey also produced, arguably, Stouffer's single most-famous finding of the war from the hundreds he reported, summarized in Table 1.1. Stouffer had grouped answers to a question about how a White soldier would feel about being in a company having both White and Black platoons by another variable: whether the soldier was in a semi-integrated company (and hence directly exposed to an all-Black platoon), in a regiment or a division that contained a semi-integrated company (less exposure to an all-Black platoon), or in a division in which there were no semi-integrated units at all (no exposure to an all-Black platoon).

The findings were quite dramatic. Analysis revealed the closer the actual contact White soldiers had with an all-Black unit, the more accepting they were of semi-integration. As can be seen in Table 1.1, 60% of White soldiers in semi-integrated companies said they would accept this arrangement, compared to ever declining percentages of acceptance with progressively less exposure: 51% of those in an otherwise White regiment, and 38% of those in an otherwise White division. Notably, only 11% of White soldiers serving in a division with no semi-integrated companies, that is, ones having no exposure to them, said they would be receptive to the arrangement. Could exposure to Black soldiers under battlefield conditions be reducing White prejudice? Commenting upon this finding, Stouffer termed it "illuminating" and stated that semi-integration could be a step toward how successful full integration of Black soldiers into White units might take place, especially under combat conditions.

Unfortunately, with the end of the war in August of 1945 and the downsizing of the Army, Stouffer's findings were quickly forgotten and the assignment of troops to either all-White or all-Black units resumed. Most Black troops returned home to a segregated landscape that did not look substantially different than it had before their wartime service.

However, the story does not end here. Social upheavals during the war had raised frustrations with Jim Crow laws and the war too had heightened political expectations among Blacks, particularly those who had served. In 1945, War

Table 1.1 Attitudes of White Soldiers toward Serving in Semi-Integrated Units, by Exposure to All-Negro Platoons[a]

Q. "Some Army divisions have companies which include Negro platoons and White platoons. How would you feel about it if your outfit was set up something like that?"					
White Soldier's Prior Exposure to All-Negro Platoons		*Attitudes of White Soldiers[b]*			
		Would Accept (%)	*Prefer Not to (%)*	*Would Reject (%)*	*(%)*
Negro platoon in otherwise White company	80	60	33	7	100
Semi-integrated company in White regiment	68	51	29	20	100
Semi-integrated company in White division	112	38	38	24	100
No negro unit in White division	1450	11	27	62	100
Number of White soldiers in sample	1710				

a *Stouffer et al., 1949, The American Soldier*, adapted from Chart XVII, p. 594.

b Categories from original are collapsed here: *would accept* = "would like it" and "just as soon have it as any other setup"; *prefer not to* = "rather not, but it would not matter too much"; and *would reject* = "would dislike it very much". Sample is from U.S. Army Division in Europe, 1945.

Secretary Robert Patterson appointed the Gillem Board to propose policy for the efficient use of African American troops. Its report six months later was a big surprise. It advised recruiting Blacks in proportion to the population, assigning qualified Black soldiers to all military duties, using Black and White officers interchangeably, and, as a long-term goal, fully integrating the armed forces. These recommendations were ignored.

On 26 July 1948, President Harry Truman dropped a bombshell, Executive Order 9981. This declared "equality of treatment and opportunity for all persons in the armed services without regard to race, color, religion, or national origin."[15] Reaction and opposition to the order within the military was virulent. Top military brass chose to interpret the directive's language as requiring equality within segregated units consistent with a separate-but-equal principle.

Little changed until an unexpected event reopened the door. On 25 June 1950, the North Korean Army, supported by China, suddenly invaded South Korea. President Truman quickly received authorization from Congress to commit troops and materiel. The Korean War would rage on in fits and starts over the next three years, and the U.S.A. contributed almost 400,000 troops to the effort. Early in the war, Black enlistments exceeded their quota, but many languished in reassignment areas. The war did not go well, and the Army almost immediately faced dire manpower shortfalls. It quietly initiated the practice of assigning Black troops as replacements in all-White units on an individual rather than a unit basis, but only to units whose battalion commanders agreed to the arrangement. Some battalion commanders jumped at the chance to bring their units to full strength, while others went short of replacements rather than accept Black ones.

Johns Hopkins University sociologist, Leo Bogart, conducted Stouffer-style surveys in Korea when there were both all-Black units and units integrated to varying degrees.[16] He found that the individual-assignment system worked better than either the segregated or semi-integrated systems of World War II. Most beleaguered White soldiers, especially in combat units, pragmatically preferred Black soldiers to no replacements. After this war, the plan to revert to segregated units was scrapped.

After the war, a second wave of sociologists expanded the scope of military studies as they analyzed their experiences. Collectively, these World War II and post-war studies covered an impressive array of topics: resocialization in early adulthood, small group processes, influence and persuasion, attitude-behavior linkages, group cohesion, and race relations. Interest in 1946 was stout enough to merit a special edition of the discipline's leading journal, the *American Journal of Sociology*, under the rubric of "*Human Behavior in Military Society*."[17]

Further, in 1954 social psychologist Gordon Allport formulated a classic treatise on prejudice and procedures to reduce prejudicial attitudes, famously known as the "contact hypothesis."[18] It stated that *working together as equals on a team to achieve some common goal* is one of the surest ways to reduce formerly held, even bitter, prejudices. Faced with the problem of ending discriminatory practices, the focus, Allport argued, should be on changing behaviors first and allowing changes in attitude to follow.

The Japanese American Experience in World War II

Immigrants from Japan had begun arriving in Hawaii and the west coast (primarily California) in the late 1800s. In Hawaii, they became the ethnic majority and enjoyed a sense of superiority. In California, they found work as gardeners and truck farmers, where they encountered acerbic *anti-Asian resentment*. Railroad barons there already relied heavily upon Chinese immigrants as low-wage workers. The attack on Pearl Harbor sent anti-Japanese American sentiment through the roof and shockwaves through Japanese American communities.

Already one of America's *problem minorities* – not White, not fully accepted, concentrated in low-level occupations, and sequestered in ethnic residential enclaves – war with Japan called into question their very status as Americans. Their leaders in California formed the Japanese American Citizens League (JACL). JACL pledged allegiance to the U.S.A. and offered help in identifying "subversives." On a personal level, most were careful to speak English with correct diction, some avoided eating rice in public, and others tried to Americanize their names or pass as Chinese.

All this changed on 19 February 1942, when President Franklin Roosevelt issued Executive Order 9066. The Order shifted jurisdiction over "suspicious aliens" from the Justice to the War Department. The very next day, Lieutenant General John L. DeWitt, commander of the Western Defense Command, designated areas along the Pacific Ocean as *security zones* from which all persons of Japanese descent might be excluded. The order implicated some 120,000 Japanese Americans, almost all of whom were U.S. citizens.

Japanese Americans were asked to leave the security zones voluntarily, but widespread public hostility made it difficult to relocate anywhere. In March of 1942 the military began forced removal of them from their homes in California, Oregon, and Washington into ten relocation camps. The "Japanese American Evacuation and Relocation Study" (JERS), an overlooked World War II study, was conducted under the auspices of the War Relocation Authority (WRA). The project was formulated and directed by sociologist Dorothy Swaine Thomas. One of her research assistants, Tamotsu Shibutani, would go on to become a well-known sociologist in his own right. We shine our Spotlight on their work.

Spotlight on Sociological Thinking and Research

Dorothy Swaine Thomas and Richard S. Nishimoto, *The Spoilage: Japanese–American Evacuation and Resettlement during World War II*,[19] and Tamotsu Shibutani, *The Derelicts of Company K: A Sociological Study of Demoralization*[20]

Dorothy Swaine Thomas (her maiden name, and later after her marriage to W. I. Thomas, her married name as well) was born in Baltimore, Maryland, in 1899. She lived with her maternal grandmother and then an uncle, doing so well in school she won a scholarship to the elite, all-women Barnard College. There she had the good fortune to study with W. F. Ogburn and Franz Boas, premier professors at the time in sociology and anthropology. After graduation from Barnard, she studied at the London School of Economics and obtained her doctorate there in 1924. Her dissertation, "Social Aspects of Business Cycles," contained state-of-the-art multivariate analyses. In this respect, Dorothy Swaine Thomas was emblematic of a new wave of women in sociology.[21]

Faculty positions were mostly nonexistent for women, so Thomas filled temporary research positions until 1926 when she became a research assistant for a
famous sociologist, W. I. Thomas. The two soon were intimately involved (she was
27 and he was 63) and they married in 1934. They were a sociological odd couple
embodying two contrasting styles: she, statistical analyses of measurable trends and
behaviors, and he, rich descriptions of subjects' inner worlds. Their collaboration
soon reflected changes in their perspectives. Thomas became a fan of what she
came to call the *behavior document* – the depiction of the total situation and its
effects on behavior, now to include perceptions and attitudes. This was the mindset
she carried into JERS.

Richard Nishimoto was born in Tokyo in 1904.[22] His parents immigrated to the
U.S.A. when he was a small child. He remained in Japan with relatives until he was
17, where he attended a private Episcopal school, acquiring exposure to American
ways. Upon joining his parents in San Francisco, Nishimoto graduated from high
school and completed an engineering degree at Stanford University. He worked
vegetable stands until World War II. In 1942, he and his family were sent to the
Poston Internment Camp in Arizona, where Thomas hired him as a research
assistant.

The Spoilage begins with a demographic assessment.[23] Of about 120,000 interned
Japanese Americans, about 70% were Nisei, that is, citizens born in the U.S.A. This
group was young – most were under the age of 20 – and did not speak Japanese
fluently. Nisei moved easily between American and Japanese ways. Most of those
born in Japan, Issei, were more than 50 years of age and more set in Japanese ways.
Internees' delinquency rates and numbers on relief rolls were lower than for Whites.

The camps were located in remote "badlands." A camp had 8,000 to 15,000
internees housed in wooden, barrack-style buildings covered with tar paper. Sheet-
rock walls inside marked off completely open, small "family apartments." Each
apartment had a stove for heat and each family member received a cot, mattress,
and blanket. None had a kitchen, water faucet, or toilet – these were provided
communally. The barrack compound itself, enclosed with barbed wire and guarded
by Military Police in guard towers, also contained buildings for a WRA administra-
tor and staff, a mess hall, meeting rooms, a canteen, and, eventually, school rooms
for children's education. Outside the wire, there were fields where internees were
expected to raise crops for their own consumption and for sale in markets elsewhere. For this, WRA paid wages ranging from $12 to $19 a month.[24]

Internees struggled mightily. At first, the norm was to gamely make the best of a
bad situation. JACL leaders in the camps suggested they strive to become a model
minority. With WRA bungling and mismanagement, the mood quickly shifted
from resignation to outright anger. Food shortages, delayed payment of wages, and
poor living conditions were constant grievances, and, too, the abject unfairness of
what had taken place fully set in. Those favoring JACL approaches – usually older
Nisei – were shunned as *inu* (literally, "dogs"). Relations between internee spokes-
men and WRA administrators degenerated into episodes of edgy confrontation.
JERS researchers were regarded as potential snitches and research assistants
recorded their field notes surreptitiously.

In June of 1942, the War Department instructed the Selective Service System to
discontinue the induction of Japanese American males. Nisei males, again, almost

all citizens, were reclassified as 4-C, that is, undesirables ineligible for military service.[25] However, before the year was out, the military began facing manpower shortages and the War Department floated the idea of a Nisei combat unit. General DeWitt opposed it, stating infamously: "A Jap's a Jap ..., American citizen or not."[26]

General DeWitt aside, the WRA and War Department sought a blanket solution to obtaining initial security clearances. They prepared two "loyalty questionnaires," one for Nisei males 17 years of age or older bearing the emblem of the Selective Service System, and one with the WRA logo for Nisei females and all Issei.[27] The questionnaires crucial questions were Q. 27 and Q. 28. Q. 28 on both forms asked: Will you swear unqualified allegiance to the United States of America ... and forswear any form of allegiance and obedience to the Japanese emperor ...? Q. 27 asked about willingness to serve in the military. The questionnaires provoked considerable consternation. Having already been stripped of most of their possessions and confined as undesirable aliens, were they now really being asked to declare themselves "loyal" or "disloyal"?

Instructions noted that Nisei males "loyal to the U.S." (answering Q. 28 affirmatively) but unwilling to serve in the military at this time (answering Q. 27 negatively) "probably will be taken into the military service in due time" anyway. In 1943, the War Department put together the 100th Infantry Battalion with recruits from both Hawaii and the relocation camps. Later a larger Nisei unit, the 442nd Regimental Combat Team was formed and the 100th Infantry placed within it. The 442nd went on to some of the worst fighting in Italy and France and became one of the most highly decorated of American fighting units in World War II.[28] Following their motto, "Go for Broke," they were intensely motivated to prove their loyalty as Americans and their mettle in combat.

Meanwhile, back at the internment camps, the WRA sought to reduce expenses, and large numbers of "loyals" were out-processed during 1944 and 1945 into the work-leave program. For many this provided a meager, but viable, new start after internment.[29] The smaller number of "disloyals" were sent to the Tule Lake Relocation Center (in California), now redesignated as a Segregation Center. There, matters deteriorated badly. A core of fed-up internees began to take a rebellious and strident pride in their Japanese-ness. The authorities responded with martial law.

Tamotsu Shibutani was born in 1920 in Stockton, California. Like most Nisei, he took advantage of public schooling. He graduated from high school at the top of his class and moved on to the University of California at Berkeley. He was a graduate student in Sociology there when the Japanese navy bombed American vessels docked at Pearl Harbor in 1941. Shibutani and his family were among those sent initially to Tule Lake Relocation Center. Thomas, who knew him from his university days at Berkeley, hired him there as a research assistant.

The War Department reinstituted the draft requirement in 1944 for draft-age Nisei in the internment camps. The irony was not lost on suddenly draft-eligible Nisei. Hence, their entry into the military took place under circumstances quite different than those of the 442nd and its volunteers. Nonetheless, most seized the chance, and some, Shibutani among them, were assigned to Company K, a non-combat training unit. Through his work for JERS he had acquired skills in observing and summarizing conversations and events in detail, a practice he now continued as an inductee. He mailed his notes to trusted friends and family at regular intervals for safe keeping.

The men of Company K became disenchanted early in their enlistment. Their officers were White, and an assignment to a segregated unit for these officers generally meant they previously had screwed up elsewhere. Their equipment and supplies, like those dispensed to all-Black units, were discards, and billeting and mess facilities often were those deemed unfit for others. There did not seem to be any plan for how to use this newly formed Nisei unit, so they were assigned jobs as janitors or were given pointless busy work. Finally, they were sent to Ft. Snelling, Minnesota, where the Military Intelligence Language School was to prepare them to be translators destined for the Pacific Theater.

At Ft. Snelling Company K descended into near total disarray. Only a few of the Nisei soldiers would fall out for morning formation, a job they rotated among themselves. Large numbers of them went AWOL from any work assignment, again something they rotated among themselves to give an outward appearance of minimum compliance. They challenged orders from their superiors, refused to perform in class, trashed their substandard barracks, and misbehaved on their free time. More than one visiting officer asked in dismay, "Where do soldiers like this come from?"[30]

One explanation might be riff-raff theory – the "good" Nisei already had volunteered for the 442nd. Shibutani sought to answer this question by offering a *sociological theory of demoralization* applicable in a variety of settings. Like the research of Edward Shils and Morris Janowitz (described later in this chapter), he underscored the importance of *primary groups* – a soldier's immediate, intimate circle of fellow soldiers – in motivating positive performance. But, what if the basic needs of the soldiers are unmet and the primary group is helpless to remedy the situation? In such instances, Shibutani argued, the primary group becomes an instrument for striking back.[31]

Victory in Japan (V-J Day) arrived on 14 August 1945 before Company K completed interpreter school. They deployed instead as part of the post-World War II Occupation of Japan to Camp Zama in the Tokyo area. Like the all-Black units who were part of the Occupation, they were assigned the camp's janitorial chores.

Thomas's *behavioral document* and Shibutani's *participant analysis* reveal the diverse reactions of Japanese Americans to the internment saga. Following the war, the camps were dismantled and most Japanese Americans quietly tried to put their lives back together. Some internees from the Tule Lake Segregation Center renounced their U.S. citizenship, and a few asked to be deported to Japan. In 1988, Congress passed the Civil Liberties Act, signed into law by President Ronald Reagan. The Act issued a formal apology to Japanese Americans who were incarcerated during World War II and extended to those still alive $20,000 apiece in reparations.

In 1948, Thomas became the first female professor at the University of Pennsylvania's Wharton School. She served as president of the American Sociological Association in 1952, the first woman to do so. Nishimoto was not so fortunate. Following the war, he was not able to obtain steady academic employment and, toward the end of his life, worked as the night watchman in a small San Francisco hotel. Shibutani obtained his doctorate in Sociology at the University of Chicago and became a well-known professor at the University of California at Santa Barbara. Still, he could not bring himself for almost 25 years to confront and write about what he had experienced during the war. With the encouragement of friends and colleagues, he tackled the project in the 1970s, the story of Company K.

Morris Janowitz and *Unit Cohesion*

Just as Stouffer's *The American Soldier* stands out as foundational research of what would become *military sociology*, one other study merits that status among the wave of sociological research reported after World War II, that of Edward Shils and Morris Janowitz, the latter a graduate student in Sociology at the University of Chicago.

Janowitz studied at Chicago under the tutelage of an array of well-known scholars, including controversial psychoanalyst Bruno Bettelheim, political scientist Quincy Wright, and acclaimed sociologist Edward Shils. Janowitz had served with Shils during the war in the Office of Strategic Services (OSS, the forerunner of what we now call the Central Intelligence Agency). Their OSS team had collected and scrutinized just about any information they could get their hands on to track the status of German forces.

Following the war, Shils had convinced Janowitz to join him at the University of Chicago. Shils was an associate professor and already renowned as a talented teacher and brilliant academic. He taught courses not only in sociology but in social philosophy, English literature, and the history of Chinese science. These talents would win him the Balzan Prize in 1983 for his many contributions to sociology – an honor given in fields in which the Nobel Prize is not awarded. We place our Spotlight on a 1948 journal article by Shils and Janowitz that has become one of the all-time classics in military sociology.

Spotlight on Sociological Thinking and Research

Edward A. Shils and Morris Janowitz, "Cohesion and Disintegration in the Wehrmacht in World War II"[32]

It may seem a bit odd to focus a Spotlight on the second author of a journal article, but in this case there is a very compelling reason. While Shils was highly renowned for his work in many other areas, Morris Janowitz is the one who went on to become the founding father of military sociology. The son of Polish immigrants, he was born in Paterson, New Jersey in 1922. At the time, Paterson was a major industrial city with some of the country's oldest textile mills producing silk (hence its nickname, the "Silk City"), and factories manufacturing firearms and locomotives. During Janowitz's boyhood, it was the site of bitter labor strikes and ethnic conflicts, events that spurred his interest later in reducing collective violence in public affairs.

Bespectacled and bookish, Janowitz early on developed a reputation as a serious student. He excelled in high school and then at the Washington Square College of New York University where he obtained a degree in Economics in 1941. His first job was with the Library of Congress' Experimental Division for the Study of

War-Time Propaganda, followed by one in the War Policies Division of the U.S. Department of Justice as a propaganda analyst. In 1943 Janowitz was drafted and put to work by the military in the OSS's Psychological Warfare Division (PWD), where he met Shils as part of General Dwight Eisenhower's staff in London.[33] He later was sent to France as a field intelligence officer where he interrogated German POWs. The experience, he later would recount, solidified his identity as a social scientist with an enduring interest in sociology.

Reunited at the University of Chicago following the war, the two revisited their German-POW data with a more discerning theoretical eye. In "Cohesion and Disintegration in the Wehrmacht," they sought an answer to an intriguing question. By 1943, two years before the *Wehrmacht*'s final collapse, victory by Germany no longer seemed likely. Yet the German Army (*Wehrmacht*) had been able to sustain itself as a formidable fighting force until the bitter end. Why so? On the Eastern front, the German invasion of Russia had turned into a ghastly fiasco and, combined with the first Allied advances on the Western Front, a crushing rout was in the making (see our extended discussion of this in Chapter 8). Yet even when defeat was imminent in 1945, fighting effectiveness remained unexpectedly high. How had the *Wehrmacht* sustained such high levels of it?

A popular explanation was that German soldiers were motivated by exceptionally strong commitments to Nazi ideology and political beliefs. Shils and Janowitz doubted this was so, or at least thought it did not provide a sufficient explanation. Hard-core Nazis congregated in elite Waffen-SS and paratroop divisions.[34] In contrast, conscripts and volunteers, the vast majority of whom were not Nazi Party members, populated the regular army (*Heer*) units. POW interviews indicated that German soldiers in these units considered themselves German patriots rather than representatives of Nazism, and thus seldom discussed politics or strategic aspects of the war among themselves.

Shils and Janowitz began by considering the ways the *Wehrmacht* could have disintegrated. Soldiers, singly or in groups, could have deserted, actively or passively surrendered, or engaged in acts of resistance within their own ranks. By and large, these did not happen until the last days of the war. The interviews pointed to a sociological account of the German Army's hardiness in the face of pending defeat.[35] The sharing of responsibility for survival and of intense experiences in combat converted squads and sections making up German units into *primary groups* – ones in which the soldiers had developed strong, durable, emotional bonds with each other. Conversely, this group solidarity also signaled the group's capacity to control the behavior of its members. The ability of the Germans to soldier on, Shils and Janowitz argued, persisted so long as these tightly connected primary groups, coinciding with the boundaries of a squad or section, were able to meet basic needs.

The *Wehrmacht* went to great lengths to promote such group solidarity. Care was taken to maintain homogeneity within units by putting into them soldiers from the same region of residence and of the same ethnicity. While most German units fought tenaciously to the bitter end, some did indeed fold. Those units most likely to crack were late-war, hastily assembled groups or ones with randomly mixed, heterogeneous, ethnic compositions – those in which Austrians, Czechs, and Poles were thrown together, for example. Throughout the war, the *Wehrmacht* also strived to keep units intact over long periods of time. For example, units that were decimated, where possible, were withdrawn from the front, reinforced, and sent through retraining together before recommitting them to battle.

Secondary group considerations – organizational, cultural, political, and strategic issues – proved relevant only when they happened to reinforce primary group concerns and needs, or conversely, when the primary group broke down. German hierarchical structure entrusted stricter, more extensive control of squads and sections to junior officers and senior noncommissioned officers (NCOs) than in the U.S. military. These leaders were expected to combine stern discipline with fatherly benevolence and, by and large, POWs spoke highly of their junior officers and NCOs. Interestingly, interviews also revealed the average German soldier generally had high regard for Adolf Hitler, whom most saw as a kind of father figure rather than a demagogue who headed up the Nazi Party. Still, though the Nazi Party regularly disseminated propaganda to regular army units, it largely was ignored.[36]

The U.S. military's PWD also tracked the effectiveness of its own propaganda efforts. During the last year or so of the war, the PWD saturated adjacent areas occupied by the German troops with leaflets and aimed alluring radio broadcasts at them. Tactical propaganda encouraged German soldiers to desert or surrender. Leaflets, for example, typically contained a coupon – designated a safe conduct pass – that could be clipped out and carried in one's wallet. The pass assured those who carried and presented it fair treatment should they desert or surrender. Strategic propaganda in contrast focused on bigger issues. These usually conveyed messages such as "The Nazi cause is not worth fighting for" or "Surrender, the German Army is about to be defeated."

Interviews suggested that the strategic messages largely fell on deaf ears. Almost no German soldiers deserted or surrendered in response to such information, and POWs recalled the themes in these propaganda messages much less often than they could the safe conduct pass information. In fact, interrogators discovered that German soldiers often had the pass in their possession. There even appears to have been a minor black market in which the coupons were sold or traded. Despite this, the pass usually was not used except by individual soldiers who had become isolated from their units or in cases where an entire squad or section had been separated from its larger unit.

"Cohesion and Disintegration in the Wehrmacht" had an immediate and visible impact. For starters, it presented information that could be interpreted positively about a recent wartime enemy. Understandably, most Americans felt strongly (and negatively) about the Nazi Party and the *Wehrmacht*. But here was an example of talking about an emotionally charged issue rather dispassionately. The article also singlehandedly established the theoretical validity of *unit cohesion* – the preferred label for *primary group solidarity* in the military lexicon.

Still, the work was not without criticism. For one, the supporting data came from interrogations of German POWs, who at the end of the war were released roughly in the order in which they were well-behaved, model prisoners. Since the most hard-core Nazis among them tended not to cooperate, it is likely that those who did consent to be interviewed were those less motivated by political ideology or those who smartly suppressed any allegiance to Nazi ideology. In anticipation of this, Shils and Janowitz had noted that special units of hard-core volunteers in the *Wehrmacht* had performed valiantly, but so did conscripted units if *primary group cohesion* was stout.

However, historian Omer Bartov's study of German soldiers on the Eastern Front emphasized the role of indoctrination by the Nazis.[37] He attributed the unwillingness of German soldiers to surrender to Russian troops to the Germans' fear the Russians simply would execute them.

Military Sociology Becomes a Discipline

The impressive pile of studies generated by World War II and post-War sociologists and social scientists provided the basis and motivation for a new subfield. What was missing was someone to bring it all together. Morris Janowitz became that person.

Soon after completing his Ph.D. at the University of Chicago, Janowitz accepted a position in the Sociology Department at the University of Michigan where he remained for ten years. While there, he organized a group he called the "Inter-University Seminar on Armed Forces and Society" (known as the IUS for short). Under the auspices of the IUS, Janowitz regularly brought social scientists interested in the study of the military to Ann Arbor to present and discuss their work. These mini-conferences resembled sessions of the *multidisciplinary research teams* of World War II, a format Janowitz had particularly enjoyed.[38] When he moved back to the University of Chicago in 1961, the IUS migrated with him and became the central professional organization for those who study the military.

Because the World War II studies and IUS sessions typically were *multidisciplinary*, there is potential for some confusion about the label "military sociology." Is it just sociological or a broader social scientific enterprise? Note that Janowitz himself skirted this issue by giving his seminar and the consequent organization a non-disciplinary name: "Armed Forces and Society." We use the term "military sociology" to designate a discipline that encompasses sociological and other social science disciplines – psychology, history, anthropology, and political science – which share an interest in and bring relevant perspectives to the study of war and the military. We reserve the term "sociology of the military" for the portion of "military sociology" that is strictly sociological.

Before leaving these early years of the field, we should note two important books which grew out of the University of Michigan IUS get-togethers, one by Harvard University political scientist, Samuel Huntington, *The Soldier and the State*,[39] and the other by Janowitz himself, *The Professional Soldier*.[40] These bear mention because they reset the focus of *military sociology* that, to this point, had concentrated on the *conscripted soldier*, the "grunt." These books moved the center of attention to the *professional officer corps*. Huntington and Janowitz disagreed on significant issues, but they agreed that the key questions revolved around the proper form of *civil–military relations* for a democracy and, correspondingly, the *desired ethic of its professional officers*.

The endgame of the Vietnam war brought a development that would again dramatically change the course of military sociology: the advent of the All-Volunteer Force (AVF) in 1973. Janowitz was an early opponent of the change. He worried about a military driven by monetary incentives to ensure market value for military personnel and noted the positive benefits a society accrues through compulsory, universal military service.[41] Amidst all this change, the time was ripe for professional journals devoted to these burning issues. In 1973, Professor George Kourvetaris at Northern Illinois University founded the *Journal of Political and Military Sociology*, and the IUS successfully established its own multidisciplinary journal, *Armed Forces & Society*, in 1974. This latter journal quickly became the sounding board for theoretical debates of the day.

The sociologist, however, who carried the torch of concerns about the AVF most persistently was Charles Moskos, Jr. at Northwestern University. His assessment, "From Institution to Occupation: Trends in Military Organization," known informally as the "I/O thesis," appeared in *Armed Forces & Society* in 1977.[42] Moskos worried that market contingencies facing an AVF would convert military service into "just a job" chosen by economically minded applicants. While this might work during peacetime, he felt it could pose serious problems during wartime. Soldiers motivated by pay may lack sufficient inspiration to persevere in combat. How long would it take, he wondered, for them to reach the conclusion, "You can't pay me enough to do this!"[43]

Moskos's I/O thesis proved enormously influential and set the research agenda in military sociology for the next two decades. It quickly gained visibility in military and some academic circles, in no small part because Moskos himself sought media publicity and published versions of his original article in several journals, including *Parameters*, the official journal of the Army War College. Consequently, it sparked considerable debate over his thesis's validity. It also resonated with U.S. military leaders who sought guidance for building effective manpower models for recruiting and sustaining an AVF. Other nations, confronting the issue of how to maintain a military in the Cold War and post-Cold War eras, soon provided substantial cross-national comparisons.

To this point, the university home for military sociology had been wherever Morris Janowitz was – first, the University of Michigan and then the University of Chicago. However, Janowitz was nearing the end of his academic career and the notoriety of the I/O thesis drew attention to the work of Moskos at Northwestern, where he attracted funding and graduate students. The Sociology Department at Northwestern, however, did not build a military sociology program around its well-known professor. Rather, that occurred at the University of Maryland under the leadership of two other luminaries of American military sociology, David R. Segal and Mady Wechsler Segal.

David Segal and Mady Wechsler Segal both earned Ph.D.s in Sociology at the University of Chicago, where Janowitz had served as their mentor. Both ultimately secured faculty positions in the University of Maryland's Department of Sociology, and the Segals became the center of an exceptionally active nucleus of sociology professors with interests in peace, war, and the military. This critical mass of sociologists at Maryland made it possible for the Sociology Department to not only offer courses, but a specialty area in Military Sociology within the graduate program. Over the years, more than 40 master's students and more than 20 Ph.D. students have received their degrees at Maryland with a focus on Military Sociology. Currently, seven permanent military and civilian professors at West Point, the Naval Academy, and the Air Force Academy hold Ph.D.s in Sociology from Maryland's Military Sociology program.

The Segals, in conjunction with their colleagues and many graduate students at Maryland, developed an international reputation for theory-based, empirical research with practical applications for the military. David Segal was among those who had provided a critique of Moskos's I/O thesis.[44] His work in this area culminated in the 1989 book, *Recruiting for Uncle Sam*.[45] Mady Segal meanwhile established herself as the foremost social science authority on women in the military and military families. Her 1986 theoretical assessment of family dynamics within military settings, "The Military and the Family as Greedy Institutions," remains one of military sociology's most frequently cited articles.[46]

Meanwhile, European military scholars, many of whom to this point had been IUS participants, formed their own consortium in 1986, the European Research Group on Armed Forces and Society (ERGOMAS). Motivation to do so came in part from discontent. Though the IUS had become increasingly multidisciplinary (drawing mostly from sociology and political science but welcoming other of the social sciences as well), it had remained only nominally international in scope. U.S. military issues dominated talk and discussion at IUS conferences, overshadowing both similar and unique military concerns elsewhere.[47]

ERGOMAS organized working groups for addressing issues in European militaries and emphasized comparative, cross-national perspectives. Sterling examples of this mandate are found in the many compilations of scholarly works by the Italian general turned sociologist, Giuseppe Caforio, Jacques van Doorn of Erasmus University Rotterdam, political scientist Gerhard Kümmel at the Social Science Institute of the Bundeswehr , and Christopher Dandeker of King's College London.[48] Charles Moskos, president of the IUS during the 1980s, wisely and happily embraced ERGOMAS, and many of its social scientists continued their participation in the IUS as well. Research and collaboration within the IUS hence developed a more robust international dimension.

The end of the Cold War in 1991 provided new avenues of research and prominence for the discipline. The collapse of the Soviet Union left most Western militaries without a strategic focus. Social scientists with an interest in military affairs probed the theoretical and practical issues confronting "post-modern" militaries. Foremost among these were cross-national studies addressing the assessments of subnational security threats, downsized defense budgets and military manpower, and the rise of military operations other than war. The attack on 11 September 2001 by *al-Qaeda* on the World Trade Center and the Pentagon accelerated the pace and interest in military affairs. Under the superb editorship of Patricia Shields, Professor of Political Science, Texas State University, *Armed Forces & Society* has documented for the past 20 years the extent to which military-oriented social science has taken up the challenge of studying the global, structural, and demographic correlates of terrorism and related issues in the post-9/11 environment.

In the upcoming chapters, we take up virtually all the topics mentioned above in greater detail.

Questions for Discussion

1   Department of War research units in World War II typically were multidisciplinary, including on the same team, for instance, a sociologist, political scientist, a historian, and maybe an anthropologist. What would you say are the advantages of this? Are there any disadvantages?

2   The U.S. military in World War II was racially segregated, and White soldiers strongly approved. Yet, when semi-integrated units were installed out of desperation, most White soldiers came to accept the change. Speculate for yourself: Why/how might semi-integration have changed their minds?

3   Japanese Americans in World War II reacted in a variety of ways to the loss of their homes and the forced internment in "relocation" camps. Visualize yourself in their shoes. How might you have responded upon internment? Upon being faced with the loyalty questionnaire? Upon being drafted?

4   One of the great contributions from World War II studies is the identification of primary groups and unit cohesion as prime contributors to military effectiveness. Would it be fair from this to assert that soldiers do not fight and die for country so much as fight and die for each other? Explain.

Notes

1   For a survey of the field, see James Burk and David R. Segal, 2011, "Editors' Introduction," pp. xx–xlvii in David R. Segal and James Burk (eds.), *Military Sociology*, vol. 1, Thousand Oaks, Calif · Sage Publications.
2   Albion W. Small, 1916, "Fifty Years of Sociology in the United States (1865–1915)," *American Journal of Sociology*, XXI (May): 721–864.
3   Ben Shephard, 2015, "Psychology and the Great War, 1914–1919," *The Psychologist*, 28 (November): 944–946.
4   Thomas W. Salmon, 1917, *The Care and Treatment of Mental Diseases and War Neurosis ("Shell Shock") in the British Army*, New York: War Work Committee of the National Committee on Mental Hygiene.
5   Hans Pols and Stephanie Polk, 2007, "War and Military Mental Health: The US Psychiatric Response in the 20th Century," *American Journal of Public Health*, 97 (December): 2132–2142.
6   Samuel A. Stouffer, Edward A. Suchman, Leland C. DeVinney, Shirley A. Star, and Robin M. Williams, Jr., 1949, *The American Soldier: Adjustment during Army Life*, Princeton, N.J.: Princeton University Press, pp. 5–27; Samuel A. Stouffer, Arthur A. Lumsdaine, Marion Harper Lumsdaine, Robin M. Williams, Jr., M. Brewster Smith, Irving L. Janis, Shirley A. Star, and Leonard S. Cottrell, Jr., 1950, *The American Soldier: Combat and Its Aftermath*, Princeton, N.J.: Princeton University Press.
7   Robin M. Williams, Jr., 1989, "*The American Soldier*: An Assessment, Several Wars Later," *Public Opinion Quarterly*, 53 (Summer): 155–174.
8   Joseph W. Ryan, 2010, "Samuel A. Stouffer and *The American Soldier*," *Journal of Historical Biography*, 7 (Spring): 100–137, p. 109.
9   Ibid., pp. 105–106.
10 As we show in Chapter 6, the terms "White" and "Black" are socially constructed terms loosely related to racial categories based upon skin color. Generally, the term "Black" refers/referred to a person of African descent and "White" to one of European descent. Incidentally, the labeling of public facilities in the Jim Crow era used the terms "White" and "Colored."
11 Stouffer et al., op. cit., pp. 486–489.
12 James Gavin, 2009, *Stormy Weather: The Life of Lena Horne*, New York: Atria Books; http://mts.lib.uchicago.edu/artifacts/index.php?id=defender and https://www.nps.gov/features/malu/feat0002/wof/Lena_Horne.htm.
13 Stouffer et al., 1949, op. cit., pp. 502–507.
14 Ibid., p. 592.
15 Presidential directive 9981 may be found at www.trumanlibrary.org/9981.htm.
16 Leo Bogart, 1969, *Social Research and the Desegregation of the U.S. Army*, Chicago: Markham Publishing Company. The assignment of Black troops to White units on an individual basis were viewed by many military planners as temporary measures to meet manpower demands during the Korean war. However, the research by Bogart and his associates documented the military effectiveness of these stop-gap policies.
17 Elizabeth S. Clemons (ed.), 1946, "Human Behavior in Military Society," Special Edition, *American Journal of Sociology*, 51 (March): 359–487.
18 Gordon W. Allport, [1954] 1979, *The Nature of Prejudice*, 25th Anniversary Edition, with new Introduction by Kenneth Clark, new Preface by Thomas Pettigrew, New York: Basic Books.
19 Dorothy Swaine Thomas and Richard S. Nishimoto, 1946, *The Spoilage: Japanese–American Evacuation and Resettlement during World War II*, Berkeley and Los Angeles: University of California Press.
20 Tamotsu Shibutani, 1978, *The Derelicts of Company K: A Sociological Study of Demoralization*, Berkeley and Los Angeles: University of California Press.
21 Robert C. Bannister, 1998, "Dorothy Swain Thomas: The Hard Way in the Profession," published originally (in German) in Claudia Honegger and Teresa Wobbe (eds.), *Frauen in der Soziologie (Women in Sociology)*, Munich: Oscar Beck, pp. 226–257, http://www.swarthmore.edu/SocSci/rbannis1/DST.html, part 2.
22 *Densho Encyclopedia*, Richard S. Nishimoto biography, http://encyclopedia.densho.org/Richard_S._Nishimoto/.
23 Thomas and Nishimoto, op. cit., 1–5.
24 Ibid., 26.

25   Shibutani, op. cit., p. 49.

26   Ibid., 18.

27   Ibid., 54–60.

28   The 442nd Regimental Combat Team was awarded seven Presidential Distinguished Unit citations, one Congressional Medal of Honor, 47 Distinguished Service Crosses, 350 Silver Stars, 810 Bronze Stars, and more than 3,600 Purple Hearts. For an interesting historical account, see Masayo Duus, 2006, *Unlikely Liberators: The Men of the 100th and 442nd*, Honolulu: University of Hawaii Press.

29   See Dorothy Swaine Thomas, 1952, *The Salvageables: Japanese American Evacuation and Resettlement*, Berkeley and Los Angeles: University of California Press.

30   Shibutani, op. cit., p. 2.

31   Ibid., Chapter 9. Paul Savage and Richard Gabriel argue similarly concerning the breakdown of discipline among units in the Vietnam war, as evidenced by fraggings (assassination of officers by enlisted men), drug use, and desertion. Savage and Gabriel, 1976, "Cohesion and Disintegration in the American Army: An Alternative Perspective," *Armed Forces & Society*, 2 (Spring): 340–376.

32   Edward A. Shils and Morris Janowitz, 1948, "Cohesion and Disintegration in the Wehrmacht in World War II," *Public Opinion Quarterly* 12 (Summer): 280–315.

33   While assigned to this job in London, Janowitz was awarded a Purple Heart. A German rocket landed near the building where he worked and knocked his typewriter off its table. The typewriter landed on Janowitz's foot, breaking a bone in his ankle.

34   *Waffen-Schutzstaffel* (Armed Protective Squadron) was created as the armed wing of the Nazi Party in Germany.

35   Shils and Janowitz, op. cit., pp. 283–288.

36   Ibid., pp. 304–306.

37   Omer Bartov, 1985, *The Eastern Front, 1941–1945: German Troops and the Barbarisation of Warfare*, New York: Macmillian.

38   James Burk, 1993, "Morris Janowitz and the Origins of Sociological Research on Armed Forces and Society," *Armed Forces & Society* 19 (Winter): 167–185.

39   Samuel Huntington, [1957] 1981, *The Soldier and the State: The Theory and Politics of Civil-Military Relations*, Cambridge, Mass.: Belknap Press of the Harvard University Press.

40   Morris Janowitz, [1960] 1971, *The Professional Soldier: A Social and Political Portrait*, New York: The Free Press.

41   Morris Janowitz, 1976, "Military Institutions and Citizenship in Western Societies," *Armed Forces & Society* 2 (Winter): 185–204.

42   Charles C. Moskos, Jr., 1977, "From Institution to Occupation: Trends in Military Organization," *Armed Forces & Society* 4 (October): 41–50.

43   Charles C. Moskos, Jr. and Frank Wood (eds.), 1988, *The Military: More than Just a Job?* Washington, D.C.: Pergamon-Brassey.

44   David R. Segal, 1986, "Measuring the Institutional/Occupational Change Thesis," *Armed Forces & Society* 12 (Spring): 351–375.

45   David R. Segal, 1989, *Recruiting for Uncle Sam: Citizenship and Military Manpower Policy*, Lawrence, Kans.: University of Kansas Press.

46   Mady Wechsler Segal, 1986, "The Military and the Family as Greedy Institutions," *Armed Forces & Society* 13 (Fall): 9–38.

47   See the ERGOMAS founding statement at www.ergomas.ch/images/stories/documents/1986-1988.pdf. Discontent also stemmed from the expense of flying from Europe to the U.S.A. for conferences.

48   Gerhard Kümmel and Andreas D. Prüfert (eds.), 2000, *Military Sociology: The Richness of a Discipline*, Baden-Baden, Germany: Nomos Verlags-geschaft. Giuseppe Caforio (ed.), 1998, *The Sociology of the Military*, Cheltenham, UK: Edward Elgar Publishing; 2003, *Handbook of the Sociology of the Military*, New York: Kluwer Academic/Plenum Publishers; 2009, *Advances in Military Sociology: Essays in Honor of Charles C. Moskos*, Bingley, UK: Emerald Group Publishing.

Recommendations for Additional Reading/Viewing

Gordon W. Allport, [1954] 1979, *The Nature of Prejudice*, 25th anniversary edition, with new Introduction by Kenneth Clark, new Preface by Thomas Pettigrew, New York: Basic Books.
(All-time classic in social psychology; theoretical and empirical appraisal of prejudice and the contact hypothesis)

James Burk, 1993, "Morris Janowitz and the Origins of Sociological Research on Armed Forces and Society," *Armed Forces & Society*, 19 (Winter): 167–185.
(An appraisal of Janowitz's role in founding military sociology by one of his Ph.D. students, now Professor Emeritus and former President of the IUS)

Elizabeth S. Clemons (ed.), 1946, "Human Behavior in Military Society," Special Edition, *American Journal of Sociology*, 51 (March): 359–487.
(Special edition with Editorial Forward plus 20 articles; sociologists' observations on their experiences in World War II)

Rolf Forsberg, 2006. *Go For Broke: Japanese Americans in World War II*, Hosted by Senator Daniel K. Inouye and narrated by George Takei.
(Full length documentary, available for viewing at www.youtube.com/watch?v=7_WG96MAirI and for purchase at www.amazon.com/Going-Broke-George-Takei/dp/B005IV550G)

Gerhard Kümmel and Andreas D. Prüfert (eds.), 2000, *Military Sociology: The Richness of a Discipline*, Baden-Baden, Germany: Nomos Verlags-geschaft.
(An assessment of the state of military sociology at the turn of the century, assembled by two of its chief European practitioners)

2 Sociology and Military Sociology

Reader's Guide: Social scientists use theories to organize their thinking. Untested theories are only plausible explanations, which inspire them to collect data to verify each theory's validity. Social science thinking takes place at three levels of analysis: the macro-level, which focuses on whole societies or social institutions; the meso-level, which moves the analysis to that of organizations or local communities; and the micro-level, which homes in on individuals and what they are thinking, experiencing, and doing. Military sociology operates at all three levels: to understand militaries and war, it helps to know something about what makes societies, organizations, and individuals tick. The first two Spotlights showcase a classic study of suicide and a theory of killing in modern wars, both couched at the macro-level. We conclude with two multi-level studies presented in tandem: the first asks: Would we have less war if more leaders were women? The second asks: Is there a link between how a society treats its women and its likelihood for initiating war? Both employ impressive datasets to answer these questions.

Here is a very simply stated theory of sorts: *first, sunrises, later, sunsets*. These few words impart much information and explanation. It accounts for why we have sunlight during the day and darkness at night: the sun rises, travels across the sky, and sets on the opposite horizon. As it appears visually to humans, Earth is stationary and flat. It is a *common sense* and satisfying explanation. But how good an explanation is it? It implies, for instance, that if one set sail in a ship westward from, say, Spain, at some point that ship would plunge off the edge of the Earth. However, the Portuguese explorer, Ferdinand Magellan, led a naval expedition of the Spanish Crown that sailed to the west in September of 1519. After three years of adroit seamanship and harrowing experiences – Magellan himself was killed in a skirmish with natives in the Philippines – one of his ships, navigating consistently in a westerly direction, arrived back in Spain on 6 September 1522.[1]

About the same time, Polish astronomer Mikolaj Kopernik (Nicolaus Copernicus) advanced a *heliocentric model of the solar system* in which the sun is at the center and stationary, and the Earth and other planets orbit around it. A ball-shaped, orbiting Earth provides an alternative account for "sunrise" and "sunset." It also explains many other details: why there are seasons (the Earth's orbit is elliptical, not round), the number of days in a year (the length of time it takes the Earth to complete an orbit), and so on. Copernicus was able to demonstrate these with precise calculations and supporting evidence. Here, we have an early example of *science*, that is, a way of knowing that

DOI: 10.4324/9781003282549-2

systematically tests plausible theories with real-world data. One theory is considered better than another if it can accurately and demonstrably account for more phenomena.

Military sociology and its implied social science disciplines – sociology, political science, history, psychology, and anthropology – follow these guidelines in forming and evaluating theories. We will be presenting theories from this range of disciplines though we – the authors of this book – are sociologists and usually work using that lens. Sociological theories are pitched at one of three different levels of analysis: the *study of societies and institutions*, the *study of social groups*, and/or the *study of individuals* within these structures. The other social sciences roam these levels as well, except for psychology which tends to confine itself to the study of mental processes. Some social psychologists though would object to this latter characterization as too limiting.

The Study of Societies and Institutions

Sociologists define a *society* as the "coordinated effort by people sharing some common identity," usually historically based in a distinct geographical location, "to survive and flourish as a collective." *Identities* are socially constructed, and achieving a *common identity* is not always a simple, straightforward process.

Nigeria, for example, represents a case common to societies with a colonial history. Its current boundaries were drawn a century or so ago by British colonizers who lumped together tribal members with no prior national history (see our discussion of this in Chapter 6). Today, Nigerians tend to identify themselves based on their tribal affiliations – notably Yoruba, Igbo, Fulani, or Hausa – and only under certain circumstances as fellow Nigerians. Or, consider the former Czechoslovakia. Formed in 1918 as a fragment of the disintegrating Ottoman Empire, it peaceably divided itself into the Czech Republic and Slovakia in 1993, in large part because Czechs and Slovaks admitted they did not consider themselves members of a common nation.

Another significant part of the definition of society refers to "coordinated efforts to survive." Survival needs would include producing the material things essential for keeping societal members alive (food, clothing, and shelter), figuring out how to make rules and resolve disputes, providing security so that people can carry out everyday life, arranging for the procreation and rearing of children (the main source of new members), and providing explanations for "big questions," such as, what is the meaning of life and what happens to us when we die? Members of a society construct an elaborate array of customs and, in most cases, laws to govern the meeting of these needs.

Sociologists use the term *social institution* to designate a cluster of rules, roles, and relationships designed to meet some specific survival need. The *economy* is the social institution in which the essentials for staying alive are produced and distributed, and *polity* is the one in which the rules for governance, dispute resolution, and internal security are carried out. Sex is regulated and children produced and reared in the *family*; and *religion* is the social institution for approaching the natural and supernatural in search of ultimate meaning. In simpler societies, one institution may serve a dual purpose, while in more complex ones there may be additional social institutions. For instance, what we call the *military*, the social institution vested with defense from external threats (and carrying out offensive ones as well), does not appear until rather recently in human history.[2]

The social institutions of a society are strongly interconnected, both in form and substance. For starters, it usually does not work very well for a society to meet only some of its basic needs and not others. For example, it is difficult for an economy to function on a day-to-day basis if security constantly breaks down. During recent wars in Afghanistan and Iraq, farmers and shopkeepers have had to close their stalls for days or weeks at a time as episodes of violence raged in the streets – disrupting both the lives of individuals and the sustainability of Afghan and Iraqi societies.

When breakdowns like this occur, people scramble to construct alternatives to address unmet needs. The inability of a society to provide internal security may be countered by families or neighborhoods forming their own gangs or militias to ensure their safety. To finance these activities, gangs and militias might develop a subterranean economy of their own or muscle into existing ones. In this respect, thriving black markets in contraband goods and services are indicators that the formal economy does not work very well, or that certain subgroups are cut off from access to the formal economy. As these examples suggest, failure to meet basic needs satisfactorily puts the long-term survival of a society at risk, in part by encouraging alternatives that may contribute to further instability or even outright collapse.

While there are many options to choose from in meeting each of a society's basic needs, not all options fit together equally well. The substance of any particular social institution constrains the possibilities within a society's other institutions. As we show in Chapter 11, a large, *extended family* system in which married spouses have many children and multiple generations and kin live together in the same compound may make perfect sense in an *agrarian economy*. Here, the family is both the basic unit of production within the economy and the basic unit for the reproduction and rearing of children. Having lots of children means more workers for the kin-group, and relying only on trusted relatives to perform critical tasks serve both the family and the economy very well.

In contrast, modern *industrial economies* require significant separation between economic and familial institutions. As large-scale factories and corporations come to dominate an industrial economy, the hiring of capable strangers works better than giving jobs to relatives, and highly mobile *nuclear families* of parents and their immediate offspring are freer to move about to obtain better opportunities. Now, having children introduces consumers into the household who tend to remain unproductive until early adulthood, an expensive proposition, so fewer generally is better. Thus, modern industrial economies tend to erode traditional family ways, and traditional family ways tend to restrain the growth of industrial economies.

The Study of Groups

The terms "society" and "social institution" are important elements of *social structure*, that is, social arrangements that take on a life of their own beyond the individuals within them. Sociologists term this the *macro-level* in recognition of their exceptionally large size and the complexity of coordination within them. At the interim or *meso-level* are two additional elements of social structure to consider: *formal (organizations)* and *informal groups*.

A *social group* is simply two or more people who interact with each other regularly (as opposed to an encounter, say, between an Uber driver and his fare for whom the interaction usually is fleeting). Some groups are small and the considerations that guide their interactions over time are quite *informal*. A group of friends, for instance, share ideas or *norms* about how friends should behave to be friends, and this sets guidelines for how they are to act if they expect to remain friends. Though such norms are important for maintaining any informal group, they usually are not written down – members just carry them around in their heads. Nor are the *sanctions*, positive and negative, for supporting or violating these norms put into writing. Informal norms and sanctions tend to work because people living in the same socio-cultural context already share some commitment to them as good ideas about how to behave.

As groups grow in size, people find it necessary to write out the norms and to specify each person's relative place and task within the group. Because this often begins with someone saying, "Let's get organized!" the resulting *formal groups* are called *organizations*. German sociologist Max Weber penned the classic works in the late 19th and early 20th centuries on a version of organizations, *bureaucracies*, and their impersonal features.[3]

Top managers within bureaucracies, Weber noted, typically develop organization charts listing positions from high to low along with corresponding standard operating procedures (SOPs). These designate who is supposed to do what, who oversees whom, and how members are expected to carry out their assigned tasks. Such formalizing can make operations more efficient. However, organizations tend to overdo it, developing more and more rules, procedures, and oversights to the point where they often are badly impersonal and quite inefficient. Weber saw this as the dominant organizing principle in modern societies and worried about the implications, referring to its end-product as "the Iron Cage." Because of their iron-cage tendencies, organizations typically contain two dimensions: formally arranged people and formally designated procedures on the one hand, and, on the other, informally defined groups whose norms may or may not match the SOPs.

Sociologist Sanford Dornbusch, then at the University of Washington, provided the following example from his days as a cadet at the Coast Guard Academy during World War II.[4] Here, the theme is two sets of rules, one formal, one informal. In his story, a first classman (in civilian-college parlance, a senior) directs one of the swabs (a freshman) to fetch him some candy. This order is problematic since candy is not allowed in the cadet area. Still, so ordered, the swab goes and gets some candy but is unexpectedly accosted on the way back by a supervising officer, who gives the swab 15 demerits for having candy in his possession. The first classman who gave the order cannot undo the demerits. Therefore, he informs other first classmen of the situation and, collectively, they allow the swab 15 demerits worth of misconduct to make up for it.

Lesson: formal and informal structures often diverge, and when they do the SOPs and informal norms may operate at peculiar cross-purposes. At that point, which of the two set of norms should prevail? In this case, the swab learned here that the informal ones are the way to go.

The Study of Individuals as Social Beings

The *micro-level* focuses on individuals in person-to-person interaction with others. Think for a moment about the daunting task facing each of us: how to figure out the social labyrinths in which we move day-to-day and to negotiate social life within that maze. *Social structure*, much like scaffolding, outlines how behavior is arranged in each bit of social space in that society. *Culture*, the substance of a society's way of life, details the content of these arrangements from the language to be spoken, to the appropriate affect and costumes, to the rules for doing any activity.

The complexity of all this can be quite intimidating. Even usually savvy adults visiting a foreign country or unfamiliar parts of their own society can encounter difficulties adjusting their perceptions, feelings, and behaviors. There is even a term, *culture shock*, for the deep, visceral reaction people may experience when abruptly taken out of their socio-cultural comfort zones. So how does one learn all this? The short answer is by hanging out and interacting with other humans.

Sociologist George Herbert Mead devoted a good deal of his writings to this. He argued it is done through developing the *mind* and *sense of self* by interacting with others.[5] He observed infants and small children to document their acquisition of language and ability to visualize the content of the social world. In the first stage, a baby imitates the sounds and gestures of others, slowly discovering that it is possible to communicate with larger people if sounds and gestures are organized in appropriate sequences and done in a very particular way. This is an exciting but laborious process requiring a couple years of practice and repetition.

Stage two consists of play in which a three-to-five-year-old child, by himself or with other children, *pretends to be the significant others* in his life. If daddy does the cooking in the household and likes baseball, his little boy may want to set up a play-kitchen and have a special costume to wear while slapping around a ball with a stick. If only mommy cooks, a little boy may conclude that pretending to do kitchen work should not be in *his* repertoire.

The child enters a socially mature state around the age of eight when she can hold conversations not only with others, but with herself about herself as others would see her if she were to act a particular way, an ability Mead called *taking the role of the other*. She may rehearse, for example, that if she goes to school unprepared, the teacher will give her a bad grade and her parents will be upset, so she diligently completes her homework; or, if she has valued delinquent pals at school, she may decide not to do her homework because she anticipates they will think more highly of her as a fellow deviant. At this point, the youngster has reached a pivotal understanding of *how the game of life is played* in her society by someone like her.

Learning to *take the role of the other* and behave appropriately wherever one finds oneself is a process that continues throughout life. Consider what happens when a recruit joins the military. Civilians do not automatically know how to be a soldier, sailor, airman, or Marine, because this thrusts them into bits of social space in which they have little or no prior experience. They must learn the content of military spaces and acquire the mindset, costumes, and skills to carry out the required social roles. Basic training in the U.S. military is an eight-week crash course in which the recruit is stripped of her civilian identity and given a new military one. As with other occupations and professions, this is only the beginning of an extended process of acquiring a new identity.

In this respect, sociologists plying this perspective would be in agreement with the lines from Act II, Scene VII, of William Shakespeare's *As You Like It*:

> All the world's a stage,
> And all men and women merely players:
> They have their exits and their entrances,
> And one man in his time plays many parts.[6]

Theory and Military Sociology

We now have quite an array of tools on the table for approaching the study of war and the military. We therefore begin with a *macro-level* definition of military sociology: the *study of the military as an institution and of the interconnections between it and the society in which it is located*. But social scientists also do studies of the military at the *meso-* and *micro-levels*. Hence, the *study of the military as an organization* and *of individuals within the military*, respectively, should be added to the definition. Finally, of course, studies of militaries ultimately raise issues about their reason for existence: the circumstances of their use of lethal violence.

The plan is to review three studies that cast light on the military and war. Only one of the three is by a military sociologist (Malešević – and even he might not consider it his primary identity), two of the three studies were done by sociologists (Durkheim and Malešević), and one of the studies truly is the work of a multidisciplinary team made up of two political scientists (Hudson and Caprioli), a psychologist (Baliff-Spanvill), and a geographer (Emmett). In all three studies, the level or levels of analysis are germane to the theorizing and research. Such is military sociology.

A Classic Study

Theorizing at the macro-level means focusing on social arrangements larger than the individual. Those arrangements (and groupings) constitute the social contexts in which individuals live and, in the case of this study, die. Suicide would seem to be a totally individual phenomenon, the taking of one's own life. However, the incidence of suicide has distinctive and enduring characteristics – males do it more than females, the relationship with age is curvilinear (it is higher among younger and older persons), more prevalent in rural than urban areas, lower for those in the military but much higher among veterans, and so on. However, in 2010, it was apparent that the rate of suicide among U.S. military servicemen was notably higher than in 2000, an increase commonly attributed to depression and distress in the All-Volunteer Force during wartime – a *micro-level* explanation.

Drawing on the work of Émile Durkheim, author of this book, Wilbur Scott, and his colleague at the U.S. Air Force Academy, psychologist George Mastroianni, expanded the discussion by pointing to *macro-level* considerations.[7] We will return to their thought-piece, but first shine our Spotlight on Durkheim's classic study of suicide.

Spotlight on Sociological Thinking and Research

Émile Durkheim, *Suicide: A Study in Sociology*[8]
   David Émile Durkheim was born in Épinal, France, in 1858. His father, grandfather, and great grandfather were rabbis, and it was assumed Émile, a gifted student, would follow in their footsteps. Though he remained close to the Jewish community, he chose instead a secular path for his education and life work. This took him to Paris for graduate studies at the renowned École Normale Superieure. His academic interests lay in the emerging social sciences and the scientific method, both unpopular in French intellectual circles at the time. He completed his doctorate in 1882.

   Durkheim's first faculty position was a not-so-prestigious one at the Université de Bordeaux which had just established a new program for prospective school-teachers.[9] The selection of a social scientist was not without controversy. That university, like those elsewhere in France, preferred literary to scientific approaches. So, he conducted his classes in this way during the week and gave free public lectures on Saturdays on a new discipline, sociology.

Durkheim defined *social facts* – ways of acting that are shaped and constrained by the community or society in which individuals reside – as the central focus of sociology. He argued in favor of a scientific approach to sociology in his 1895 book, *Rules of the Sociological Method*. And, in his 1897 book, *Suicide*, he purposely chose the study of suicide, a seemingly individualistic act, to illustrate sociology's relevance. In 1898, he founded *L'Année Sociologique* (*Annual Sociological Review*), France's first social science journal. These efforts established Durkheim's reputation as the best-known sociologist in France, culminating in his appointment in 1902 to Sorbonne Université in Paris with the title, Professor of Education and Sociology.[10]

Durkheim sought to demonstrate sociology's unique contribution by studying suicide with an eye toward society's characteristics, not those of individuals. France already had a long history of meticulously compiling birth and death records for all its provinces in a central location in Paris. Further, statisticians there already had devised techniques for analyzing and presenting these data in tables and maps.[11] This enabled him to draw upon an extensive trove of data for testing his ideas.

A society is akin to a living organism, Durkheim argued, and the health of a society is reflected in the relative incidence of suicide and other social ills.[12] For it to survive two basic needs must be met: *integration* (its members must be sufficiently immersed in relationships with others to willingly cooperate in group life) and *regulation* (the rules of conduct must be sufficiently clear, so people know what is expected of them). Further, these must be at optimum levels. This theory, applied to the incidence of suicide, is presented in Figure 2.1.

Here are guidelines for interpreting the graph: the levels of integration and of regulation are indicated on the horizontal axis, and the incidence of suicide is listed

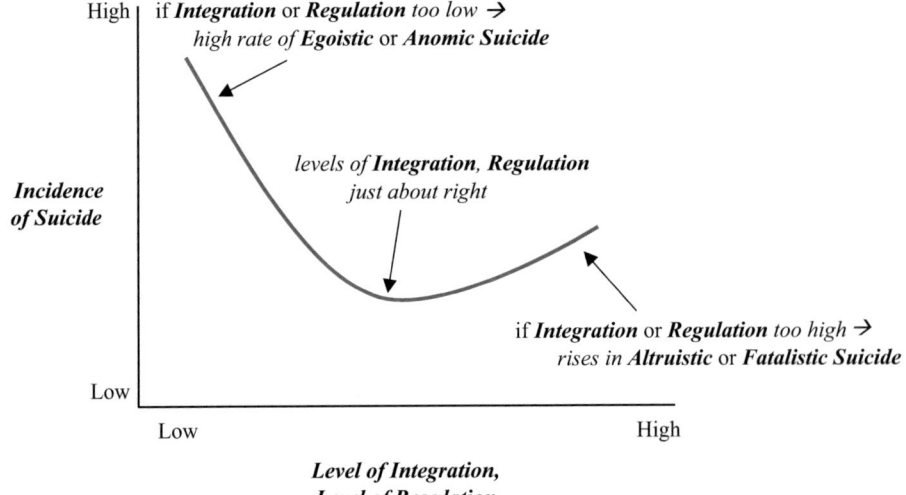

Figure 2.1 Durkheim's Theory and Types of Suicide

Source: Authors' Creation.
Note: Figure constructed from Durkheim's descriptions of the types of suicide and well-known depictions of them.

on the vertical axis. When levels of integration or regulation are too low, Durkheim predicted (and found) higher suicide rates. He designated these *egoistic suicide* and *anomic suicide*, respectively, depending upon whether they occurred in response to too little integration or too little regulation.

Why might these low levels be problematic? Beginning with integration, a healthy society is one whose members are sufficiently immersed in group life.[13] Societal groups can both sustain and constrain their members, but these important tasks are left undone when individuals lack close relationships with others. Among the negative consequences is an increase in suicide. Durkheim noted, for instance, the suicide rate among Protestants in 19th-century Europe (then, primarily Lutherans and Calvinists) was higher than among Catholics. The difference, he argued, was not because Protestants were more accepting of suicide – they were not – but because Catholicism at the time more effectively integrated its adherents into collective communities.

Similarly, when regulation is too low, individuals are left adrift because there is insufficient consensus about what the rules are.[14] As with low integration, low regulation represents an instance in which individuals are not sufficiently constrained. In the former, societal groups have lost their ability to influence isolated individuals; in the latter, individuals are not held in check because the normative order has lost its coherence. Both instances represent breakdowns in the control that groups must exert over individuals for a society to unite and persist over time.

Because integration and regulation perform chores that enhance the vitality of a society, a rise in their levels decreases the incidence of suicide – an indication of a "healthy" society. However, more and more is not always better. There is a point beyond which optimal gains in societal health begin to decline. If there is too much integration or regulation, there is another spike in suicides, this time of the *altruistic* and *fatalistic* varieties.[15] In the case of the former, the intensity of group ties has become counterproductive. Durkheim explained:[16]

> Now, when a person kills himself, it is ... because it is his duty. We thus confront a type of suicide differing by incisive qualities from the preceding one ... [Egoistic suicide] occurs because society allows the individual to escape it ... [altruistic suicide] ... because society holds him in too strict tutelage.

Durkheim found that the suicide rates of soldiers in European countries sometimes were higher than for those among civilians of the same age. His careful review of these suicide rates showed that they were highest among elite troops

> best suited to [the military's] needs ... The [higher incidence] is then caused ... by the sum total of states, acquired habits or natural dispositions making up the military spirit. ... In short, a soldier's principle of action is external to himself, which is the quality of the state of altruism.[17]

A special case of altruism may be found in a type of suicide occurring in combat which Durkheim terms "heroic suicide." Sociologist Jeffrey Reimer, then with Tennessee Technological University, studied Congressional Medal of Honor citations from 1863 to the present, focusing on the 125 cases in which the soldier specifically "chose to sacrifice his life for his comrades."[18]

The modal example of heroic suicide studied by Reimer is one of a soldier throwing himself over a grenade to save the lives of others around him. His analysis of these cases supports the basic hypotheses advanced by Durkheim about altruistic suicide. Heroic suicide was more common in elite than non-elite units, in cohesive than non-cohesive units, and among those in leadership positions than among followers.

Durkheim devotes much less attention to fatalistic suicide, where regulation becomes suffocating. As an example, he suggests the conditions experienced by those held in slavery. Important for our purposes here, Durkheim was able to establish his typology of suicides through a careful sequencing of hypotheses drawn from his theory and, then, through data presented as tests of these hypotheses.

An important point stands out. While *committing suicide* is an individual act, the *incidence of suicide* is a group characteristic, raising the question: Why does the incidence vary from one group or society to the next? A purely micro-level answer, say a theory of psychological depression, does not suffice. Not to make light of what it portends for individuals, Durkheim shows variations in the incidence are indicative of the relative health of societies themselves.

Returning to the issue of increases in suicide in the U.S. military, an Army investigation of the problem in 2009 revealed that less than 40% of recent military suicides had served a tour of duty in Afghanistan or Iraq, and even smaller percentages had exhibited prior behavioral miscues such as going AWOL (absent without leave), Article 15s (administrative punishment for infractions), or civilian legal problems. Mastroianni and Scott consider how breakdowns in military group cohesion, confusion about the rules of engagement in irregular warfare, and growing public opposition to the wars in Afghanistan and Iraq might have created conditions where egoistic and anomic suicides were more likely.[19]

Theorizing about Violence and War in the Modern Era

Sociologist Randall Collins recently advanced a *micro-level theory of violence*, zeroing in on the characteristics of *violent encounters* rather than violent individuals as the basic unit of analysis. Even very violent individuals, Collins notes, are nonviolent most of the time. Collins draws violent-encounter data from three distinct sources: video-tape recordings of real-life violence, reconstructions of violent events by the very participants who took part in them, and ethnographies of violent events by trained social scientists and sophisticated journalists. He compares the substance of violent encounters across the range of settings in which they occur – from battlefields to city streets to sports arenas and dysfunctional households – in search of common, universal features. He concludes "this vast array can be explained by a relatively compact theory."[20]

Collins argues that violent episodes are marked initially by *confrontational tension*. People become edgy at the first hints that violence may be in the offing because humans are biologically disposed to cue into other humans. Encountering one another, humans note facial expressions and details of body language that prime them to do what humans ordinarily do best with others – *interact*. But, if these cues yield a sense this will not be a pleasant encounter, emotional energy is held in check and, with the release of adrenalin, is expressed as awkward tension and possibly fear.

This tension does not by itself lead to violence. Rather, *posturing* – words, gestures, and other actions designed to stand off the adversary – is a common next step and, in most cases, the episode moves no closer to outright violence than that.[21] However, the presence of certain variables – prior experience in resolving threats violently, orders from an authority figure to proceed, the presence of others encouraging violence, and/ or a highly dehumanized view of the adversary – may serve to overcome this hesitation so that *confrontational tension* is followed by a *burst of violence*. How violent will the violence be? That depends upon the extent to which still other variables further increase confrontational tension, for example, the *length of time anticipating violence* and *how helpless potential victims* are (greater helplessness ironically increases the ferocity of the violence).

Consider an example, provided by Collins, from Phillip Caputo's experience while a Marine platoon commander during the Vietnam war. In this scenario, his platoon approaches a village, Ha Na, after having been fired upon and briefly pinned down. However, once there they find no men of fighting age, only defenseless villagers:[22]

> Then it happened. The platoon exploded. It was a collective detonation of men who had been pushed to the extremity of endurance. I lost control of them and even myself. … [W]e rampaged through the village, whooping like savages, torching thatch huts, tossing grenades into the cement houses we could not burn. In our frenzy, we crashed through hedgerows without feeling the stabs of the thorns. … We shut our ears to cries and pleas of the villagers. One elderly man ran up to me, and, grabbing me by the front of my shirt, asked, "*Tai Sao? Tai Sao?* Why? Why?"
>
> "Get out of the goddamned way," I said, pulling his hands off. I took hold of his shirt and flung him down hard, feeling as if I were watching myself in a movie …. We passed through the village like a wind; by the time we started up Hill 52, there was nothing left of Ha Na but a long swath of smoldering ashes, charred tree trunks, their leaves burned off, and heaps of shattered concrete.
>
> … The platoon snapped out of its madness almost immediately. Our heads cleared as soon as we escaped from the village into the clear air at the top of the hill …. The change in us, from disciplined soldiers to unrestrained savages and back to soldiers, had been so swift and profound as to lend a dreamlike quality to the last part of the battle.

Collins's theory explains many things well. It especially gives an indication why a significant amount of violent encounters end, oddly enough, in either nonviolent episodes of posturing or in orgies of violence. He refers to the latter as *forward panics*. In such cases, prolonged instances of *confrontational tension* combine with the *relative helplessness* of the victims to produce especially explosive instances of violence.

However, accounting for wars and the immense violence within them also requires *macro-level* theorizing. A small number of scholars, starting with political scientist Quincy Wright at the University of Chicago, devoted great attention to war itself after the horrors of World War I. Wright established a multidisciplinary program there and directed graduate student research on war-related topics. His own magnum opus, *A Study of War*, was published in 1942.[23]

Though U.S. sociologists studied the military during and after World War II, the study of war itself never really caught on in sociology as it did in political science. An exception is Yugoslavian-born Siniša Malešević's attempt to make sense of his country's fragmentation and descent into savage ethnic-religious warfare. We now place our Spotlight on his macro-level sociology of war and violence.

Spotlight on Sociological Thinking and Research

Siniša Malešević, *The Sociology of War and Violence*[24]

Siniša Malešević's place of birth and boyhood experiences provide a personal point of departure for his analysis of modern war. Born in Bosnia-Herzegovina in 1969, he recalled being raised in a multi-cultural environment of Orthodox Serbs, Bosnian Muslims, Catholic Croats, and Jews. The ethnic violence and ethnic cleansing during the wars of Yugoslav succession (1991–1995) stood in stark contrast to his happy childhood memories of playing with kids of different ethnic backgrounds.[25]

Reviewing the slaughter in the Napoleonic wars (1803–1815), World Wars I and II, and the wars of Yugoslav succession, Malešević asks several questions: What is distinctive about war in the modern era? What causes war in these times? And, in an era espousing liberty, democracy, justice, and equality, why are otherwise peaceful citizens at times such willing killers of each other? The first two are macro-level questions, the latter is a micro-level one.

Malešević begins with an observation: despite all the warfare in human history, killing does not come naturally or easily to humans. U.S. Army Lt Col. Dave Grossman's well-known books (*On Killing* and *On Combat*), for instance, argue that inter-species killing is common, while intra-species killing is rare.[26] Grossman contends intra-species killing is dysfunctional from an evolutionary point of view, that is, the prospects of survival over the eons are grimmer for a species given to killing its own members than for species that do not do so. Among humans, however, the inhibition against the wholesale killing of other humans can be rather easily overcome. The techniques for doing so are well known: the killing needs be orchestrated by an accepted leader, done in the company of comrades, and involve a cause in which adversaries are marginalized as "others."

The answer then to the "why" and "how" of people as willing killers of each other contains a paradox: we are social creatures, and it is our very sociality that makes modern humans both compassionate altruists and enthusiastic killers. Malešević states:[27]

> We fight and slaughter best when in the presence of others – to impress, to please, to conform, to hide fear, to profit, to avoid shame and for many other reasons too. And it is these very same social ties that make us equally and often simultaneously martyrs and murderers.

Modern warfare combines two long-time historical trends, depicted in Figure 2.2: one, the organizational capacity to marshal and deliver lethal violence (bureaucratization of coercion) and, two, a supporting ideology strong enough to justify the use of this lethal capacity (centrifugal ideologization).

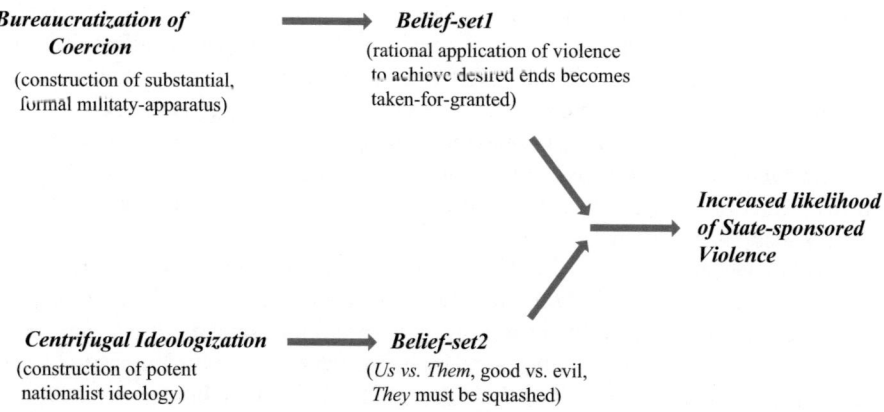

Bureaucratization of Coercion

(construction of substantial, formal military-apparatus)

Belief-set1

(rational application of violence to achieve desired ends becomes taken-for-granted)

Increased likelihood of State-sponsored Violence

Centrifugal Ideologization
(construction of potent nationalist ideology)

Belief-set2
(*Us vs. Them*, good vs. evil, *They* must be squashed)

Figure 2.2 Malešević's Theory of War and Ethnic Violence
Source: Authors' Creation.
Note: Constructed from Malešević's description of his theory.

Malešević's starting point, the "bureaucratization of coercion" is an unusual term. Precisely because bureaucracy is a form of social organization rationally constructed for purposes of efficiency, it is capable of very impersonal qualities. Its emphasis on orderly rules to routinize activity and on position-to-position rather than person-to-person linkages can convert its members into capable cogs in a well-oiled machine. This is abetted in modern societies by technological innovations increasing both the machinery of modern militaries and the lethal capacity of modern warfare (trains, ships, trucks, tanks, aircraft, machine guns, artillery, bombs, nuclear weapons, etc.). The resultant modern military, so organized and equipped, has the potential to unleash massive amounts of death and destruction. Curiously, at the same time, its bureaucratic nature allows for a rather callous, if not cavalier, approach to the killing and mayhem. In the bureaucratic world, Malešević reminds us, the mechanization of lethal violence depersonalizes the link between killer and adversary.

Having the ability to do so, however, does not always mean this capacity will be used. Paralleling the bureaucratization of society is some reverence for the principles of a common humanity. Hence, the wanton use of lethal military force requires the development of an ideology that neutralizes these values.[28] Malešević's answer begins with a modern society's need for some ideological glue to tie a diverse citizenry together "into ... entities able and willing to support war and other coercive causes when necessary."

Primary among these ideologies is *nationalism*, a "shared perception that 'our nation' is morally and ideologically right and that 'our' actions are universally justifiable."[29] A visceral sense of identity and belonging based upon community (a special, distinctive group), territory (attachment to a special place), culture (a shared language and way of life), history (a shared past and legacy), and an imagined future together may all figure in nationalism. Such notions of "a people" are the stuff of human history. What is distinctive in the modern era is how this sense of "a people" meshes, or not, with the power of the nation-state.

While vitally important, such feelings may lie politically inert in the absence of struggle – a situation challenging the well-being of the group and calling for some collective response. Malešević is careful to note that nationalism by itself does not lead to collective violence and war. It does however have the easy capacity to provide the imagery, language, and appeal for labeling exactly what is wrong and who, that is, which "other," is responsible for what has gone awry. Hence, while there are positive functions associated with nationalism, its downside is a propensity to serve as war's instigator and fellow traveler.

To kill in warfare, both at the individual and collective levels, it helps if the enemy can be defined as especially blameworthy or even undeserving of human respect. Such *dehumanization* is useful because humans in modern societies are socialized to cherish and treasure human life, or at least the lives of certain humans. In war, nationalism provides a way out of this dilemma by declaring fellow group members as morally just and the enemy as "monsters and animals ... whose actions prove their intrinsic inhumanity."[30] Ideology thus allows citizens to become willing, if not righteous, killers, or at least fervent supporters of such action on their behalf.

Man/Woman, the State, and War

Political scientist Kenneth Waltz has posited a multi-level treatise on warfare.[31] He begins at the micro-level with *man*, by which he meant *human nature* rather than males, proceeds to *the state*, the political apparatus of a large society, and on to *war*, inter-state relations of the violent variety. Waltz first reviews two polar-opposite views of human nature – an *optimist's* view: humans are pacifists by nature, life in society introduces strife and warfare; and, a *pessimist's* stance: humans are brutish and life in society is necessary to constrain the mayhem. In either case, he contends, human nature is whatever it is, so the *analysis of warfare needs to take place at the macro-level*. Importantly, the *domestic character* of a state's political apparatus, Waltz contends, is most certainly reflected in *how that state conducts business* in the international arena.

Waltz's use of the word "man" in the title of his 1959 book eventually prompted lively responses from a host of feminist scholars, who took the choice of the term, intended or not, to mean "males." Political philosopher Jean Bethke Elshtain, for whom Waltz was one of her professors during her doctoral program at Brandeis University in the early 1970s, reviewed these rejoinders 40 years later in a commentary, "Woman, the State, and War."[32]

Radical feminists, she states, have argued that men historically have had their say and, given frequent and tragic warfare under their watch, now must be relegated as completely as possible to the sidelines. On the other hand, *liberal feminists*, so-called because of their classical liberal-like emphasis on individual rights, have demanded equal opportunity for women in all spheres of public life. A common thread running through these divergent critiques is that the political worlds of which Waltz spoke, and the consequent likelihood of warfare, would be very different with the participation of women.

Elshtain herself was a frequent critic of *ideological feminism*, a political narrative unwaveringly requiring "the storyline of history be the history of men oppressing women."[33] Unlike radical feminists, she did not see male aggression as the sole driving force of conflict in human societies. A more dominant role, she argued, is played by *sacrifice*, the willingness to give of oneself for the collective, a role also played historically by women in wartime as nurturers and mothers of soldiers, and sometimes as the very symbol of the collective itself. The reduction of warfare in her view thus calls for the active participation of citizens – men and women – devoted to a more nuanced concept of justice: a *just society*, especially one attentive to elevating the truly dispossessed, leads to a *just world*.

Elshtain challenges feminist scholars to empirically establish their claims. Political scientist Mary Caprioli for years has devoted her research agenda to that end, and she and a team of researchers headed by Valerie Hudson have continued her earlier efforts.

Spotlight on Social Science Thinking and Research

Mary Caprioli, "Gendered Conflict"[34] and Valerie Hudson et al., "The Security of Women and the Security of States," Chapter 4 in *Sex and World Peace*[35]

Mary Caprioli is Chair of the Department of Political Science at the University of Minnesota-Duluth and Director of its International Studies program. She begins her research with a micro-level observation: many public opinion polls and studies have found that women are less likely than men to favor war and, in almost all realms of foreign and domestic policy, to prefer less belligerent policies than do men. Hence, the institutional willingness to use force as a matter of policy – a macro-level variable – should decline as women obtain greater social, economic, and political equality.

Caprioli tests the hypothesis that increased gender equality produces lower levels of militarism. To do so, she drew data from the Military Interstate Dataset (MID) for the years 1960 to 1992. MID contains data on incidents of 2,187 international disputes for 159 states during this time, allowing for classification of each state's response to each incident on a five-point scale, ranging from "used no military action" through "use of threats of action," "display of force," "use of force," to "war."

To explain variation in militarism, Caprioli includes measures of social equality: the fertility rate (the number of children per women 15–44 years of age – lower is better), political equality (the length of time between women's suffrage and the international dispute, and the percentage of women in society's legislative body), and economic equality (percentage of women in the paid labor force). Control variables include the number of contiguous states, the number of alliance partners, relative wealth, and the extent of the features of democracy. The inclusion of these control variables allows for the elimination of them as potential competing explanations.

Caprioli found that increased levels of gender equality – social, economic, and political – are significant predictors of reduced militarist responses. One important implication of her work is that foreign policy initiatives aimed at reducing incidents of international conflict might do well to focus on supporting programs and organizations devoted to improving gender equality.

Valerie Hudson is Professor and Chair of the George H. W. Bush School of Government and Public Service at Texas A&M University. She has a long history and record of research on national security matters and issues affecting the security of women. Psychologist Bonnie Ballif-Spanville is Professor Emeritus, Brigham Young University, where she specialized in the comparative study of domestic violence worldwide. Geographer Chad Emmett is Associate Professor at Brigham Young University and has served as the Director of the Middle East/Arabic Studies Program. And, we already have met Mary Caprioli. Since 2001, this team has been the driving force in putting together the WomanStats Database (www.womanstats. org), the most comprehensive available database on the status of women cross-nationally, containing over 320 variables for 170 nations.

Hudson and her associates define sex as the biological differences between women and men, and gender as socially defined differences between them in a society. A gendered perspective is one that considers both sex and gender in accounting for some phenomenon. They define inequality as the relative power a man or woman has in each society, an indication of the amount of subordination.[36]

Standard macro-level theories accounting for warfare – democratic peace thesis, scarcity of resources, wealth-gap hypothesis, Huntington's clash of civilizations, etc. – are gender neutral. In contrast, Hudson and associates' point of departure is at the micro-level: how are a society's women treated relative to men in everyday life? When women are badly subordinated and incur violence in everyday interactions, the authors theorize, this pattern permeates structures of the larger society and establishes a blueprint of generalized violence and exploitation at the societal level, in turn, affecting that society's posture toward other nations. Thus, they hypothesize: does the security of women affect the security of states?

Specific data for testing such hypotheses are not available in the growing number of datasets on the status of women, hence the team's devotion to constructing WomanStats. In it, they include specific measures of the physical security of women[37] (PSOW), namely, reports of domestic violence, rape, marital rape, and murder as well as laws pertaining to these and whether these laws are enforced. Also included are measures of a preference for male children (the relative cultural valuation of male and female life) and variations in the sex ratio. Usually, the sex ratio at birth is about 106 male babies for every 100 females, so if a country has a sex ratio for babies of, say, 130, this most likely is indicative of gender-selective abortion or female infanticide. Finally, measures are included for maternal mortality, compliance with laws regarding the trafficking of females, discrepancies in educational levels by gender, and governmental participation by women.

For the dependent variable, the security of states, Hudson and her associates employed standard measures to enhance comparability with other studies.[38] The three were: a measure of state peacefulness (GPI, the Global Peace Index), degree of behavioral deviancy (SOCIC, the States of Concern to the International Community), and NR, the Neighboring-Country Relations scale. They also included control measures for each of the competing nongendered theories in the international relations literature for state security. The team then conducted appropriate bivariate and multivariate statistical analyses of all these variables.

The list of findings is long and instructive but can be correctly summarized as follows. At the bivariate level, PSOW, as hypothesized, is positively and strongly correlated with each of the three measures of state security, GPI, SOCIC, and NR (as are several other variables). Multivariate analyses therefore should give an indication if the cluster of security of women measures explains state security better than the control variables, that is, the conventional nongendered theories of state security.

Here, to cut to the quick, they show the PSOW variable is a more significant predictor of state security than are any of the control variables. Hudson and her associates conclude:[39]

> [Our] results indicate that *if a scholar or policy maker had to pick one variable* – level of democracy, level of wealth, prevalence of Islamic culture, or the physical security of women – to assist him or her in predicting which states would be the least peaceful or of concern to the international community or have the worst relations with their neighbors, *the physical security of women would be the best choice.*
>
> [Emphasis added.]

These findings by Caprioli and by Hudson and her associates merit a high level of credibility because of their theoretical reasoning, devotion to data collection, inclusion of control variables, and use of multivariate statistical analyses. The basic theory holds up well. Simply stated, *more secure women, more peaceful world.*

Questions for Discussion

1 Visualize yourself reporting for military basic training (any era, any military). In a sentence or two for each, how would a micro-, a meso-, and then a macro-level theory approach an explanation for what led you to sign up. With which level of explanation do you feel the least comfortable? Why?

2 Using either *integration* or *regulation*, explain to an intelligent but uninformed listener how the relative health of the military, measured at the meso-level, has relevance for suicides in the military, that is, what do you know about suicide having meso-level information that you would not otherwise know?

3 Malešević notes that inter-species killing is common but intra-species killing is rather rare. Yet, humans, especially in modern societies, are able to kill other humans rather easily and on a very large scale. How is this so? Are you persuaded by his explanation?

4 Take a careful look at the WomanStats database (www.womanstats.org), especially the codebook, to get a first-hand feel for the data. What are your impressions? After seeing their data and revisiting their conclusions, in what ways would you challenge "gender-neutral" theories of war?

Notes

1 See, for instance, Lawrence Bergreen, 2004, *Over the Edge of the World: Magellan's Terrifying Circumnavigation of the Globe*, New York: HarperCollins Publishers.
2 See any introduction to a sociology textbook for a more in-depth discussion. We are partial to Rodney Stark, 2006, *Sociology*, 10th edition, Boston, Mass.: Cengage Learning.
3 Max Weber, [1915] 1947, *The Theory of Social and Economic Organization*, translation by A.M. Henderson and Talcott Parsons, New York: Free Press.

4 Sanford K. Dornbusch, 1955, "The Military Academy as an Assimilating Institution," *Social Forces* 33 (May): 316–321.
5 George Herbert Mead, 1934, *Mind, Self and Society*, edited by Charles W. Morris, Chicago: University of Chicago Press.
6 The monologue identifies seven stages of life: infancy, childhood, adulthood (lover, soldier, wise person), old age, and dementia/death, http://shakespeare.mit.edu/asyoulikeit/asyoulikeit.2.7.html.
7 George R. Mastroianni and Wilbur J. Scott, 2011, "Reframing Suicide in the Military," *Parameters*, 41 (Summer): 6–21. Featured in Mark Thompson, "A Scary New Way of Looking at Military Suicides – In the Mirror," *Battleland*, 10 November 2011, http://battleland.blogs.time.com/2011/11/10/a-scary-new-way-of-looking-at-military-suicides-%e2%80%93-in-the-mirror-2/
8 Émile Durkheim, [1897] 1951, *Suicide: A Study in Sociology*, trans. by John A. Spaulding and George Simpson and with Introduction by George Simpson, New York: The Free Press of Glencoe. Available more recently as Emile Durkheim, [1897] 1979, *Suicide: A Study in Sociology*, The Free Press, Kindle edition.
9 Most of this biographical information is drawn from Robert Alun Jones, *Emile Durkheim: His Life and Work, 1858–1917*, http://durkheim.uchicago.edu/Biography.html.
10 Durkheim's stellar career came to an early end. After most of his prize students in sociology were drafted into the military and killed in 1916, Durkheim withdrew from academic life and died of a stroke in 1917.
11 Michael Friendly, 2007, "A.-M. Guerry's Moral Statistics of France: Challenges for Multivariable Spatial Analysis," *Statistical Science*, 22 (3): 368–399.
12 Durkheim, 1951, op. cit., pp. 145–151, 297–325.
13 Ibid., Chapters 2 and 3.
14 Ibid., Chapter 5.
15 Ibid., Chapter 4.
16 Ibid., pp. 219–220.
17 Ibid., pp. 237–239.
18 Jeffrey W. Reimer, 1998, "Durkheim's 'Heroic Suicide' in Military Combat," *Armed Forces & Society*, 25 (Fall): 103–120. The Congressional Medal of Honor, instituted during the Civil War, is the nation's highest award for heroism in combat. The act of heroism must be "above and beyond the call of duty."
19 Mastroianni and Scott, op. cit., pp. 12–19.
20 Randall Collins, 2008, *Violence: A Micro-Sociological Theory*, Princeton, N.J. and Oxford, U.K.: Princeton University Press, pp. 1–8.
21 The phrase, "fight or flight," often is tossed about as the only two options at this point. As Collins and others have shown, posturing is a more common response that either of these two.
22 Phillip Caputo, 1977, *A Rumor of War*, New York: Ballantine Books, pp. 287–289, quoted in Collins, 2008, op. cit., p. 87.
23 Quincy Wright, 1954 [1942], *A Study of War*, second edition, Chicago: University of Chicago Press.
24 Siniša Malešević, 2010, *The Sociology of War and Violence*, New York: Cambridge University Press.
25 Siniša Malešević, personal communication, 2 May 2012.
26 Dave Grossman, 1996, *On Killing: The Psychological Cost of Learning to Kill in War and Society*, Boston: Little, Brown and Company; Dave Grossman and Loren W. Christensen, 2004, *On Combat: The Psychology and Physiology of Deadly Conflict in War and in Peace*, Millstadt, Ill.: Warrior Science Group.
27 Malešević, op. cit., p. 3.
28 Malešević, op. cit., p. 130, argues that the very necessity of this stands as proof that humans are not by nature violent and warlike.
29 Ibid., p. 142.
30 Malešević, op. cit., pp. 142–143.
31 Kenneth Waltz, [1959] 2001, *Man, the State, and War: A Theoretical Analysis*, New York: Columbia University Press.
32 Jean Bethke Elshtain, 2009, "Woman, the State, and War," *International Relations* 23 (June): 289–303.

33 J. Daryl Charles, 2006, "War, Women, and Political Wisdom: Jean Bethke Elshtain on the Contours of Justice," *Journal of Religious Ethics* 34 (June): 239–269, p. 251.

34 Mary Caprioli, 2000, "Gendered Conflict," *Journal of Peace Research*, 37 (January): 51–68.

35 Valerie M. Hudson, Bonnie Baliff-Spanvill, Mary Caprioli, and Chad F. Emmett, 2012, *Sex and World Peace*, New York: Columbia University Press (Kindle edition).

36 Ibid., location 227.

37 Ibid., location 1182–1331.

38 Ibid., location 2248–2262.

39 Ibid., location 2298.

Recommendations for Additional Reading

Miguel A. Centeno and Elaine Enriquez, 2016, *War & Society*, Cambridge, UK and Malden, Mass., USA: Polity Press.

(A readable and very useful sociological introduction to war with considerations at all three levels of analysis)

Dave Grossman, 2009, *On Killing: The Psychological Cost of Learning to Kill in War and Society*, New York: Back Bay Books.

(Grossman's theory and methods have been criticized, but this is a popular introduction to a study of "killology")

Valerie M. Hudson, Donna Lee Bowen, and Perpetua Lynne Nielson, 2020, *The First Political Order: How Sex Shapes Government and National Security Worldwide*, New York: Columbia University Press.

(Hudson's initial work sparked surprise and interest; this follow-up study has even better data and stouter findings)

Jeffrey W. Reimer, 1998, "Durkheim's 'Heroic Suicide' in Military Combat," *Armed Forces & Society*, 25 (Fall): 103–120.

(An analysis of a sample of U.S. Medal of Honor winners who gave their lives to save others)

3   Biological and Cultural Bases of Warfare

Reader's Guide: Is war prevalent because humans are naturally warlike or because cultural elements lead them down that path? Recent developments in genetics and socio-biology have moved far beyond the traditional nature vs. nurture debate, and we bring the reader up to date. Much can be learned too from the incidence and intensity of warfare among hunting-gathering, horticultural-herding, agrarian, and industrial societies. Variations suggest strong social and cultural elements. Another feature of war merits special examination: societies vary in how they do gender, and in how they do war, but almost always do war with men. Our first Spotlight thus reviews the substantial literature on gender and warfare and how each impacts the other. The second Spotlight examines the social organization of a small slice of warfare – a warship in World War II. The third addresses military culture. Though there are distinct differences between civilian and military cultures, there is no one military culture, but many in response to several factors, none more dramatic than whether a military unit is in combat or in garrison.

Is human behavior innate or learned? Most sociologists would answer without much hesitation: it is learned! Their assuredness stems from the often-rancorous nature vs. nurture debate that raged at the turn of the 20th century. On the *nature* side was physician-turned-psychologist William McDougall. He argued that instincts drove human behavior and complex behaviors could be explained by combinations of instincts.[1] On the *nurture* side was physicist-turned-anthropologist Franz Boas.[2] He contended that humans were "blank slates" who learn as members of society how to act – hence the immense variations in human behavior from one society to the next. In the years before World War II, the debate was resolved in favor of the nurture side, at least as far as social scientists were concerned, and sociologists mostly have moved on without any further discussion.

The nature vs. nurture debate had its origins in the fascination with race in Western Europe and the New World. European colonizers encountered African and New World societies previously unknown to them, societies that appeared to them less complex than Western European ones. Why were these societies different? During the 1850s the French writer and diplomat, Arthur de Gobineau, provided an explanation based on the racial appearances of peoples.[3] His musings attributed differences among human societies to the color of their inhabitants (see our further discussion in Chapter 6).

About 20 years after the appearance of Gobineau's work, a self-taught English ethnologist and archeologist, E. B. Tylor, provided a rebuttal. Referring to his travels and

DOI: 10.4324/9781003282549-3

fieldwork among the Anahuac Indians of Mexico, he portrayed types of societies as survival adaptions in response to their different environmental settings.[4] Further, current primitive societies, he argued, look like what advanced societies once were and, given the human capacity for adaptation, progress by primitives to more advanced stages was inevitable.

On the American front, a similar argument was advanced in the 1870s by Lewis Henry Morgan, a Connecticut lawyer.[5] Morgan had befriended nearby Iroquois and Seneca tribes during his legal defense of their land claims. He learned about their kinship structures to sort out how treaties might apply to current tribal members, and later worked as a field researcher among Plains tribes for the Smithsonian Institution. Morgan laid out a theory of *social evolution* and identified *technological innovation* as its driving force. He noted that American Indian societies already represented points all along the continuum: the Comanche and Apache were hunter-gathers, the Cherokee and Chickasaw were farmers, and the Aztecs and Mayas, independently of the Old World, had developed empires.

However, it was Franz Boas who rather single-handedly put anthropology on the map as a scientific and academic discipline. His decades of fieldwork with the Baffin Island Eskimos and the Kwakiutl Indians of British Columbia, along with his founding of a prestigious Department of Anthropology at Columbia University, earned him the title of "Father of Modern Anthropology." Taking on the Gobineaus of the world, Boas argued that *race* was a useless concept. It is not the "human mind" that explains culture, he stated, but the other way around: human capabilities and behaviors are consequences of the cultures in which people are raised.[6]

We have devoted space to all this because sociologists carry the memory of this contentious history and its unsavory political baggage. This has left many of them very suspicious of explanations that human behavior has an extensive biological basis. We show how this might be recast to incorporate recent evolutionary and genetic thinking into sociological ways of looking at the world.

Nature via Nurture

In 1997, sociologists Richard Machalek and Michael Martin examined the 20 best-selling textbooks used in Introductory Sociology courses.[7] They did so with an eye toward what the texts have to say about biology and culture and about an emerging subfield in evolutionary biology, *socio-biology*.[8] From these (and other sources), they summarize the Standard Social Science Model (SSSM): (1) human social behaviors are learned; (2) the biology of the brain shapes human behavior *only* by endowing humans with the capacity for culture; (3) there is insufficient variation in the human genome to produce the full range of social behaviors found within and across cultures; (4) culture, not biology, explains variation in behavior within and between societies; and (5) features of human societies are not reducible to either psychological or biological variables.

Most sociologists endorse SSSM quite fully, Machalek and Martin contend, because they see, as the only alternative, the antithesis of these points, that is, human social behaviors are innate, genes wire the human brain to express unalterable patterns of behavior, and so forth. This latter position might be termed the Exclusive Genetic Determinist Model (EGDM).[9] The question is: How well do SSSM and EGDM stack up as theories in the light of the evidence?

In the 1930s, biologists combined ideas from Charles Darwin's evolutionary theory with those of Gregor Mendel's experimental genetics to produce a neo-Darwinian theory of evolution. The synthesis increased the power of Darwin's theory by providing his principle of evolutionary change, *natural selection*, with the genetic mechanisms through which it might take place. This neo-Darwinian version has become the dominant paradigm within the biological sciences. Even sociologists, comfortable in the position that social evolution was a different matter, accepted its authenticity. The discomfort came later as geneticists, evolutionary psychologists, and socio-biologists extended this thinking into the social realm.[10] Most daunting for sociologists is the worry that genetic considerations might wrongfully supplant social explanations for human behavior.

In any case, it is helpful to focus on the distinction between *genotype* (the chromosomal content of a gene or gene set) and *phenotype* (a consequent trait or response associated with a gene or gene set). Importantly then, the *norm of reaction* refers to the amount of variability in the phenotype(s) associated with any genotype. Some genotypes have a very narrow norm of reaction, that is, their chromosomal content consistently produces a single, identifiable phenotype. For example, eye color is a response to the interaction of several genes, and specific combinations invariably produce brown (or blue) eyes. However, for other genotypes the norm of reaction can be quite wide, that is, a gene or set of genes can produce a range of phenotypes. This is especially so when, in the words of socio-biologist E. O. Wilson, there is "joint influence of heredity and environment."[11]

The norm of reaction most likely is an unfamiliar concept to most sociologists, and its absence tends to paint them into the SSSM corner. Incorporating the concept, however, provides a basis for conversations between sociology and socio-biology. What might be some points of convergence? Wilson, a myrmecologist (a biologist who specializes in the study of ants), has underscored the primacy of culture in accounting for human behavior.[12] Other points of convergence are: human behavior relies upon learning; to the extent biology influences human behavior, the influence is not fixed; and, societies are complex social systems that cannot be adequately explained only by biology or psychology.

However, there are some sticking points. Sociologists are likely to think the human genome enables *equipotentiality*, that is, humans' marvelous ability to adapt means that virtually any adaptation (phenotype) is possible and viable. This would mean members of a culture, potentially at least, could just as easily acquire adaptive behaviors *a*, *b*, or *c*, or *m*, *n*, or *o*, or even *x*, *y*, or *z*. In rebuttal socio-biologists espouse *prepared learning*, that is, human genotypes predispose them to acquire some behaviors (phenotypes) more easily than others. This suggests an inborn propensity to learn *a*, *b*, or *c* swiftly and decisively, while *m*, *n*, and *o* also are possible but have steeper learning curves, and *x*, *y*, and *z* are nonstarters. Prepared learning thus does challenge the notion that "anything goes."

The Bio-Social Basis of War

In 1941, the *American Journal of Sociology* published a special issue on "Theorizing about War." In it, British anthropologist Bronislaw Malinowski[13] – then at Yale University – noted there were two competing schools of thought. At stake was the question of whether war is at base a biological phenomenon or a social-organizational one. On the latter side were theorists who believed in "the *primeval pacifism* of man."[14] Malinowski himself argued that *endemic warfare* is a recent phenomenon, first appearing in agrarian societies only 5,000 or so years ago. Johns Hopkins University biologist, Raymond Pearl, expressed the counter-position. The *will to live*, he maintained, "shared by plant and animal, amoeba and man," is a biologically rooted impulse underlying all aggressive and ultimately warlike behavior.[15] Here, warfare is seen as an endemic feature of human life spanning millions of years, obviously preceding *Homo sapiens*.

Wilson has provided a new starting point, most recently in his book, *The Social Conquest of Earth*.[16] Wilson notes that species which form large, integrated societies persisting across generational lines – the characteristic of *eusociality* – are rare. Among the millions of vertebrate and invertebrate species only a handful have done so, most notably ants, wasps, and termites, and humans. He contends *eusocial* species have a tremendous competitive advantage over those that operate either as solitary or loosely coordinated individuals and, consequently, are the very ones who dominate the earth.[17] There are of course critical differences between how and why bees and humans form and carry out *eusocial* arrangements.

Among bees, the queen bee carries sperm, acquired in an encounter with a male bee, with her in a little sac as she flies to a new territory. She subsequently can fertilize herself from the sac (in the forever absence of her male "partner") to produce castes of daughter-worker bees who carry out the hive's division of labor. New generations of worker bees remain in the hive, creating and replicating complicated arrangements for foraging, feeding, defending the hive, and so forth. Such "bee civilizations" are built by *instinct* in the truest sense of the word: the altruistic worker bees are a phenotypical extension of the queen bee with an exceptionally *narrow norm of reaction*, and so execute their tasks, in Wilson's words, like "little robots." This arrangement among ants, bees, termites, and wasps has evolved over millions of years, with natural selection favoring those "queen bees" who over eons developed progressively *eusocial* colonies.

Human evolution is, again in Wilson's words, "fundamentally different." The closest ancestor of modern-day humans (*Homo sapiens*) is *Australopithecus*, only one of several hominid species from which humans could have evolved. In a nutshell, *Australopithecus*'s advantage, and then that of *Homo sapiens*, was increasing *eusociality* built upon natural selection taking place at two levels. At the individual level, natural selection favored those who foraged, thrived, and therefore reproduced more successfully, a process that genetically privileged *individuals* acting greedily in their own self-interest. However, effectively cooperating *groups* of individuals also staved off external threats, foraged, and reproduced more successfully than groups of humans who were less cooperative. Natural selection thus has favored *dominant individuals* at one level, and *effectively cooperating clusters* of them at another. The genetic expression at the phenotypical level thus is creative rather than robotic. *Eusociality* in successful human groups reached its most intense expression as *Homo sapiens* developed culture and language.[18]

There are many implications, but our interest is how all this impacts war. Wilson states:[19]

> Group-selected traits typically take the fiercest degree of resolve during conflicts between rival groups. ... To form groups, drawing visceral comfort and pride from familiar fellowship, and to defend the group enthusiastically against rival groups – these are among the absolute universals of human nature and culture. ... The instinct that binds them together is the biological product of group selection. People must have a tribe.

Thus, Wilson is saying, here are the *a, b,* and *c*'s of our biological heritage: we are decidedly *eusocial* rather than solitary creatures who readily, compulsively even, form groups; we do so even though we must subordinate our own desires somewhat to make the group life we crave work well; to ignore this balance might be an *m, n,* or maybe even a *z*; and, this drive to form robust groups (and societies) necessitates strong in-group/out-group boundaries. This may lead to a range of outcomes, from friendly competition to lethal conflict between resultant groups.

War by Type of Society

Anthropologist Keith Otterbein has traced the origins and history of warfare in human societies.[20] He begins by defining war as "armed combat between political communities." There are advantages to this definition. It applies to organized violence in all types of societies between political entities using weapons. It is specific enough to distinguish violent skirmishes among isolated individuals from organized encounters of such and is broad enough to include clashes between the armies of two nations and the shooting of arrows at each other by members of two conflicted kinship groups.

His "waves of warfare" theory is depicted in Figure 3.1. The key independent variable in Otterbein's model is the *type of society* – hunting-gathering, horticultural-herding, agrarian, or industrial – which he uses to explain changes in the *intensity of warfare*. He argues that pre-agrarian warfare, while not as lethal as later wars, nonetheless was often deadly. Archeological evidence, he argues, does not support the contention that pre-agrarian humans were reluctant to kill other humans. This is especially so of big-game hunters 50,000 years ago. The weapons and social coordination required to hunt woolly mammoths successfully could also be used to kill rival clans or groups of strangers who encroached upon favored hunting areas, and, Otterbein says, they often were. This line of warfare declined with the overhunting of big game, making possible, or even requiring, a heavier reliance on gathering, the hunting of smaller game, and then gardening.

Ultimately, these changes set the stage for the transition some 10,000 years ago to the full-time cultivation of plants and the domestication of animals as sources of food. *Endemic warfare*, Otterbein observes, is not compatible with the development of horticultural-herding endeavors, which require greater stability to complete cultivation cycles. Over time, certain horticultural-herding societies became increasingly complex, first in the Fertile Crescent – present-day Syria and Iraq – then, principally through diffusion, in other areas. The new agrarian societies were marked by increases in the scale of agriculture, population, and hierarchies of authority to oversee the production and distribution of goods. Writing – the etching of words on tablets or parchment – developed about this time, though there is debate: Did the need for written records to keep track of all this come first or was writing subsequently applied to record-keeping?

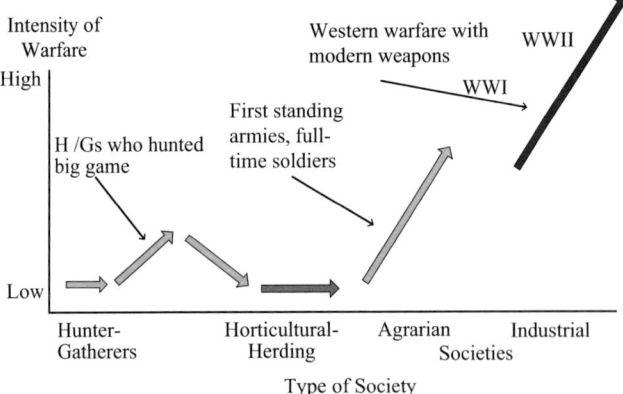

Figure 3.1 Otterbein's Waves of Warfare

Source: Authors' Creation.
Note: Figure constructed from Otterbein's description of his theory.

Agrarian societies also enjoyed enough surplus production to pull some people, typically men, out of agricultural work and reassign them to other tasks, principally the manning of a standing army. Such societies soon became quite capable of plundering their neighbors to acquire goods and slaves. Ironically then, the origins of *civilization* – societies marked by social and political complexity and rudimentary literacy and science – also are associated with *endemic warfare* and *slavery*. Finally, industrial societies continued the war-faring trend. They are not more savage than agrarian ones but have conducted warfare with monstrously more potent weapons.

War and Gender

There is an intriguing puzzle associated with gender and warfare. Societies tend to impose upon their members stringent gender phenotypes, although exactly what these gender traits and roles consist of varies significantly from society to society. And, as we have seen, warfare of some sort is a feature of virtually all known societies, although exactly what it consists of and how it is done varies considerably across time and societies. However, almost without exception, warfare has been the *near-exclusive domain of males*, that is, societies have devoted considerable time and effort to ensuring that their fighting ranks literally were "manned." Except for warfare, there is almost no activity apart from childbearing that is done almost exclusively by females or males everywhere. We turn our spotlight to a comprehensive treatment of this puzzle.

Spotlight on Social Science Thinking and Research

Joshua Goldstein, *War and Gender: How Gender Shapes the War System and Vice-Versa*[21]

The child of parents who were both microbiologists, Joshua Goldstein grew up on the campus of Stanford University where he "picked up a feeling for the world of natural science." This would come in handy later in life when, as a political science professor at the American University in Washington, D.C., he would tackle the puzzle of war and gender. Providing some answer, Goldstein notes, draws upon a large, complex body of literature spanning the disciplines of biochemistry, anthropology, psychology, sociology, political science, and history. He advises the reader that he has never served in the military and that he begins the study with an admiration for feminist theories critiquing standard answers to the puzzle. He promises, however, to follow the science wherever it takes him.

Goldstein begins by defining his terms. It is common to use sex to denote the biology of being female and male,[22] and gender to indicate culturally created definitions of femininity or masculinity. He argues this creates an arbitrary dichotomy and gives the impression sex is rigid while gender is flexible. While notions of femininity and masculinity do vary tremendously across cultures, he asserts it is more helpful to think of it this way: "biology provides diverse potentials, and cultures [creatively] limit, select, and channel them," that is, the norm of reaction here is not as narrow as often portrayed. He uses gender to refer to masculine and feminine roles produced by the dynamics of biology and culture, and sex to sexual behaviors.

War is lethal intergroup violence, and the war system refers to how societies organize themselves to participate in war.

This much we know about physical violence: whether interpersonal or intergroup, it is much more likely to be carried out by males, especially younger adult ones, than by females. This has led to the contention that testosterone may be the culprit. This hormone is secreted by the testes and ovaries in humans (and other mammals), though at higher levels in males than females (and in younger adult males than older ones). Higher levels of testosterone in males of primate species (the order Primates of the class Mammalia includes lemurs, tarsiers, monkeys, apes, and humans) produce secondary sex characteristics, that is, taller stature, heavier musculature, a deeper voice pitch, and so forth, than found in females of the same species. Might testosterone levels explain why warfare has been the near-exclusive domain of men?[23]

Some find studies of laboratory rats suggestive. For example, male rats fight less often when castrated (hence, reducing the release of testosterone), but once again become feisty when injected with testosterone. Female rats given testosterone exhibit more aggression and at levels usually observed in male rats. There are some issues though with generalizing these findings. Similar studies among primates do not show the same findings, and castration (usually by accident) in human males does not necessarily reduce aggression. Finally, there is evidence in humans for reverse causality, that is, acting aggressively can result in an increase in testosterone. Bottom line: the relationship between testosterone and aggression in primates is much more complicated – has a wider norm of reaction – than the rather narrow one found in laboratory rats.

A related possibility is that the XY chromosomal pairing (the genetic initiator of cellular changes associated with being biologically male) might trigger more aggressive behavior than the XX pairing (the gene pairing for being biologically female). Physical anthropologist Richard Wrangham at Harvard University contends so, basing his claims on his studies of chimpanzees. Chimps, humans' closest genetic primate relatives, are not the cute, peaceful little primates they sometimes are thought to be.[24] Male chimps are quite aggressive. Unified groups of them work together to kill neighboring groups of chimps to protect territory and acquire females – certainly qualifying as lethal intergroup violence – and routinely assault females and other males in their own group when frustrated or sexually aroused. Wrangham argues for a genetic basis for this in chimps and, by extension, in humans.

Bonobos also share an extremely close genetic relationship with chimpanzees and humans. The DNA (deoxyribonucleic acid, molecules containing genetic "instructions") in humans, chimps, and bonobos is 96 to 99% identical (depending on the comparison criteria). However, chimps and bonobos exhibit very different gendered behaviors. Bonobos, male and female, are more playful than aggressive. Goldstein describes them as "radically promiscuous" and females typically are bisexual. When squabbles arise over food or other issues, male bonobos usually defer to the females, who often have sex with each other to "smooth conflicts" within or between groups. These varying behaviors among primates who are genetically so closely related suggest no consistent explanation for warfare is likely to be found here.

Although warfare has been the near-exclusive domain of men, instances of women excelling in combat negate the argument they have been excluded because they are not capable. Two recent, more fully documented cases are the army of the Dahomey Kingdom (present-day Benin) in 18th and 19th- century Africa, and the Soviet military in World War II. In the former, women served in gender-segregated units making up from one-tenth to one-third of the total army.[25] The women's corps dressed and behaved like their male counterparts, were armed with muskets and swords, and shared regularly in combat. They were respected for bravery in battle and their cruelty with captives often exceeded that of the men. The Dahomey army ended in defeat at the hands of the French colonial army in 1892 after several bloody battles.

In the case of the Soviets,[26] their army early in World War II took a savage beating at the hands of the invading German blitzkrieg (see our discussion in Chapter 8). Out of necessity rather than by design, women were mobilized to fill the ranks. While they performed many roles, hundreds of thousands of them were included in combat missions as snipers, machine gunners, anti-aircraft gunners, and pilots and bombardiers in tiny all-women air force units, and so on. Although some of the lore may have been amplified for propaganda purposes, the consensus is that women proved themselves and gained "acceptance and even admiration" from Soviet military men who were "initially skeptical or hostile" to the idea.

So, the puzzle remains. Goldstein turns to culturally imbued socialization processes. Until the 19th century, human societies required very high birth rates to offset very high death rates (the *demographic imperative*) – one reason societies historically have made a big deal out of gender. Meeting this imperative requires a certain amount of gender polarization, that is, the organization of social life around female–male differences to emphasize childbearing. Until the age of seven or eight, testosterone levels in little boys and girls are nearly similar. As they reach ten or so, differences in the production of testosterone intensify secondary sex characteristics – physical traits noted above associated with being male and female.

In "primitive" societies where warfare is less advanced, there are advantages to being bigger, more muscular, and not being frequently pregnant. Societies develop rationales such as these and others to account for gender polarization, the amount of which almost always exceeds that required to meet the demographic imperative. This polarization often is achieved by laying out different routines for little girls and boys, marked by higher levels of rough play for boys than for girls and rites of passage for boys and girls, converting the former into frequent warriors and the latter into frequent mothers. This is not achieved lightly. A concerted effort and mix of potent positive and negative sanctions are required to pull this off. The result is gender roles that often are the antithesis of each other.[27]

There is cross-cultural variation in exactly how the above scenario plays out, but generally the more gender polarization, the more warlike. The direction of causality very likely runs the other way as well, that is, societies marked by the relative absence of war require less gender polarization. Conversely, the Sambia of New Guinea and Yanomamo of Brazil and Venezuela are both notably, if not violently, sexist and extremely given to vicious inter-group violence. Among the former, boys are taken from their mothers at the age of seven or so and tutored exclusively by older males to become warriors. This period of tutelage is marked by homosexual activity believed to enhance masculinity. In the latter, families practice female infanticide until the first son is born.

Finally, industrial societies, with their very low birth and death rates, require much less gender polarization, but do so to some extent, largely out of custom. Notably, post-industrial societies are the very ones who have resorted to gender-integrated all-volunteer forces in past decades (more about this in Chapters 5 and 7).

Goldstein's analysis now is twenty-some years old but still stands as a comprehensive treatment of the topic. It alerts us that the issue requires a multi-level approach spanning several disciplines and that, while some answers are better than others, there are no easy ones.

Military Culture(s)

In the social sciences, the term *culture* most simply refers to all the "stuff" members of a society need to know to live within it. This begins with having an intimate knowledge of the social settings in which one circulates. People must become very adroit at moving day-in and day-out from one social space to another, each time changing costumes, affect, and actions to match requisite expectations. How do they know how to do so?

The term *socialization* denotes the process humans go through to gain requisite mastery of these details. This begins in the first year of life. The acquisition of language by infants is central to their someday having the knowledge and skills to become fully functioning human beings. This is not something that happens spontaneously but is developed by imitating and interacting with other humans. By age one the little person has begun organizing the vocal sounds to catch the attention of others, and by two or three pretending to be the others in her or his life. Youngsters reach a crucial level of social maturity as they gradually can *take the role of the other*, that is, are able to visualize the expectations and roles at play in each social setting.

Sometimes, socialization is not simply a matter of imitation but of *reward-and-punishment* (in the case of being parented), *direct instruction* (in the case of going to school), or *role acquisition* (in the case of learning how to act in a new setting). Basic training in the military has been studied and commented upon extensively as an especially intense example of all three.[28] Within minutes of arriving at the induction station, *socializing agents* in the form of drill instructors rudely begin stripping new recruits of their civilian selves and instilling in them a new *military self*. Perhaps nothing signifies this transition in the popular view more than the ceremonial cutting of the hair. Shortly after arrival, male trainees' heads are shorn with a few theatrical whisks of an electric clipper, scattering stylish locks about the floor in pathetic little mounds. What once was before has been cast aside by the military in the flick of a wrist. Women must surrender some hair as well.

To speak of *military culture* implies a distinctive *occupational culture* that differs from those found in civilian life. There are important cues new members of an organization might use to decipher what that organization's culture is all about. Incoming "plebes" and "doolies," basic trainees at the U.S. Naval Academy and U.S. Air Force Academy

(USAFA), respectively, are informed from day one of the countless rules covering everything imaginable. These are spelled out in exhausting detail in the *Midshipman Regulations Manual/Daily Schedule* and the *Cadet Sight Picture*. But even if plebes and doolies were to commit all these to memory, it still would not guarantee they would know how to act. Social life is more complicated than that.

The newbie might first begin by noticing the *symbols and artifacts.*[29] The Naval Academy, for instance, sits on the Severn River and backwaters of Chesapeake Bay. Hemmed in by water and the city of Annapolis, Maryland, the campus takes up a mere 340 acres. Its most striking features are massive, white stone buildings, many designed by Beaux-Arts architect Ernest Flagg at the turn of the 19th century. The Beaux-Arts style combines noble-looking, classic Greek-and-Roman exteriors with baldly utilitarian interiors. Bancroft Hall's frescoed facade and sweeping staircases give way to dorm rooms, charitably described as spartan, housing all 4,000 midshipmen in a single building. The style says military grandeur and tradition are at the heart of service in the Navy.

In contrast, USAFA sits at 7,500 feet above sea level on almost 18,000 acres of national forest land to the north of Colorado Springs, Colorado. The message here is that of a new military tradition at the cutting edge of technology. Built in the late 1950s, its edifices are modern-looking, metal-and-glass, decidedly functional boxes, all set in a rectangle around the Terrazo, a small compact area within which cadet life takes place. The only non-box-like building is the chapel. Designed by Walter Netsch, it is an architectural marvel, soaring 150 feet into the air with 17 metal-and-stained-glass spires in the shape of what appear to be jets pointed skyward.

Another way to access culture is to consider an organization's *myths and stories.* Participants interpret what policies "really mean" based on what happens in *defining moments* rather than on what the policies formally say. One might assume, for instance, that midshipmen and cadets who most enthusiastically embrace the rules would be the ones most highly regarded. Not so – idealistic newcomers quickly learn there is such a thing as the appropriate amount of enthusiasm. Midshipmen and cadets have firm ideas about just how gung-ho one should be. At USAFA, cadets who are thought to overdo it are labeled *tools* to signify their status as mindless stooges and detested ass-lickers whose exuberance makes others needlessly look bad.[30]

Finally, organizations may be known by their *rituals.* While military academies have many, marching in close-order drill is a common image when people think of the military. Midshipmen and cadets line up in perfectly squared formations at morning formation, just before the noon meal, or whenever told to do so. They then move in unison by striding lockstep with each other. Students at Columbia University or the University of Colorado – only miles from West Point or USAFA, respectively – have not been observed walking about in this manner.

One topic that fascinated those ruminating over their days in the World War II military was the dramatic resocialization required of them to pull off the transformations from civilian-to-soldier and soldier-back to-civilian. Indeed, the whole process of reinventing oneself in early adulthood – something sociologist Erving Goffman called *developing one's official self* – became one of the enduring themes in the sociological literature from this era.

We now place our spotlight on a description of wartime duty by one of the kingpins of post-World War II sociology, George C. Homans.

Spotlight on Sociological Thinking and Research

George C. Homans, "The Small Warship"[31]

George Casper Homans was born in Boston, Massachusetts, to a patrician family in 1910, the great-great-grandson of John Quincy Adams, the sixth president of the U.S.A., and great-great-great-grandson of John Adams, 2nd president of the U.S.A. Befitting his social background, he attended the prestigious St. Paul Preparatory School in Concord, New Hampshire, then Harvard University where he earned the baccalaureate degree in English in 1932. Upon graduation, Homans became a junior fellow in Sociology at Harvard's Society of Fellows[32] and, in 1939, became an Instructor in Sociology in the Center for Advanced Studies. An accomplished yachtsman and naval reserve officer, he served on active duty during World War II for four and a half years, two of those as a junior officer on a "small warship" whose task was to provide an escort for American troop ships through submarine-infested waters.

After the war, Homans returned to his position at the Center. His wartime commentary reflected his theoretical thinking in this part of his academic career, which culminated in the book, *The Human Group*.[33] Homans contended that a group (two or more interacting individuals) is the most basic unit in society and therefore the most elementary unit of analysis. He assumed groups somehow fit together to form society, but that was not a matter of concern. Interaction within a group arose, Homans continued, from an *external system* – the environment that sets the conditions for its existence. An *internal system* also develops, and the two systems make up the total system. The sociologist's job is to observe what group members do (activity), how each member acts relative to the others (interaction), how members feel about the group (sentiments), and what codes of behavior they adopt (norms).

Homans's description of the warship emphasizes the ship at sea as an entity unto itself under the absolute control of its captain. Though the captain and officers have the authority to issue orders, exclusive reliance on this mechanism of control is unwise. Homans thus offers observations on good leadership by describing the kinds of *exchange relationships* likely to result in voluntary compliance. After all, he notes, "morale is just a fancy word for willingness to cooperate."[34]

Good leaders first must be able to hold up their end of the exchange. At sea, the sailors are totally dependent upon the demands, rules, and schedules provided by the officers in charge. Their lives literally are in their officers' hands. The sailors' first question then is: Do the officers have the technical skills and organizational acumen to operate the ship and take care of the details? Sailors thus look for tell-tale signs of whether the officers are competent or not. Homans tells of sailors standing on the deck watching new officers guide the ship into port – can they do so expertly, or do they knock some of the paint off the hull before getting it in position alongside the pier – and of monitoring chow (military slang for "food") and berthing schedules to see if they are efficiently executed. Lack of expertise is a warning sign to them that sailors may have to take some matters into their own hands.

Secondly, there must be a balance of tasks. The ship consists of many subgroups of sailors – those working on equipment and guns on deck (deck apes), those in the boiler and engine rooms (snipes), and so forth. At times, some of these groups must work especially hard while the demands on others at that moment are slight.

Assignment of tasks and privileges, Homans notes, must be done in such a way that each group feels the officers are aware of the fluctuations in task demands and balance out the burden over time fairly.

Homans's third observation concerns the regulations. Like the Army Regulations, the Navy's formal rules are highly detailed and, if applied mindlessly, can prove to be oppressive and counterproductive. If one expects the sailors to comply willingly, Homans notes, officers must at times "protect them from the Navy" by deciding that some rules, though in the books and backed by law, are not worth following. Done in a judicious manner, this stance tends to increase compliance for those rules the captain chooses to enforce. Mindless enforcement of pointless rules is one of the most despised of some officers' habits.

Finally, Homans addresses the problem of communication. The organization chart, listing the authority structure of the ship from high to low, provides a model of how information is supposed to flow from top to bottom. Generally, the downward flow of information, he states, though often incomplete and garbled by the time it reaches the lowest sailors, still is more robust than the flow of information upward. The use of fresh water on a ship, for instance, is a critical issue. There is a finite amount of it for a range of uses from drinking and showering to cooling and cleaning the equipment. Efficient, voluntary compliance is critical. However, exactly how the ship's water policies are affecting sailors at every level is not something the officers will be aware of, unless they make special effort to get frank, honest feedback.

In the years after the publication of "The Small Warship," Homans developed one of sociology's main micro-level theories, *social exchange theory*.[35] Its focus is on individual actors and the choices they make based on the content of relationships with others. These relationships are marked by *reciprocity*, that is, individuals tend to respond to others in kind, meaning they make assessments of the tone, content, and purpose of each interaction and react positively or negatively in accordance with these assessments. Homans contended that social life could be understood as the sum of these *social exchanges*.

So far, we have proceeded as if *military culture* were a single entity distinct from that of civilian sectors. However, there is no single military culture. The cultures of military organizations vary both within any military and among the militaries of different societies. For an understanding of how this works, we shine the spotlight on the collaborative work of three social scientists.

Spotlight on Social Science Thinking and Research

Joseph Soeters, Donna Winslow, and Alise Weibull, "Military Culture," in *Handbook of the Sociology of the Military*[36]

Our three researchers originate from the Netherlands, Canada, and Sweden, respectively. Each has expertise in some dimension of military culture through work with the military of his or her home country. Joseph Soeters holds a doctorate in Social Sciences from Maastricht University. A specialist in how to measure effectiveness in both civilian and military organizations, he has traveled widely to study Dutch and multi-national military operations in Bosnia, Kyrgyzstan, Afghanistan, Lebanon, Liberia, the Congo, and Bolivia.[37] He is Professor of Management and Organizational Studies at the Royal Netherlands Defence Academy and Tilburg University.

Donna Winslow began her career as an aid worker with indigenous tribes in New Caledonia before completing her doctorate in Anthropology at the University of Montreal. Her fascination with the military began in the early 1990s when she was invited by the Canadian government to lead an inquiry into the grisly torture and death of a Somali national at the hands of a Canadian Airborne Regiment on a peacekeeping mission in the town of Beledweyn.[38] Her study implicated misplaced loyalties associated with the culture of the unit. Winslow continued her research on military culture as Chair of the Department of Anthropology at Vrije University in Amsterdam and, upon her return, as consultant for the Auditor General of Canada and the U.S. Army Training Command.

A longtime professor and sociologist at the Swedish National Defence College, Alise Weibull has focused on the sociology of education and the military. She is especially known for her research on the education and training of officers for the Swedish military, and on the changing culture in the Swedish Air Force as it moved from an adjunct unit within the Army to a stand-alone entity.

Soeters, Winslow, and Weibull identify three themes in studies of military culture. The first emphasizes the disparities between military and civilian occupational cultures, the second, variations within military culture because of differences among the Army, Navy, and Air Force, and among job specializations within these. Finally, the third details instances in which subcultures within the military depart from the values and expectations of the larger organization (as we saw in Winslow's study of Canadian paratroopers in Somalia). Of the three themes, the first is the one found most often in the literature. The basic contention here appears obvious: because it is entrusted with the legitimate use of lethal violence, the military must be organized and conduct its affairs differently than civilian organizations. The profession of arms thus includes practices that highlight warfighting, its accoutrements, and accompanying values of service, discipline, and obedience.

While these distinctions yield useful insights, the second category provides a nuanced understanding. Variations among the services reveal something about the realities of their operating environments. Ground combat by the Army is a totally different creature than warfare encountered by the Navy at sea. Killing in the Air Force is done by officers – pilots – from far away, while the killers in the other services are enlisted personnel fighting their adversaries in close quarters. Weibull's study of the Swedish Air Force showed that air staff, pilots, and mechanics progressively valued expertise over rank and tradition once they no longer were a component of the Army. The result was a "social free-zone" among them in which professional knowledge and skills were great equalizers and diminished the hierarchy inherited from the Army days.[39] One might expect a similar trend among Army, Navy, and Air Force medical personnel whose work requirements differ radically from those of their fellow soldiers, sailors, and airmen closer to the "tip of the spear."[40]

Soeters, Winslow, and Weibull contend that the characteristic producing the most variation in military cultures is proximity to combat or danger. They use the terms "cold" and "hot" organizations to capture the phenomenon.[41] *Cold (military) organizations* are combat units that are in garrison, support units in secure environments in or out of the combat zone, and sections of combat organizations that by their work always are removed from the fight. Cold units take on many bureaucratic features for which the military is famous, or, more correctly, infamous. Days may be consumed by routine training exercises and make-work activities meant to keep everyone busy. Leadership may lapse into more traditional, less transactional, and coercive modes and trivial infractions may take on bigger significance than usual. Often, the staff of any unit or a military subculture whose work is removed from the tip of the spear might more closely resemble an office or work group found in a civilian organization.

Hot (military) organizations occur when warfare or simply dangerous operations become reality. Thus, the culture of a combat unit, by function at the very tip of the spear, will fluctuate depending upon whether it is a hot or cold state. Similarly, a service occupation – truck driver, for instance – may ordinarily be cold but become a hot one in an area where convoys may be ambushed. Hot organizations have a tempo and flexibility not found when cold, and the organizational posture may encourage more initiative. Routine activities and training take on a greater sense of urgency and purpose. Threats of punishment are less effective motivators when real hostilities and attendant risks are taking place, so leaders may display more transformational and inspirational styles. There often is a greater sense of togetherness in hot organizations and unit cohesion, or the absence of it, becomes a barometer of likely success or at least the ability of the group to endure.

Finally, military culture varies across societies. Soeters provides a personal example from a research venture with an Irish–Swedish stabilization force in Liberia.[42] He found their outpost, Camp Clara, divided into Swedish and Irish halves with little cooperation between them. On his arrival, the Swedes warned Soeters of communicable diseases and pointed out a special port-a-john reserved for those with diarrhea. Before meals, an officer was stationed near the mess hall door to ensure all washed their hands twice, once with water and a second time with alcoholic gel. The canteen provided excellent food – dinner the first night was deer steak with red currants. To the chagrin of local workers on the outpost, leftovers always were trashed immediately. In Africa, most people do not have the luxury of wasting leftovers. Likewise, old equipment and worn furniture were packaged up and sent back to Sweden for disposal. In this spirit, Swedish troops on patrol were cautioned to limit their contact with the locals to prevent corruption and disease.

The Swedish also restricted their interaction with the Irish contingent, who operated quite differently. While Swedish troops observed the two-beer per day limit, Guinness flowed freely on the Irish side. Soeters noted there were hand-washing facilities at the entrance of the Irish canteen, but no one seemed to use them. Food entrées centered around overcooked potatoes. When it was time to do joint patrols, Irish soldiers scheduled to go out often reported ill, so the Swedish and Irish patrolled separately. The Liberians however loved the Irish, whom the Swedish conceded "have the gift of gab." The former took an interest in the Liberians and their sporting events, allowed the kitchen help to take leftover food home, and interacted frequently with the locals while on patrol, including occasional flings, in Soeters's words, with "black Venuses at the entrance of the camp."

On the last day of his visit, Soeters asked a Swedish general about the gap between Irish and Swedish soldiers. "The Irish have a problem with abortion, but not with prostitution. The Swedish ... have a problem with prostitution, but not abortion," the general summarized for him. Soeters concluded, "Just try to reconcile those points of view!"[43]

Questions for Discussion

1 The *nature vs. nurture* debate usually is framed in just that way, nature *versus* nurture. A more nuanced view, considering recent developments, is to reframe this as *nature via nurture* (or even *nurture via nature*). Explain this, making use of such concepts as *norm of reaction* and *prepared learning*.

2 Joshua Goldstein has tackled a tough question: societies vary in how they do gender and how they do war. So why has warfare historically been done essentially by males? He explores several possibilities. Which explanation did you find most interesting? Were you persuaded one way or another?

3 George Homans suggests that social life can be understood as a sum of social exchanges. Do you think leadership can (or should) be conceptualized in this way? If so, what can leaders do to facilitate voluntary compliance?

4 Joseph Soeters and his associates argue that, while there is a noticeable gap between civilian culture and military culture, there is no one "military culture" (just as there is no one "civilian culture"). What stood out to you in their analysis? Why is there no one military culture?

Notes

1 William, McDougall, [1908] 2001, *Introduction to Social Psychology*, Boston, Mass.: Adamant Media Publishing. McDougall, incidentally, served in the French army as an ambulance driver during the early years of World War I and, later in the British army, he worked with shell shock victims.

2 Franz Boas, [1911] 1983, *The Mind of Primitive Man: A Course of Lectures Delivered before the Lowell Institute, Boston, Mass., and the National University of Mexico, 1910–1911*,Westport, Conn.: Greenwood Press.

3 Joseph Arthur de Gobineau, [1853–1855], 1915, *An Essay on the Inequality of the Human Races*, 4 vols., translation by Adrian Collins, New York: G.P. Putnam's Sons.

4 E.B. Tylor, [1871] 2010, *Primitive Culture: Researches into the Development of Mythology, Philosophy, Religion, Art, and Custom*, 2 vols., Cambridge, England: Cambridge University Press.

5 Lewis H. Morgan, [1877] 2000, *Ancient Society: Researches in the Lines of Human Progress from Savagery through Barbarism to Civilization*, with new Introduction by Robin Fox, Piscataway, N.J.: Transaction Publishers.

6 Franz Boas, [1921] 1961, *The Mind of Primitive Man: A Course of Lectures Delivered before the Lowell Institute, Boston, Mass., and the National University of Mexico, 1910–1911*, New York: Free Press.

7 Richard Machalek and Michael W. Martin, 2004, "Sociology and the Second Darwinian Revolution: A Metatheoretical Analysis," *Sociological Theory* 22 (September): 455–476.

8 Socio-biology is a field of study that assumes social behavior may be thought of and studied as a product of evolution. See, for instance, E.O. Wilson, 1975, *Sociobiology: The New Synthesis*, Cambridge, Mass.: Belknap/Harvard University Press.

9 Machalek and Martin, op. cit., pp. 457–459.

10 Machalek and Martin, op. cit., pp. 455–456.

11 E.O. Wilson, 1998, *Consilience: The Unity of Knowledge*, New York: Alfred A. Knopf, discussed and cited in Machalek and Martin, op. cit., p. 460.

12 Discussed and cited in Machalek and Martin, op. cit., p. 462.

13 For a summary of Malinowski's place in the history of anthropology, see George Peter Murdock, 1943, "Bronislaw Malinowski," *American Anthropologist*, 45: 441–451, downloaded from http://www.aaanet.org/committees/commissions/centennial/history/095malobit.pdf.

14 Bronislaw Malinowski, 1941, "An Anthropological Analysis of War," *American Journal of Sociology*, 46 (January): 521–550; cited in Otterbein, op. cit., p. 31.

15 Raymond Pearl, 1941, "Some Biological Considerations of War," *American Journal of Sociology*, 46 (January): 487–503.

16 Edward O. Wilson, 2012, *The Social Conquest of Earth*, New York: Liveright Publishing Corporation.

17 A single colony of African driver ants may have as many as 20 million workers. Although weight is not necessarily a sound measure of dominance, Wilson estimates that all of the ants on earth weigh roughly the same amount as the earth's 8 billion humans; ibid., pp. 111, 117.

18 Wilson's version of eusociality has elicited a firestorm of criticism from some evolutionary biologists. Richard Dawkins identified *inclusive-fitness theory* as the evolutionary rationale for eusociality. This theory argues that individuals are most motivated to serve altruistically in groups made up of close kin. Wilson's extrapolation of this principle to "the tribe" does not meet Dawkins's standard. For Wilson's rebuttal, see pp. 143–147, 166–182 in *The Social Conquest of Earth*.

19 Wilson, op. cit., pp. 54, 57.

20 Keith Otterbein, 2004, *How War Began*, College Station: Texas A&M Press. For a summary of his modes of theorizing and handling of data, see pp. 15–21. He employs a combination of theoretical induction (using archeological and ethnographic material), logical deduction, and cross-cultural comparisons in search of explanatory patterns.

21 Joshua Goldstein, 2001. *War and Gender: How Gender Shapes the War System and Vice Versa*, New York: Cambridge University Press.

22 In the U.S.A., if doctors filling out birth certificates see that a baby has a penis, he is a boy and if the baby lacks a penis, she is a girl. The official guideline is: to be a penis, the baby's genitalia must be longer than 2.5 centimeters or less than 1 centimeter to be a clitoris (both penises and clitorises develop from the same undifferentiated organ in embryos). Note there is a gap between 1 and 2.5 centimeters. In these cases, other criteria are used. Even in these cases, there may be lack of clear evidence. See Anne Fausto-Sterling, 2000, *Sexing the Body: Gender Politics and the Construction of Sexuality*, New York: Basic Books.

23 Goldstein, op. cit., pp. 142–157.

24 Ibid., pp. 184–194. Goldstein summarizes the studies on both chimpanzees and bonobos in this section.

25 Ibid., pp. 60–64.

26 Ibid., pp. 64–70.

27 Ibid., pp. 17–179, 228–238.

28 See, for instance, Louis Zurcher, Jr., 1968, "The Naval Recruit Training Center: A Study of Role Assimilation in a Total Institution," *Sociological Inquiry*, 37 (Winter): 85–98. Also, the *American Journal of Sociology*, 1945, 50 (January), devoted several articles to this topic in their special edition of "Life in the Army."

29 Edgar H. Schein, [1985] 2010, *Organizational Culture and Leadership*, 4th ed., San Francisco: Jossey-Bass.

30 Similarly, Roger Little's study of an infantry unit in the Korean War found that soldiers referred to overly gung-ho types whose idea of bravery put others at risk as *heroes* (hence, a derogatory term). The term *bums* denoted soldiers who were ostracized because they could not be counted on during combat skirmishes. Knut Pipping reports the use of similar terms and connotations in his study of a Finnish infantry company during World War II (2008, *Infantry Company as a Society*, Helsinki, Finland: National Defence University).

31 George C. Homans, 1946, "The Small Warship," *American Sociological Review*, 11 (June): 294–300.
32 The Society, established in 1933, provides (to this day) young scholars with the opportunity to pursue studies in any department free of any formal duties.
33 George C. Homans, 1950, *The Human Group*, New York: Harcourt, Brace and Company.
34 Homans, 1946, op. cit., p. 295.
35 George C. Homans, 1961, *Social Behavior: Its Elementary Forms*, New York: Harcourt, Brace and Company.
36 Joseph L. Soeters, Donna J. Winslow, and Alise Weibull, 2006, "Military Culture," Pp. 237-254 in Giuseppe Caforio (ed.), *Handbook of the Sociology of the Military*, New York: Kluwer Academic/ Plenum Publishers.
37 For his personal reflections on these studies, see Joseph Soeters, 2012, *Green about Green: A Civilian in Military Life*, available as a Kindle book through Amazon Digital Services.
38 Donna Winslow, 1997, *The Canadian Airborne Regiment in Somalia: A Socio-Cultural Inquiry*, Ottawa: Canadian Government Publishing Center; Winslow, 1998, "Misplaced Loyalties: The Role of Military Culture in the Breakdown of Discipline in Peacetime Operations," *The Canadian Review of Sociology and Anthropology*, 35: 345–367. The paras posed and took pictures of themselves posturing around the dead Somali in a manner strikingly similar to that of American soldiers years later at Iraq's Abu Ghraib prison.
39 Alise Weibull and T. Bjorkman, 1997, *I Vantan pa JAS*, Stockholm: FHS, described in Soeters et al., 2006, op. cit., p. 245.
40 The tip-of-the-spear colloquialism developed to refer to military activities distinct to combat.
41 Soeters et al., 2006, op. cit., pp. 246–249.
42 Soeters, 2012, op. cit., chapter entitled, "The Mysterious Other."
43 Ibid., Kindle location 634 of 1774.

Recommendations for Additional Reading

John Hockey, 2002, *Squaddies: Portrait of a Subculture*, Liverpool, UK: Liverpool University Press.
(Classic sociological study of a British Army unit; the author follows them from basic training to a field exercise in Canada and then deployment to Northern Ireland)

J.E. Linden, 2005, *Soldiers and Ghosts: A History of Battle in Classical Antiquity*, New Haven, CT: Yale University Press.
(A fascinating account by University of Virginia historian of how the Greeks and, later, the Romans did warfare)

Margaret MacMillan, 2020, *On War: How Conflict Shaped Us*, New York: Random House.
(A smart, informative narrative about war, culture, and society by a well-known Canadian historian)

Jonathan H. Turner, 2018, *On the Biology and Sociology of What Made Us Human*, New York and London: Routledge.
(A guide to recent developments in genetics and socio-biology by a noted sociological theorist)

Donna Winslow, 1997, *The Canadian Airborne Regiment in Somalia: A Socio-Cultural Inquiry*, Ottawa, Canada: Commission of Inquiry into the Deployment of Canadian Forces to Somalia.
(An anthropological analysis of a Canadian unit's culture and its link to a peacekeeping mission gone awry)

4 The Military as a Bureaucracy and as a Profession

Reader's Guide: The military is an unusual creature as it is simultaneously a bureaucracy and a profession inspiring its members to give their all. The two classic examples of bureaucracy are the mass-production plant (perfected in the U.S.A. by Henry Ford) and the mass, draft-based military. We review the mechanics of both, and how the U.S.A. rode a Fordist approach in production and military organization to dominance in World War II. Early military sociology thus was the sociology of the mass army. The first Spotlight underscores military sociology's shift to the study of the military as a profession, especially for the career officer, and presents two classic versions of how professionalization should be achieved. Our second Spotlight addresses the U.S. military's transition to an All-Volunteer Force (AVF) at the end of the war in Vietnam. This was not without some trepidation: Does this threaten to transform the military from a profession to "just a job"? The final Spotlight reviews the characteristics of modern corporations and AVFs, showing how each has become a post-Fordist organization.

Societies have gotten better at doing war as their capacity for more complex social organization has increased. The end of the Middle Ages (around AD 1500) saw a series of innovations portending an uptick in battle potency. Ingenuity in the use of gunpowder produced the combustive properties needed for firing cannons, muskets, and handguns. Most military historians attribute the big changes in warfare to such technical developments. However, sociologist James Aho plucks a succinct quote from one of sociology's founding fathers, Max Weber, to introduce a supplementary set of considerations: "It was discipline and not gunpowder which initiated the transformation."[1]

Weber's insight stemmed from his recognition of the primacy awarded to *rational authority* in secular (but not sacred) matters in the 16th and 17th centuries. Nowhere was this change in thinking more evident than among northern Europe's Protestant military innovators: Oliver Cromwell of England, King Gustavus Adolphus of Sweden, Prince John Maurice of Nassau, the Netherlands, and, later, Frederick the Great of Prussia. Their thinking centered on "what past experience [had shown] to be the most practical ... regardless of moral scruples or honorable custom."[2] Second, they were part of a revival in studying and imitating "the greatest army of antiquity," the Roman Legion. The result was a shift in the type of military they were able to field and a substantial increase in killing efficiency. The Thirty Years War (1618–1648) between Catholics and Lutherans is estimated to have resulted in somewhere between 4 and 8 million deaths.

DOI: 10.4324/9781003282549-4

Weber's theory of bureaucracy appeared in *Economy and Society*, a multi-volume work he completed during World War I.[3] His point of departure was a phenomenon of power. Both the modern state and industrial economies rely extensively upon a hyper, legal-rational apparatus Weber labeled *bureaucracy*. *Legal-rational* authority creates efficiency through the application of impersonal, universalistic standards. Tasks are identified and associated with specific positions, and these positions are connected in a hierarchy by lines of communication and control. People become incumbents of these positions by having the requisite knowledge and experience. Finally, detailed rules and regulations govern all aspects of organizational life. *Rationality* thus makes bureaucracies successful on the one hand and dehumanizing on the other, as anyone who has ever been treated like a faceless number can attest.

Bureaucracies and World War II

The most frequently mentioned prototypes of bureaucracy are the *mass-production manufacturing plant* and the *modern mass-army*. Both forms of organization reached their apex during World War II and the following decade of the 1950s, an era where their ability to marshal vast amounts of material and participants into a machine-like organization provided unparalleled dominance.

The mass production plant was not an exclusive invention of the U.S.A., but American innovations stand out. Legend, now disputed, had it that the American inventor, Eli Whitney, perfected the use of *interchangeable parts* toward the end of the 18th century while carrying out a contract to manufacture muskets for the Department of War.[4] The technique called for breaking down a product into its myriad of constituent parts so that its final manufacture could consist simply of assembling identically made replicas of these parts into a whole. This allowed lesser-skilled workers to put together a product of quality more rapidly than skilled master craftsmen making it from scratch start-to-finish. It also provided a means for easily repairing a product once it left the factory, since new parts could be inserted to replace broken ones. The manufacturing successes in the American gun industry were copied during the 19th century, most notably in the assembly of sewing machines, farm implements, and bicycles.[5]

The person who took manufacturing to another level is Henry Ford. At the turn of the century, Ford headed one of several groups of mechanic-inventors making horseless carriages – automobiles – to order and from scratch. Ford however had a vision. After several fits and starts, he established the Ford Motor Company (FMC) in 1903 with a goal:

> to produce [a] light ... car with ... ample horsepower ..., capable of carrying its passengers anywhere that a horse-drawn vehicle will go without the driver being afraid of ruining his car, ... and whose price [was] within reach of the millions who cannot yet afford automobiles [made to order].[6]

Ford and his engineers solved one problem after the other over a decade of trial-and-error. He settled on the Model T and directed that it was the only car his company would make. He also insisted that parts be made with such precision that line workers would not have to do any fitting, and hit upon the idea of placing the chassis on a moving conveyer belt which moved down a line of stationary workers. The sum of these innovations (and borrowed best practices) increased production dramatically. FMC produced about 5,500 Model Ts in 1908 at a retail price of $850 apiece. By 1916, it turned out 585,000 at a retail price of $360.[7]

Ford's technical successes did not come without human cost. Boring tasks, lack of discretion, and low pay produced soaring levels of worker dissatisfaction and turnover. In 1913, turnover among FMC line workers was almost 400% and the company that year had to hire 50,488 men to maintain a labor force of 13,623.[8] Ford tried several remedial strategies, including raising the minimum wage to $2.34 per day and hiring newly arriving immigrants, many unable to speak English but happy to find work. Still, these measures fell short until Ford instituted the unheard-of wage for line workers of $5.00 a day.[9]

Meanwhile, with World War I still fresh in mind, the U.S.A. had adopted a firm isolationist stance in international affairs. While the U.S.A. led the world in per capita production of cars and trucks for civilian use, its military was neglected and in disrepair.[10] All that changed with the bombing of Pearl Harbor in 1941. Overnight federal agencies began the conversion of American industry to military purposes. The Army Service Forces, a new agency, centralized the authority needed to direct and manage the Army's logistical chain for an active-duty force that grew from about 240,000 to a peak of almost 12 million by the end of the war.

Ford's perfection of mass production led the way. World War II was a war of ships, tanks, planes, and mechanized weapons, and the U.S.A. produced more of these than anyone else.[11] Its assembly lines churned out 15 million rifles, carbines, and sub-machine guns, almost 3 million machine guns, 700,000 rocket launchers, 170,000 mortars and howitzers, and a billion rounds of ammunition of all sorts. Among the big-ticket items, it made 2.5 million trucks, 96,000 tanks, 22 aircraft carriers, 750 destroyers, cruisers, and escort ships, 8 battleships, 200-plus submarines, and more than 300,000 training aircraft, transports, fighters, and bombers.[12]

To appreciate the scope, consider the production of the B-24 *Liberator* bomber, of which almost 19,000 were made during World War II. A four-engine, heavy bomber, it was made up of more than 550,000 parts and took over 700,000 rivets to hold together. By comparison, a Model T had about 15,000 parts. Most WWII B-24s were made at FMC's Willow Run plant near Ypsilanti, Michigan. At first, it took several days to make one bomber. However, by 1944, it took only 55 minutes and Willow Run was churning out almost 650 of them per month.[13] Again, worker dissatisfaction and high turnover were endemic. Still, Henry Ford, despite war-time labor shortages, was adamantly opposed to hiring women – the famous Rosie-the-Riveters of WWII – until quite late in the war.

The U.S.A.'s successes in WWII at first were taken after the war as a sign that organizations, whether civilian or military, are at their best when very *Fordist*. However, flaws in this thinking, especially its autocratic, top-down management style, inspired the quest for alternatives. In manufacturing, this took the form of more collaborative styles of management. During the 1950s and 1960s, Japanese industrialist, Eiji Toyoda and American university professor, W. Edwards Deming, collaborated at the Toyota Corporation to develop an alternative system known as *lean production*.[14] This relied on managers, engineers, and line workers working together in teams to create constant improvements on the assembly line – a system that soon vaulted Toyota to the forefront of world automobile production.

The first American automobile manufacturer to hit the wall because of the consequent loss of market-share was the FMC. Hence, FMC's board (Henry Ford died in 1947) formed a working agreement in the 1970s with Japan's Mazda Corporation, which had copied Toyota's lean-production regimen. In 1981, FMC's new Chief Executive Officer, Donald Peterson, and his executives visited Mazda's production line in Hiroshima, Japan. Incredible as it may sound, 36 years after that city was leveled by an atomic bomb to end

World War II, Ford executives were walking an auto production line there, asking Japanese managers, "show us how you make cars."[15]

The Study of the Military as a Profession

The military too was not without its problems during and after the war. Many of these stemmed from its bureaucratic nature, which made it both an effective organization in some respects and one difficult to endure in others. Many WWII studies by sociologists, especially those who served in the ranks, portrayed the military from the point of view of the common soldier who, on the one hand, was committed to winning the war, but who otherwise chafed under the weight of endless, seemingly mindless, rules and regulations. Few aspired to remain in the military after the war. Their goal was simply to win the war and to return to civilian life as soon as possible.

In the 1960s, political scientist Samuel Huntington and sociologist Morris Janowitz shifted the focus in military studies from enlisted soldiers – informally called the "sociology of the grunt" – to the officer corps. We shine the spotlight on their path-breaking works.

Spotlight on Social Science (Huntington is a political scientist, Janowitz a sociologist) Thinking and Research

Samuel P. Huntington, *The Soldier and the State: The Theory and Politics of Civil-Military Relations*[16] and Morris Janowitz, *The Professional Soldier: A Social and Political Portrait*[17]

Samuel Phillips Huntington was born in New York City in 1927, so he turned 18 years of age just as World War II ended. Still, he did a stint in the U.S. Army after completing his baccalaureate at Yale University in 1946. Following his hitch in the army, he received an M.A. in Political Science from the University of Chicago and a Ph.D. from Harvard University in 1951, where he then joined the faculty.

It was during his early years at Harvard that he wrote ***The Soldier and the State*** (*TS&TS*). In part, this was in response to his unease over the insubordination of General Douglas MacArthur during the Korean War and the general's subsequent firing by President Harry Truman. It raised the question: In a free, democratic society, what should be the relationship between the military and civilian authority?[18] In *TS&TS*, Huntington sought to provide a framework for how to think about this question.

The central "civil-military *problematique*," to borrow a phrase from one of Huntington's protégés, Peter D. Feaver, is: How does a society guard itself from its own military?[19] A society's military is entrusted to protect it from external foes through the legitimate use of lethal violence, should the situation call for it. However, a military large enough and powerful enough to do so also is large and powerful enough to turn on its host society, or at least to become its own arbiter over when and how to use lethal violence. Huntington's answer to this paradox, and that of Janowitz a few years later, sparked a debate that became a major theme in military sociology.

The basics of Huntington's theory are depicted in Figure 4.1. It begins with two antecedent variables that set the stage: the nature and level of external threats facing a particular society (*functional imperatives*) and the internal thinking and legal

Functional Imperatives
(threats to society's security)

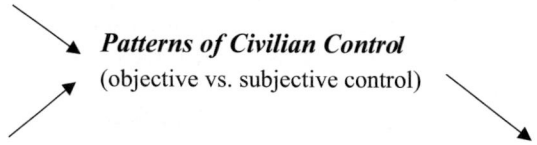

Patterns of Civilian Control
(objective vs. subjective control)

Societal Imperatives
(ideologies, legal frameworks)

Military Preparedness
(military's ability to meet external threats)

Figure 4.1 Huntington's Theory of Civil–Military Divergence

Source: Authors' Creation.
Note: Adapted from Peter D. Feaver and Erika Seeler, 2009, "Before and After Huntington: The Methodological Maturing of Civil-Military Studies," pp. 72–90 in Suzanne C. Nielsen and Don M. Snider (Eds.), *American Civil-Military Relations: The Soldier and the State in a New Era*, Baltimore, Md.: The Johns Hopkins University Press, pp. 80–82.

framework within that society (*societal imperatives*). He assumed, in the case of the U.S.A., that the former has varied over time but that the latter has remained relatively constant. American history has been one of a healthy suspicion of military forces. Large armies were commonly seen as threats to liberty, democracy, economic prosperity, and peace – a formidable list of potential harms.[20] As a result, this produced a tight *pattern of civilian control* and a rather low level of *military preparedness*, that is, the U.S.A. historically has maintained a rather small standing army that was expanded in size only in time of war.

The U.S. Constitution does not provide for *civilian control of the military*, mostly because the founders did not envision a military made up mostly of professionals. Rather, they preferred a "well-regulated militia" consisting of amateur officers and soldiers who were citizens first and soldiers second. The safeguard against the *problematique* thus lay in the composition of the military itself. Still, there were Constitutional provisions for the separation of powers and oversight: Congress was given the prerogative to decide when to go to war and control over the purse strings.[21]

Given this history, it is paradoxical, Huntington notes, that 10 of the 33 U.S. presidents between 1776 and 1950 were military generals. By comparison, only one general officer has been prime minister of England since 1789. The anomaly was made possible by an American custom among military men (and they were all men in those days) of switching "one hat for another" when running for office: a general, as office-seeker, "abandoned his military trappings," evidence to Huntington that the two roles are incompatible in the U.S.A.

After World War II, the military for the first time in U.S. history maintained a substantial portion of its wartime size and capacity in the post-war years. In part, this occurred because of a growth in the U.S.A.'s stature as a military power, and, in part because of the advent of the Cold War with the Soviet Union. The shift, Huntington argued, calls for a recalibration in how to protect U.S. society from its own military.

At the time of Huntington's writing, the social scientific study of professions was well underway.[22] A *profession*, as a subcategory of occupations, is one requiring specialized knowledge and formal certification by a supervisory group that sets and enforces standards. The concept of a profession carries a loftier status as a "calling" or vocation than does a mere occupation. Professions also incur a sense of corporateness, an awareness they are an entity apart from ordinary laymen. Traditionally, all this referred to fields of law, medicine, academe, and religion, that is, to judges and attorneys, physicians, professors, and the clergy.

Huntington thus spoke of the military as the *profession of arms*, one entrusted with the awesome responsibility of exercising lethal violence when so authorized.[23] The management of violence incurs much more than the mastery of technical skills. It calls for a way of life requiring unusual discipline and integrity. These traits should not be mandated of those outside the profession of arms in a free society, where notions of individual rights and freedoms encourage other modes of thinking and activity. Because of this, Huntington reasoned, the military should *stand apart*, that is, be well-insulated from civilian institutions.

The second part of Huntington's solution calls for *objective civilian control* of the military. While the focus of the military rightfully should be upon the management of lethal violence itself, that of civilian leaders should be on political and foreign affairs that set the contours of defense policy. As a political scientist, Huntington's bias was to ensure this division of labor through a *formal chain of command* that clearly subordinated uniformed military officers to civilian politicians, specifically, the President as Commander-in-Chief, followed by the Secretary of Defense and the Secretaries of the Army, Navy, and Air Force.

We already met Morris Janowitz in Chapter 1 as the founder of military sociology in the U.S.A. His classic work, *The Professional Soldier* (*TPS*), followed Huntington's. It too addressed the military as a profession but did so in an entirely different way. While Huntington articulated a structural arrangement to best solve the civil–military *problematique*, *TPS* described "the professional life, the organizational setting, and the leadership of the American military" in the 20th century and offered a normative solution.[24]

Janowitz's research for the book focused on the highest-ranking career-military officers. The analysis drew on historical and documentary materials, biographical data covering the social backgrounds and careers of more than 760 generals and admirals appointed between 1910 and 1955, and opinion data from a survey of 550 staff officers while assigned to the Pentagon. From the latter group, he conducted further, intensive interviews with 113 of them.

Janowitz's findings led him to different conclusions than Huntington's. In the 19th century, Janowitz noted, feudal traditions of the officer corps, coupled with the military's duties, contributed toward a *divergence* between civilian and military institutions (see top of Figure 4.2).[25] In those days, U.S. military duties revolved around direct-combat occupational specialties. More than 90% of enlisted jobs in the military during the Civil War, for example, were directly related to combat. Warfare implied mass armies clashing head-to-head on conventional battlefields. Command and control tended to be simple and hierarchical, so the military's organizational structure resembled a triangle, with lowest-echelon foot soldiers being the most numerous, followed by descending numbers all the way up the ranks

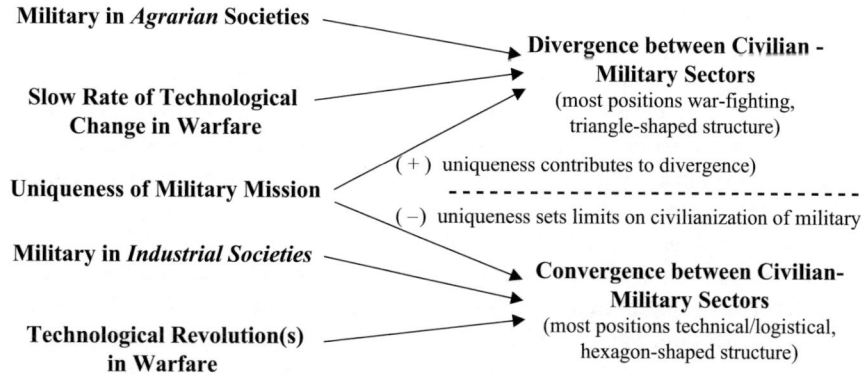

Figure 4.2 Janowitz's Theory of Civil–Military Convergence
Source: Authors' Creation.
Note: Figure constructed from Janowitz's description of his theory.

to the top general or admiral. The officer corps controlling this organization came from the upper echelons of society and were educated at elite military academies to become *officer gentlemen* and *heroic leaders*.

In contrast, the trend in the mid-20th century has been toward *convergence* (see bottom of Figure 4.2), principally because modern weaponry is in the form of air-craft carriers and destroyers, long-range fighters and bombers, intercontinental bal-listic missiles, and nuclear warheads. By the 1950s less than 30% of enlisted military jobs were directly related to combat, and middle layers consisted of highly skilled specialists in logistics, maintenance, and administration. Military organizations now more closely resembled an octagon than a triangle. This personnel shift, and increased contact with supporting civilian entities, meant those in command and supervisory positions in the military needed skills in team building, consensus-making, and nego-tiation. The *management of violence*, Janowitz noted, ironically now requires bureau-cratic structures and roles more like those of the civilian sectors. These render the *stand-apart officer corps* envisioned by Huntington unfeasible and undesirable.

So, what is feasible and desirable? Janowitz began with a portrait of the officer corps' current state. Career officers in the 1950s military were found to have more diverse class origins than in the past and came disproportionately from middle class, rural, and Southern backgrounds.[26] Similarly, the sources of commissioning had expanded. While graduates of the military academies remained overrepre-sented in the upper echelons of the military, those commissioned through civilian university Reserve Officer Training Corps programs had increased. Post-baccalaureate education now often included master's or even doctoral degree pro-grams at civilian universities. Janowitz viewed these changes as desirable.

Janowitz also detected two dominant approaches to military policy, the *absolut-ist* and the *pragmatic*.[27] The former implies a black-and-white view of foreign policy dilemmas that favors military over diplomatic solutions, including heavy-handed ones that one might associate with *total war*. The latter approach reflects more of a "shades of gray" perception of foreign policy issues, endorses restraint in the appli-cation of force, and favors consideration of a range of options over purely military solutions. Interestingly, Air Force officers were more likely to be *absolutists* and those in the Army and Navy to be *pragmatists*.

Like Huntington, Janowitz has some recommendations. The spectrum of military power ranges from weapons of mass destruction at the top end to limited, specialized capacities at the other. The sheer destructiveness of the nuclear option foreshadows future reorganization of the military toward that of a *constabulary force*, that is, one likely to operate most often at the middle and lower levels of the spectrum. Hints of such were anathema to absolutists but more acceptable to pragmatists.

In any case, modern officers must be especially attuned to the social and political consequences of military actions. This calls for adjustments in both military organization and education. Such a military must find a comfort level with executing a full range of taskings, a balance among likely officer roles this will entail, and a new authority continuum consonant with mission and personnel. Officer education must concern itself with the "whole man" concept, that is, developing technically proficient officers who are both *modern heroic leaders* and *thoughtful military managers*.[28]

Huntington sought a direct solution to the civil–military *problematique*: create a stand-apart military subservient to civilian control. Janowitz's answer is more oblique: the officer corps is subject to civilian control because of "self-imposed professional standards … integrated with civilian values."[29] By analogy, there is an old tale in which some house mice are plagued by a cat who has been sneaking up on them at night and devouring them one by one. The mice hold a meeting and decide someone needs to put a bell on the cat's collar to warn them of his thereabouts. But an old mouse then asks, "This is all well and good, but who among us can put a bell on the cat?" If the cat is a professional officer, a Huntingtonian answer is: the house's owner must bell the cat. The Janowitzian response: the cat will put the bell on his own collar.

Professionalization in the All-Volunteer Force

Janowitz's book was especially well- or ill-timed, depending on one's point of view. Five years after *TPS* first appeared, 3,500 Marines landed in Da Nang, South Vietnam. By 1969, the number of American servicemen in-country would swell to over 600,000. Janowitz himself had been prescient in his observations concerning likely strategic outcomes in a world defined by nuclear weapons, here, the prevalence of "proxy" wars in which the superpowers fought it out through intermediaries in limited, irregular fashion.

The American effort in Vietnam yielded disappointing gains and excessive casualties. Restrictions on the use of airpower were strongly opposed by the *absolutists*[30] and over the course of the war their position gained wider acceptance among U.S. professional officers. After the war, a common complaint among them was, "we didn't lose, but were not allowed to win." Officers steeped in the *pragmatist* tradition presented a counterargument: political and military leaders had stubbornly resisted necessary adaptations.[31] Such truculence was exemplified in one indignant Army general's statement: "I will be damned if I will permit the U.S. Army, its institutions, and its doctrine to be destroyed just to win this lousy war."[32]

The war in Vietnam proved to be intensely and progressively unpopular. In WWII, opposition to U.S. participation dropped from 34% in 1941 prior to Pearl Harbor to 8%

by 1944; in the case of Vietnam, opposition increased from 24% in 1965 to 57% in 1970. This placed the military and professional officer corps at odds with a large segment of the civilian population. This deterioration did not cause but certainly contributed mightily to the Congressional decision in 1973 to convert the U.S. military to an AVF.

TPS was reissued in 1971 and editions available after 1974 contained a new prologue, allowing Janowitz to register his concerns about the AVF. Sociologist James Burk, one of Janowitz's protégés from the University of Chicago and a highly respected social theorist in his own right, explained Janowitz's opposition. He notes the importance of citizen participation in the affairs of the republic. When there is participation by a broad base of citizens in both the military and civilian sectors, and when the two sectors thus converge, there is reduced likelihood for things to go awry. Removing the citizens' obligation to serve, Janowitz argued, created an imbalance between rights and obligations. The right to safety from external threat is no longer a shared duty, and also runs the risk of a professional force less representative of the larger society.[33] However, the AVF already was a reality and there was no going back. Janowitz thus advocated some form of required national service, of which a stint in the military was one of the options.

The sociologist who most effectively voiced concerns about the AVF is Charles Moskos, who tirelessly talked, wrote, and disseminated information about its problematic aspects. We shine the Spotlight here on his work that recalibrated research in military sociology for decades.

Spotlight on Sociological Thinking and Research

Charles C. Moskos, Jr., "From Institution to Occupation: Trends in Military Organization"[34]

Charles Moskos, Jr. was born in Chicago in 1934 to parents of Greek descent who had emigrated to the U.S.A. from Albania. He went to Princeton University on a scholarship, which he supplemented by waiting tables. Upon graduation in 1956, he was drafted by the Army and served in Germany with the Corps of Engineers. After military service he completed a Ph.D. in Sociology at UCLA, writing his dissertation on the Albanian political elite. His service in the military as an enlisted man was a source of both enduring personal pride and inspiration for his professional work. Indeed, he stated in a 1993 interview with the *Wall Street Journal* that he sought simply to be a "walking tribute to the enlisted man."[35] Upon his death on 31 May 2008, his wife, Ilca, sent out the following press release: "Charles C. Moskos of Santa Monica, Calif., formerly of Evanston, Ill., *draftee of the U.S. Army*, died peacefully in his sleep after a valiant struggle with prostate cancer" (italics added).

We noted in Chapter 1 that the epicenter of military sociology in the first three decades after World War II was wherever Morris Janowitz was. This meant the University of Chicago from 1946 to 1951, the University of Michigan from 1951 to 1961, and then again the University of Chicago. Upon Janowitz's return to the latter, Michigan hired the young political sociologist, Charlie Moskos, to replace him. However, Moskos soon moved to Northwestern University located in Evanston, a northern suburb of Chicago. Burly, smart, fast-talking, and infectiously gregarious, he soon established a new locus for military sociology, notable especially as Janowitz neared the end of his career and Moskos's work on AVF surged to the fore.

Like Janowitz, Moskos focused on the *obligation-* rather than the *rights-*side of the citizen's role. Moskos, however, was more worried about the consequences within the military itself. The reliance on volunteers rather than draftees as the source of manpower, he maintained, shifts the meaning of military service from that of a calling or *profession* to that of an *occupation*. The differences Moskos expected between the military as an *institution* and the military as an *occupation* are summarized in Table 4.1 in what was called the I/O thesis.

The bedrock of an *institutional* or traditional military is the notion of duty or service. As a result, recruits and draftees are paid very little – literally a few dollars a month. Pay is better for those who stay on but still low by civilian standards. Recruits are provided extensive "in-kind" compensation, for example, "three hots and a cot," (three meals a day in the mess hall and a bunk in the barracks). Recruit and "lifer" alike are expected to adhere to norms and traditions mostly unique to the military and carry out their lives largely within its confines. Only a few are married and the spouses of those who are need to be consummate team players. Since manpower needs draw from a large pool of eligible males, participation by women is limited and highly restricted. All this is consistent with the expectation that military members will put themselves in harm's way, a sacrifice for which they cannot be adequately compensated.

An *occupational* military, Moskos argued, takes on a very different character because its manpower needs compete with those of the civilian marketplace. Hence, enlistment bonuses and lucrative wages are necessary incentives, particularly for skills that are in greater demand. These inducements are more attractive when they are substantial enough to allow service members to purchase their own food,

Table 4.1 Moskos's I/O Thesis: Projected Characteristics of Traditional, Draft-based Military vs. All-Volunteer Force[a]

	Traditional Military	*All-Volunteer Force*
Issues	Institutional	Occupational
Legitimacy	Normative values	Market value
Social regard	Esteem for service	Prestige for pay
Role commitments	Generalist, diffuse	Specific, specialist
Recruit appeals	Duty, service	Pay, technical training
Evaluations	Holistic	Quantitative
Reference groups	Within military	Outside military
Basis for pay	Rank, seniority	Salary, bonuses
Mode of pay	Much in-kind	Skill, organizational need
Legal system	Military justice	Civilian jurisprudence
Female roles	Limited, restricted	Wide, open career
Spouse	Integral part	Reduced role
Residence	Military housing	Civilian housing
Post-service status	Veteran	Civilian

a Adapted from Charles C. Moskos and Frank R. Wood, 1988, *The Military: More Than Just A Job?* Washington, D.C.: Pergamon-Brassey, p. 16.

housing, and such. Market conditions also mean that the supply of males may not be able to sustain upticks in manpower thresholds, thereby opening the door to the recruitment of women as members of the military in full standing. This creates greater convergence with the civilian sector and reduces military uniqueness. Moskos conceded an occupational military might work during peacetime, but worried it would implode during times of war when death and dismemberment became looming realities. Military planners apparently agreed. Their projections called for a reactivation of the draft during times of war.

As the term I/O thesis implies, Moskos did not mean these listings as foregone conclusions but rather as hypotheses about the likely directions on a continuum in which the AVF would push the military. He thought any negative impacts would be greatest for the Marine Corps and Army – prototypes of the institutional model before the AVF days[36] – and the least for the Air Force (whom he thought more closely resembled an occupational military).

In 1977, the seminal article on the I/O thesis appeared in military sociology's academic journal, *Armed Forces & Society*, and in *Parameters*, the official journal of the Army War College. The *Parameters* article received wide and sympathetic readership among top U.S. military leaders. Some were uneasy about the indefinite suspension of the draft and related recruitment issues, including the reliance on women as equal or near-equal military members. It was Air Force General David C. Jones who did the most to promote the work. General Jones, then Air Force Chief of Staff, purchased 13,000 copies to distribute to Air Force officers and put together a conference at the U.S. Air Force Academy on the topic with Moskos as the featured speaker. From the conference proceedings came the book, *The Military: More Than Just a Job?*[37] which featured theoretical and research articles on each of the issues in Table 4.1.

There were two important outcomes from the conference. One, it gave the I/O thesis such center stage that it displaced Huntington/Janowitz's focus on the professional officer as the pivotal research agenda in military sociology. Enlisted recruitment and life in the military reappeared on the research docket. Further, at Moskos's behest, military sociologists from other nations had attended the conference. Britain had gone to the all-volunteer format in 1962 and other European militaries were about to. It no longer was a matter of looking at what was going on in the U.S. military. Now it was a question of asking what was happening in other nations and moving toward a comparative, cross-national perspective.

It is clear from the above discussion that Moskos's I/O thesis was widely read and debated, both in military sociology and within the military itself. Its importance lay in its usefulness in framing the debate and stimulating research. We continue this discussion in Chapter 5.

Into the 21st Century

Two events in the early 1990s proved critical for sustaining the concept and reality of AVFs. Persian Gulf War I – the U.S military response to the invasion of Kuwait by Saddam Hussein's Iraqi Army in 1991 – stands as one of those defining moments. A U.S.-led coalition fielded a massive military force in the Persian Gulf and crushed the Iraqi

Army in a combined air and armored blitzkrieg, the latter lasting a mere 100 hours. It was an absolutist's dream war, one version of what some meant by "no more Vietnams,"[38] and an illustration that AVFs could be stunningly effective fighting forces.

The second event was the collapse of the Soviet Union at about the same time, marking a dramatic end to the Cold War. As the rejoicing in the streets dissipated, Western militaries began to take stock of what their mission now would be. Whatever the answer, nations not already using AVFs began converting to that format. The time was ripe for an updating of the I/O thesis, and Moskos – along with political scientist John Williams and David Segal, one of the authors of this book – provided one.[39] Under the rubric of the *post-modern military*, they contrasted the security threats of the post-Cold War era with two earlier ones, 1900–1945 and 1945–1990. In the earlier eras, the security threats were invasion of the homeland by an enemy force and, after 1945, extermination by nuclear weapons. In the post-Cold War era, the security threats were most often subnational violence or military operations against nonstate actors. Moskos and others then identified resultant organizational trends that this shift in security portends.

While the concept of the post-modern military gained quick acceptance, in no small measure because of the stature in the discipline of those advancing it, there are reasons for reassessing its usefulness. The very term "post-modern" carries with it some unwanted and unnecessary baggage. More broadly, it connotes in disciplines beyond military sociology a critique of scientific thinking. Thus, the directions of analysis suggested by Moskos and others do not flow naturally from the term "post-modern." We turn our Spotlight to a military sociologist who has advanced just such a critique and offered an alternative.

Spotlight on Sociological Thinking and Research

Anthony King, "The Post-Fordist Military"[40]

British military sociologist Anthony King earned his doctorate in Sociology at the University of Salford in Manchester, England. This is worth mentioning for several reasons, not the least of which is the convenience this university's location afforded in doing his dissertation. To collect the data for this research, King conducted participant-observation fieldwork by attending every Manchester United football match (to Americans, soccer games) at old Trafford Stadium during the 1993–1994 season. His focus was on the "lads" for whom following the Red Devils' football was a way of life.[41]

Historically, football as a spectator sport in England has been a working-class pastime. By the 1980s, the lads had elevated this pastime into a lifestyle revolving around Man U matches and "cracking," that is, getting off on the camaraderie there while drinking, singing, and fighting. This was facilitated by traveling to and from matches in throngs, standing together during matches in the arenas' packed, seatless terraces, and, if all went well, "going off" on violent rampages at the end of matches. A common explanation for the lads' devotion to cracking centered on declines in British working-class life in a post-Fordist era. If so, Man U owners and those of other teams added another layer of apprehension in the 1990s. To limit rife hooliganism and violence and to attract fans from a wider range of social classes, the owners replaced the open terraces in their stadia with cordoned off sections of more expensive assigned seating. Still, the disgruntled lads soldiered on, and King had a dissertation explaining the hows and whys.

Of course, if this was the main consequence of post-Fordism, we would not be spotlighting King's work. King developed his career in military sociology by providing insightful accounts of changes in the composition, missions, and practices of European militaries in comparison to those of the U.S.A. King's critique of Moskos and others using post-modernism to account for these changes is not that they necessarily were wrong. Rather, he argued, they overlooked a more useful concept plied by industrial sociologists: *post-Fordism*. Properly appreciated, it may account for a wide range of social phenomena of which football and the military are only a part.

Why post-Fordism? Fordism, as we showed above, is a concept derived from Henry Ford's style of organization and production in making cars that dominated economic (and other) worlds in the first 50 years or so of the 20th century. It was eclipsed by a new system of organization and production – so-called *lean production* – devised by the Toyota Corporation in the 1950s. That and subsequent changes in global markets rendered the strengths of mass production a poor fit. By the 1970s, many Western corporations, including the Ford Motor Company, abandoned or modified their Fordist thinking. Industrial sociologists have categorized the adaptations in four areas: workforce composition, outsourcing, management centralization, and corporate networking.[42]

Fordist companies relied on semi-skilled workers who manned assembly lines where large stockpiles of parts were stored at each step. In the case of cars, the line churned out a huge volume of vehicles with the same characteristics before it was reset. Finished cars were sent to dealers who negotiated the sale price with wholesale or retail customers based upon supply and demand. In 1950s Japan, Toyota did not think this arrangement would work very well for them and sought to make cars in response to pre-paid orders. This required highly skilled teams of engineers and workers who could set and reset assembly lines daily in response to customer demand. Toyota fostered a small core of these teams to whom they promised lifetime employment and cultivated a peripheral flex-force which could be called upon or stood down as needed. They also *outsourced* the acquisition and delivery of parts for the assembly line to other independent companies on a *just-in-time* basis.

While lean production assembly lines empowered workers, it also required a high degree of management control to coordinate and oversee the process. This combination, King observes, paradoxically increased *managerial centralization* while simultaneously *decentralizing the production process*. At its best, this development makes for motivated and smart execution at both levels. Taken together, these changes notably increased the rapidity, quality, and cost-effectiveness of Toyota's production line, creating what James Womack, Massachusetts Institute of Technology's director of economic research, called "the machine that changed the world."[43] Finally, the instability and competitiveness of global markets encouraged companies in the 1970s to create business alliances as a buffer. Within these *corporate networks*, companies who otherwise would be grim competitors assisted each other's development while maintaining flexibility.[44]

American and European militaries since the 1990s have undergone parallel processes. For a variety of reasons, some quite rational, others quite contradictory and self-serving, the U.S. military charted a path that Europeans – in King's research, Britain, France, and Germany – have followed out of necessity. First, just as *mass production* saw its heyday wane after WWII, so did *mass armies*. Britain adopted the AVF format in 1962 and, as we have seen, the U.S.A. followed suit in 1973, and France and Germany have done so since then. The major impact has been a significant reduction in the size of *post-Fordist militaries*. King reports that the British military now is the smallest it has been in 200 years, and the U.S. military currently is roughly one-half the size it was in 1960.

Just as productivity in the post-Fordist industrial sector soared despite reductions in mass, smaller has not meant less potent militaries. To the contrary, the technologically enhanced destructive capabilities of today's militaries dwarf those of their forebears. AVFs too have become more highly trained and professionalized. Further, Americans and their European counterparts have invested significantly in Special Operations Forces (SOFs). Indeed, King states, "[SOFs] constitute the heart of the *new military core*" (emphasis added).[45] For example, there were about 3,000 SOF soldiers in the U.S. military in the late 1970s. That number increased by 2010 to more than 30,000, despite grave reservations among senior military leaders about the consequences for conventional units.

These days the Reserves and National Guard serve as the peripheral flex-force for the Active-Duty military. In the Vietnam war, rather than activating Guard and Reserve units, the U.S.A. relied upon the draft to infuse Active-Duty forces. In fact, one popular way to avoid service during that war was to join the Guard or Reserves (and both enjoyed long waiting lists). The Reserves and Guard now have been restructured to better augment the Active-Duty military. The high operational tempo of the wars in Afghanistan and Iraq has meant that many Reserve and Guard units have deployed multiple times, both as backfills stateside for departed Active-Duty units and to the warzones as front-line support and combat units. This too has been a pattern among European militaries.

In a similar vein, the U.S. military has adopted significant *outsourcing* measures. Stateside, this was at first evident in the conversion of garrison support in food service, base maintenance, and the like (historically performed by troops themselves) to positions filled by civilians. The trend has since expanded to include select tasks in logistics, maintenance, and technical combat support. The U.S. military has shifted portions of its logistics chain to private contractors providing just-in-time delivery of supplies and parts, that is, "focused" logistics in Pentagon parlance.[46] Further, the emergence of private military companies (PMCs) has allowed certain quasi-combat roles to be carried out by civilians, comprising a kind of corporate "foreign legion." These developments, particularly when tasks include those in war zones, raise prickly questions, some legal, some practical and political.

Just as post-Fordist industrial managers moved toward more *centralization of authority*, military commands have as well. In the military this has taken the form of, to use King's word for it, *jointery*.[47] In WWII, an army and navy could operate in parallel but independently. Later this was somewhat true too of an air force. These days, diverse and irregular missions require *joint commands*. Here the roles of each of the services must be carefully orchestrated to take advantage of specialized capabilities, and new technologies make it possible to plot and monitor the process like never before. As in industry, *centralization of command* has been accompanied by *decentralization of execution*. Mass armies call for tight control to carry out orders, but highly skilled, volunteer forces can operate more efficiently when top commanders delegate tactical implementation. Developing a *commander's intent* to accompany mission statements and orders can help ensure subordinates know the parameters of their discretion.

Finally, mass armies fighting conventional wars usually prevailed if they could concentrate more troops in a pivotal area than could their adversary. King notes that *victory by attrition* – winning a war by wearing down an enemy at the point of attack – was the U.S.'s and the Soviet Union's plan of attack (undergirded by nuclear deterrence) in the early years of the Cold War. From a numerical point of view, this favored the Soviets. The U.S.A. eventually sought an alternative in a so-called AirLand Battle Strategy that, as its label suggests, called for coordinated air and land attacks deep behind enemy lines at its weakest points. Victory here is achieved by causing the enemy to collapse from the rear forward. This notion of *network-centric warfare* is the one the U.S.A. and its European allies carried into the 21st century.

This discussion shows how weapon technologies have evolved rapidly since industrialization. Still, soldiers throughout history who fought in "big wars" would have been able to follow what was happening in battle. However, whether one prefers the post-modernist or the post-Fordist concepts for discussing 21st-century militaries, something has fundamentally changed.

Questions for Discussion

1   Drawing on Max Weber's theory of bureaucracy, explain how the U.S.A.'s approach to production and organization led to military dominance in World War II.
2   Both Samuel Huntington and Morris Janowitz suggest that military professionalism is essential to civil–military relations. Having studied their positions, what do you think: Should military professionals separate themselves from civilian institutions or should they be more integrated with them?
3   Western militaries have engaged in a number of conflict-ridden operations since Charles Moskos's I/O thesis appeared in the 1970s. How have all-volunteer forces performed? Does it seem that today's servicemembers consider military service mainly as a job? Why, or why not?
4   According to Anthony King, the tendency of all-volunteer forces to "civilianize" some portion of their operations, even in war zones, creates prickly questions. Do you see the long-term consequences of this for civil–military relations a good thing or a bad thing? Explain.

Notes

1 James Aho, 1979, "The Protestant Ethic and the Spirit of Violence," *Journal of Political and Military Sociology*, 7 (Spring): 103–119, p. 104.
2 Ibid., p. 110.
3 Max Weber, 1978, *Economy and Society*, translated and edited by Guenther Roth and Claus Wittich, Berkeley and Los Angeles: The University of California Press; first printing, 1968,

New York: Bedminister Press. *Economy and Society* was first published posthumously in Germany under the direction of Marianne Weber (*Wirtschaft und Gesellschaft*, 1922).

4 Robert S. Woodbury, 1960, "The Legend of Eli Whitney and Interchangeable Parts," *Technology and Culture*, 1 (Summer): 235–253. Whitney previously had invented the cotton gin, but loopholes in patent laws kept him from cashing in on what should have been a very profitable invention.

5 David A. Hounshell, 1984, *From the American System to Mass Production, 1800–1932*, Baltimore, Md.: The Johns Hopkins University Press, pp. 6–10, chapters 2, 4, and 5.

6 Ibid., p. 219.

7 Ibid., p. 224.

8 Daniel M.G. Raff and Lawrence H. Summers, 1987, "Did Henry Ford Pay Efficiency Wages?" *Journal of Labor Economics*, 5: 57–85, p. 63. One disgruntled worker was quoted as saying, "If I keep putting on Nut No. 86 for 86 more days, I will be Nut No. 86 in the Pontiac bughouse."

9 Ibid., pp. 57–60.

10 In 1939, there was 1 car/truck for every 6 people in the U.S.A., compared to 1 for every 49 in Great Britain, 1 for every 194 in Germany, and 1 for every 1,789 in Japan (John Paxton, 2008, "Myth vs. Reality: The Question of Mass Production in World War II," *Economic & Business Journal*, 1 (October): 91–104, p. 101); Robert J. Overy, 2005, *The Air War, 1939–1945*, Washington D.C.: Potomac Books, p. 190, cited in Paxton, p. 91.

11 Paxton (op. cit., pp. 95–97) analyzed aircraft production data for Germany, Great Britain, Japan, the U.S.A., and the U.S.S.R. for the World War II years. His comparative analysis demonstrates that U.S. production of total aircraft, aircraft engines, and aircraft weight was about three times higher than any of these other countries.

12 Army Service Forces, 1993, *Logistics in World War II: Final Report of the Army Service Forces*, Washington, D.C.: Center of Military History, U.S. Army, Chart 5, pp. 25–30.

13 Paxton, op. cit., p. 94. A news/propaganda film made during WWII about the production of the B-24 at the Willow Run plant may be viewed at www.youtube.com/watch?v=iKlt6rNciTo.

14 James P. Womack, Daniel T. Jones, and Daniel Roos, 1990, *The Machine that Changed the World: The Story of Lean Production*, New York: Harper Perennial, pp. 48–70.

15 Ibid., pp. 212–214, 237.

16 Samuel P. Huntington, [1957] 2003, *The Soldier and the State: The Theory and Politics of Civil-Military Relations*, Cambridge, Mass.: Belknap Press of Harvard University.

17 Morris Janowitz, [1960] 1971, *The Professional Soldier: A Social and Political Portrait*, New York: The Free Press.

18 Huntington's use of the term "liberal" here does not correspond to how it now is commonly employed, but refers instead to a broader ideology of freedom, individualism, and other democratic values that largely are incompatible with military ideals.

19 Peter D. Feaver, 1996, "The Civil-Military *Problematique*: Huntington, Janowitz, and the Question of Civilian Control", *Armed Forces & Society*, 23 (Winter): 149–178.

20 Huntington, op. cit., pp. 156–158.

21 Huntington, op. cit., pp. 163–176.

22 David R. Segal and Karin De Angelis, 2009, "Changing Conceptions of the Military as a Profession," pp. 194–212 in Suzanne C. Nielsen and Don M. Snider, *American Civil-Military Relations: The Soldier and the State in a New Era*, Baltimore, Md.: The Johns Hopkins University Press, pp. 197–200.

23 Huntington, op. cit., pp. 7–17; the quote appears on p. 11.

24 Janowitz, op. cit., pp. lvi, 6–8.

25 Janowitz, op. cit., pp. 22–36.

26 Ibid., pp. 79–97.

27 Ibid., pp. 283–300. Janowitz identifies the contemporary origins of the *pragmatic* and *absolutist* approaches with the separation of strategies during World War II between the European and Pacific theaters and the philosophies of the respective commanding generals, General George C. Marshall and General Douglas MacArthur. The Army Air Corps and Air Force general most closely associated with the *absolutist* position is General Curtis LeMay, who devised the bombing strategy for the European and later the Pacific theaters.

28 Ibid., pp. 417–433. The "whole man" quote appears on p. 425 and again on p. 431. Janowitz argues that military education should encourage officers to become officer-intellectuals in the sense that they should bring an intellectual dimension to their work.

29 Ibid., p. 420 (emphasis added).

30 General LeMay stated in his 1965 memoir, "my solution to the problem of [North Vietnam] would be to tell them frankly ... that we're going to bomb them back into the Stone Age. And we would shove them back into the Stone Age with Air power or Naval power – not with ground forces" (General Curtis E. LeMay, with MacKinlay Cantor, 1965, *Mission with LeMay: My Story*, New York: Doubleday, p. 565).

31 This position is most eloquently made by Col. Harry G. Summers (1982, *On Strategy: A Critical Analysis of the Vietnam War*, New York: Presidio Press), who argues that the U.S.A. won militarily but lost politically because the American people failed to unite behind the war effort. The counter-position is developed, also eloquently, by Lt General H.R. McMaster (1998, *Dereliction of Duty: Johnson, McNamara, the Joint Chiefs, and the Lies that Led to Vietnam*, New York: Harper Perennial).

32 Brian M. Jenkins, 1970, *The Unchangeable War*, RM-6278-ARPA, Santa Monica, Calif.: Rand Corporation.

33 Morris Janowitz, 1983, *The Reconstruction of Patriotism: Education for Civic Consciousness*, Chicago: University of Chicago Press.

34 Charles C. Moskos, Jr., 1977a, "From Institution to Occupation: Trends in Military Organization," Ar*med Forces & Society*, 4 (October): 41–50; 1977b, "The All-Volunteer Military: Calling, Profession, or Occupation?" *Parameters* (7): 2–9; and 1986 "Institutional and Occupational Trends in Armed Forces: An Update," *Armed Forces & Society*, 12 (Spring): 377–382.

35 Douglas Martin, 2008, "Charles Moskos, Policy Advisor, Dies at 74," *The New York Times*, 5 June, downloaded from http://www.nytimes.com/2008/06/05/us/05moskos.html?_r=0.

36 Moskos had earlier published *The American Enlisted Man: The Rank and File in Today's Military* (1971, New York: Russell Sage Foundation). The study is based on his field interviews with enlisted soldiers in Vietnam – the only sociologist to do so during that war – Germany, Korea, and the Dominican Republic.

37 Charles C. Moskos and Frank R. Wood, 1988, *The Military: More Than Just A Job?* Washington, D.C.: Pergamon-Brassey.

38 "No more Vietnams" has meant different things to present-day absolutists and pragmatists. For example, to General Colin Powell it meant becoming militarily engaged only in conflicts that are amenable to the massive use of force in obtaining a military solution. In contrast, a General David Petraeus would argue for using a full spectrum of kinetic and non-kinetic assets in the many limited conflicts of the post-Cold War era.

39 Charles C. Moskos, John Allen Williams, and David R. Segal, 1999, *The Post-Modern Military: Armed Forces after the Cold War*, New York: Oxford University Press.

40 Anthony King, 2005, "The Post-Fordist Military," paper downloaded from people.exeter. ac.uk/acking/Papers/postfordistmil.doc; and 2006, "The Post-Fordist Military," *Journal of Political and Military Sociology*, 34 (Winter): 359–374.

41 Anthony King, 1997, "The Lads: Masculinity and the New Consumption of Football," *Sociology*, 31 (May): 329–346.

42 King, 2005, op. cit., pp. 8–11; 2006, op. cit., pp 360–361.

43 Womack et al., op. cit.

44 King, 2006, op. cit., p. 361.

45 Ibid., p. 362.

46 Ibid., p. 365.

47 King, 2005, op. cit., p. 37–39.

Recommendations for Additional Reading/Viewing

Bernard Boëne, 1990, "How 'Unique' Should the Military Be? A Review of Representative Literature & Outline of a Synthetic Formulation," *European Journal of Sociology/Archives Européennes de Sociologie/Europäisches Archive Für Soziologie*, 31 (1), 3–59.

(A classic analysis by the leading French military sociologist of the requisites of military organization)

Ford Motor Company, *Building the B-24 Bomber in WWII: The Story of Willow Run*. Film owned and supplied by www.PeriscopeFilm.com. Available for free viewing on YouTube.

(Grainy, black-and-white footage of the construction and operation of Willow Run during WWII)

Bernard Rostker, 2006, *I Want You!: The Evolution of the All-Volunteer Force*, Special edition with DVD, Santa Monica, Calif: RAND Corporation.
(An assessment of the performance of the All-Volunteer Force, plus extensive trove of documents on DVD)
Don M. Snider and Lloyd J. Matthews, 2005, *The Future of the Army Profession* (2nd ed.), New York: McGraw-Hill Education.
(An updated edition of 2002 classic statements about the nature of a profession, specifically in the U.S. Army)
Ori Swed and Thomas Crosbie, 2017, "Private Security and Military Contractors," *Sociology Compass*, 11 (November): e12512.
(Thoughtful analysis of who private contractors are and the issues posed for the military by the use of them)

5 Soldiers, Sailors, Airmen, Marines (and Terrorists)

Who Fights and Why They Fight

Reader's Guide: The supply of male bodies for manning (literally) a military is determined to a significant degree by a society's birth and death rates. France's disadvantage with Germany in World Wars I and II provides a classic example. That leaves the question of "who?" Historically the U.S.A.'s approach has been to have a small, core, professional military augmented in time of war by a deep military draft. The first Spotlight reviews how this approach has worked, and what have been its successes and problems. Post-World War II All-Volunteer Forces (AVFs) everywhere have had to move beyond all-male militaries to meet the numbers required to make them viable, that is, have added women as a matter of necessity, then practicality. Why military members fight is another matter. Since WWII, military sociology has pointed to the vital role of unit cohesion. The second Spotlight examines the types of cohesion and their utilities, drawing evidence from the Argentine–British war in the Falkland Islands. We conclude with a Spotlight focusing on how al-Qaeda has successfully attracted its volunteers.

The two issues posed in this chapter's title, *who fights* and *why they fight*, are related but ultimately require very different lines of explanation. Many factors play into the former, including culturally defined notions of who is service-worthy, the supply of those said to be eligible, and the motivations of those eligible to take up the call. That call usually takes the form of *coercion, material gain, notions of duty and honor*, or some combination of these. In any case, there is quite a gap between what it takes to incentivize those eligible to sign up and what is required for them to endure thereafter. Compulsion and utility are powerful incentives, but by themselves generally not sufficient for the latter. To persevere, military units and their members must have a special something, described by Leo Tolstoy in his classic novel, *War and Peace*, this way[1]:

> In warfare the force of armies is the product of mass multiplied by something else, an unknown X. ... X is the spirit of the army, the greater or less desire to fight and to face dangers on the part of all the men composing the army, which is quite apart from the question whether they are fighting under leaders of genius or not, with cudgels or with guns that fire thirty times a minute.

But first, *who fights*? In the *hunting-gathering societies* in which humans have lived for most of history, hunter-gatherers typically became part-time warriors when situations called for it. The same is the case for *horticultural-herding societies*. But, as we saw in

DOI: 10.4324/9781003282549-5

Chapter 3, all this changed around 5,000 B.C. with the development of *large-scale agriculture*. Humans in these societies had the capacity to produce more foodstuffs and other necessities than were needed to survive, and so could redirect some workers, almost always males, into other activities, like being full-time soldiers. As we saw in Chapter 4, this has reached its highest level in *modern, industrial societies* with their ability to raise *mass armies* and then *AVFs*.

Demography and the Military

Putting aside for a moment the issue of *who* is service-worthy, the numbers of young men in a society available for military service is a *demographic* matter, that is, one affected by birth and death rates. A striking historical example of this may be found in France, England, and Germany. In 1700, all three were large-scale agrarian societies with *high birth and death rates*, and hence *low levels of population increase*. France was the largest nation in Europe with 21 million inhabitants. It was one-third larger than neighboring Germany (population of 14 million), with whom it had a long history of warfare, and four times bigger than England (5 million), another sometime foe.

The first of these countries to industrialize was England in the late 18th century. Early industrialization is marked demographically by declines in the death rate (while the birth rate remains high), resulting in a population explosion. By the end of the 19th century, England's population was 30 million, six times the size it was 200 years earlier. Similar growth occurred in Germany during this time. Its population by 1900 was 56 million, almost four times bigger than in 1700.

France's trajectory was different. As industrialization matures, the birth rate then begins to decline as well, reducing the magnitude of the population explosion. France entered this stage before England and Germany because the French took more readily to advances in the use of contraceptives, crude by today's standards but still somewhat effective.[2] Another contributing factor was that French young people started marrying in their early to mid-20s instead of in their mid-teenage years.[3] As a result, France's population only doubled in size by 1900 to 40 million, an increase paltry only by comparison to England and Germany. Germany's population by 1900 was 40% larger than France's, and England's was rapidly catching up.

There was disagreement among the French themselves about whether these developments were a good or bad thing. However, that all changed with the "shattering defeat" of the French army by the Germans in the 1870 Franco-Prussian War.[4] Although the French defeat more likely was due to superior German military organization than the size of its military, Germany's superior 19th-century birth rates provided a more palatable explanation: that the Germans were outbreeding them sounded better to the French than that the Germans were outsmarting them.

Thus, *fertility fever* swept across France. In the ensuing years, more than 80 pro-natalist organizations were formed to sing the praises of having babies and large families. The French Assembly passed laws prohibiting abortion and restricting the exchange of information on contraceptive practices. French politicians introduced bills to give tax breaks and government stipends to families with more than three children, financed in part by punitive taxes on bachelors, spinsters, and childless couples. The focus however was on the demographic plight of the French military. During the 1914 elections, one pro-natalist organization distributed more than 1 million posters depicting two French soldiers being bayoneted by five German troopers.[5] However, increases in fertility take 18 years or so to impact military-conscription capacities.

World War I, 1914 to 1918, devastated France. The French fielded an army of 7.5 million compared to Germany's 11 million. A whopping 75% of French soldiers were either killed or otherwise incapacitated, while Germany's casualty rate, horrific too, was "only" 54%.[6] These battlefield losses gutted the numbers of young, adult males in France and Germany (and England as well). However, the impact was especially evident in France, which had fewer males to start out with and suffered higher casualty rates. Combined with the wartime disruption of its birth rate, France's rate of population increase dipped well below replacement level during WWI, leaving the country virtually defenseless 20 years later in World War II. That war started in 1939 and raged until 1945, but France collapsed and surrendered to Germany in 1940.

This tale illustrates a basic point: militaries need a supply of young, able-bodied adults to do their business. As a result, military planners have learned to appreciate finely tuned demographic analyses. We turn our Spotlight now on one of the best-known studies of the U.S. military and its manpower issues.

Spotlight on Sociological Thinking and Research

David R. Segal, *Recruiting for Uncle Sam: Citizenship and Military Manpower Policy*[7]

David Segal already has been introduced in Chapter 1 as the main driver, along with Mady Wechsler Segal, of what became the country's premier Military Sociology program at the University of Maryland. His *Recruiting for Uncle Sam*, which blends demographic with social-cultural analysis, is regarded as the classic statement of how the U.S.A. has perceived and dealt with its military manpower dilemmas.

David Segal and Mady Wechsler were both born in New York City, he in 1941 and she in 1945. When he turned 18 years of age, he dutifully reported to his local draft board as required in those days. Local draft boards had latitude in deciding individual cases and, since he wanted to go to college, he was instructed by them to pursue his education. So directed, he went for his baccalaureate degree to Harpur College, the liberal arts college of the newly formed State University of New York. Wechsler obtained her college degrees at Queens College of the City University of New York. She and David subsequently met at the University of Chicago, and they married soon thereafter.

David completed his doctorate in Sociology in 1967 and Mady in 1973. Their first joint employment was at the Army Research Institute (ARI) in Washington, D.C. in 1973. Morris Janowitz had been their mentor at Chicago, but this was their first actual work in military sociology. The military had just switched from a draft-based military to an AVF. The shift was not without trepidation among military planners and the Segals provided research assessments.

Segal began his work on U.S. military manpower issues during his tenure at the Army Research Institute in the early 1970s. The military was particularly open in those years to input and advice from demographically oriented sociologists. A fundamental question in those days was: Would the U.S. military be able to attract enough volunteers to fill its needs? The answer to that depended on several moving parts: What are the military's manpower requirements under likely national defense scenarios? What are the projected sizes of available pools of young men (and, eventually, young women) from which to draw volunteers? And what are the factors affecting volunteer rates among the eligibles? These are questions that Segal has studied over the years.

The first question involves two parts. It includes not only assessments of what these scenarios might be but also conceptions of what the nation's military force should look like. In the very early days of the country, there was considerable suspicion of the military. Hence, in the days of the American Revolution, military leaders relied upon armed citizens who had been guaranteed the right to arm themselves. However, the *militia system*, like later systems of conscription, crumbled under the weight of exemptions and deferments. The notion of a militia rested on the ideal of service by all able-bodied, white males. However, a blizzard of state statutes granted exemptions to categories of largely privileged citizens, so that required service in militias became the lot of those who were less well-off and less powerful.[8]

The militia system also proved untenable because citizen volunteers in the American Revolutionary War felt free to come and go. A volunteer militiaman might decide to go home for a week or two to tend his crops though an important battle was pending. The solution was to activate militias to fight for the duration of specific campaigns. At the war's conclusion, militias reverted to peace-time status, leaving only a very small professional army in place. The U.S.A. did not institute a *national draft* until the Civil War, some 90 years later. Even then the notion of a draft was considered coercive, and both the Union and Confederate armies allowed draftees to hire paid substitutes to take their places.[9]

Thus, the military operated not unlike an accordion: it expanded markedly in time of war as its core professional force was augmented by volunteers, draftees, and, if needed, state militias (now called the National Guard), and, contracted after each war as these part-timers returned to civilian life. World War I also was a period of unrest with conscription. In contrast, World War II was a period of high voluntarism, but still one in five men in the military were draftees.

Mobilizations after 1850 have been carried out in the context of declining birth rates, illustrated in Figure 5.1. Early in the 19th century, U.S. death rates began to decline for the same reasons they had in Western Europe – the country was beginning to industrialize. Soon, birth rates also were declining noticeably, from a high before industrialization of 55 births per 1,000 people, to around 48 births per 1,000 people in 1850, and on down to a current birth rate of about 13 per 1,000. The sole exception in this decline has been the ten-year baby boom after World War II.[10] Throughout this time, some form of a military draft, administered by local Selective Service boards, was used to provide the bodies needed to meet manpower requirements, higher in times of war and lower in periods between them.

This system also allowed deferments and exemptions, policies that favored the middle and higher socio-economic classes by making it less likely they would be drafted. For example, deferments commonly were given at one time or another to those in certain critical civilian occupations and to draft-age eligibles wishing to attend college. Debates over the fairness of the draft reached a head during the Vietnam war. The baby-boom generation, by then of draft age, comprised a sizable cohort from which to draw conscripts. The Department of Defense thus decided to

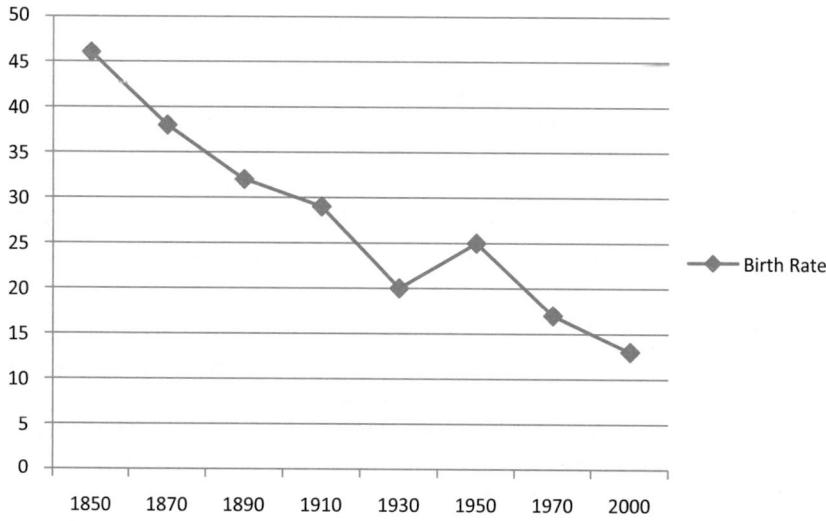

Figure 5.1 Number of Births per 1,000 of U.S. Population, 1850–2000

Source: Authors' Creation.
Note: Excerpted and constructed from data found in Michael Haynes, 2010, "Fertility and Mortality in the United States," *EH.Net Encyclopedia*, https://eh.net/encyclopedia/fertility-and-mortality-in-the-united-states./.

fight the war with the Active-Duty military augmented by volunteers and, instead of the ready Reserves and National Guard, by draftees.[11] While feasible because of the size of the baby boom cohort, this proved to be a very unpopular decision politically.

The bias in the Vietnam-era draft was complex.[12] Those located in the lower socio-economic classes were more likely to be drafted. This was so mainly because the Selective Service System gave deferments throughout the early years of the war to those who wished to attend college, a move favoring those in the upper half of the class system.[13] The consequences of not being able or willing to go to college incurred a high likelihood of being drafted and of being assigned to slots in ground combat forces. Skilled and technical military assignments incurred much less risk of injury and death. A notable exception was that of Air Force pilots and crew who flew bombing missions over North Vietnam. Especially toward the end of the war when the North Vietnamese had upgraded their air defense systems, the casualty rates for these pilots and crews were quite noticeable. Still, toward the end of the war, the Selective Service System instituted a draft lottery – literally drawing ping-pong balls inscribed with birth dates out of a hopper – to reduce class inequities.

A corollary problem was that both the premise and strategy of the war were very controversial. Vietnam was an *undeclared war*, making it an especially lethal police action with more than 58,000 American deaths. College campuses – where students faced the likelihood of being drafted as the scope of permissible deferments was tightened – became hotbeds of debate and anti-war protest. Recommendations for avoiding such strife in future engagements included securing an *official declaration of war* for anything beyond limited military actions, activating both Reserve and

National Guard units to supplement Active-Duty forces, and using a draft only as a last resort.

The Gates Commission, appointed in 1969, went a step further: it prepared a blueprint for a military composed of nothing but volunteers with entry-level military pay competitive with civilian salaries. In 1973, President Richard Nixon's Secretary of Defense, Melvin Laird, announced the formation of an All-Volunteer Force (AVF), bringing the era of draft-based militaries in the U.S.A. to an end.

Demographics of the All-Volunteer Force

In the early years of the AVF, military planners worried that they would draw volunteers heavily from the more destitute citizens in American society, because those with more education and skills would have plentiful opportunities in the civilian labor market. And, with birth rates below 20 per 1,000 and still sinking, it appeared there might not be enough male volunteers to meet manpower needs. As we shall see, the first worry has not been the reality, but the second one has come to pass. Every modern military that has gone the route of an AVF has had also to institute gender integration. This has been the case even though AVFs deliver much more military punch with much smaller numbers.

Table 5.1 summarizes the size and service composition of the U.S. Active-Duty military in ten-year intervals from 1980 to 2019 (the last officially released data at the time of writing). In 1960, the Active-Duty component of the U.S. military was about 2.5 million soldiers, sailors, airmen, and Marines. The military augmented this number during the war in Vietnam with volunteers and draftees to a total of 3 million troops. Following withdrawal from Vietnam in 1972 and the shift to an AVF in 1973, the size of the Active-Duty force was reduced by one-third, bringing it back to 2 million by 1980. To this point, we see the usual war–peace accordion effect.

Since then, despite the Persian Gulf War and the ongoing wars in Afghanistan and Iraq coming to an ungraceful conclusion as we update this chapter, roughly 1.3 million

Table 5.1 Size and Composition of the Active-Duty U.S. Military, 1960–2019[a]

Year	Army	Navy	Marine Corps	Air Force	Total	+/–
1960	873	617	171	816	**2,475**	
Vietnam War 1965– 1972						
1970	+ 52%	+ 11%	+ 53%	– 3%	**3,065**	+ 24%
All-Volunteer Force 1973						
1980	−41%	−23%	−27%	−29%	**2,051**	−33%
1990	−6%	+ 9%	+ 5%	−4%	**2,044**	−.1%
Persian Gulf War 1991–1992						
2000	−34%	−36%	−15%	−33%	**1,384**	−32%
War on Terror 2001–2021						
2010	+ 19%	−11%	+15%	−8%	**1,431**	+ 4%
2019	−17%	+ 1%	−7%	−1%		−7%
(Numerical size in thousands)	**480**	**332**	**186**	**328**	**1,326**	

a Table constructed from raw data, Table 510, Department of Defense Personnel, 1950–2010. Includes National Guard, Reserves, and retired regular personnel on extended or regular active duty. Downloaded from http://www.census.gov/compendia/statab/2012/tables/12s0510.pdf.

troops are on Active-Duty. The only service component larger in 2019 than it was in 1960 is the Marine Corps (186,000 vs. 171,000). The other services are roughly half the size they were 60 years ago. Despite an uneven performance in Afghanistan and Iraq, it still is considered a premier fighting force. The problem in these wars was not that the U.S. military was too small. We address this in Chapter 10.

So, who is filling the ranks of the AVF? First, the AVF generally attracts volunteers who are better qualified than initially projected.[14] If one visualizes the distribution of family income in terms of quintiles from the lowest 20% to the highest 20%, the middle three income quintiles have been consistently overrepresented in the AVF, and the lowest and highest quintiles underrepresented. Further, AVF recruiters consider an entry-level applicant as a *high qualifier* if he or she holds a high school diploma from a graded program of classroom instruction (or certain nontraditional ones) *and* scores above the 50th percentile on the Armed Forces Qualification Test (a nationally normed test of math and verbal skills). The goal has been for at least 60% of enlisted *accessions* – those who actually sign up for military service – to be high qualifiers. For fiscal year 2019, almost 70% of accessions were so qualified, and during the height of the wars in Afghanistan and Iraq (2009–2014), that number was over 80%.

Military planners long suspected that racial and ethnic minorities might be consistent sources of volunteers, and that has proved to be true. In fact, increasing its racial and ethnic diversity has been one of the U.S. military's goals for some time under the argument that a military should be representative of the population it defends. Since 2003, the military has reported racial/ethnic data using five racial categories (White, Black or African American, American Indian/Alaska Native, Asian, and Native Hawaiian/Pacific Islander) and two ethnic categories (Hispanic/Latino or non-Hispanic/Latino), though the latter two may overlap with a racial category. The Active-Duty enlisted accessions for these racial/ethnic groups and for gender are presented in Table 5.2.

African Americans make up 14.8% of the U.S. civilian population in the target recruiting ages, 18–24. Continuing a trend since the early days of the AVF, 18.9% of Active-Duty accessions in 2019 for all four services were Black. For the individual services (not shown in Table 5.2), African Americans made up about 24% of the Army's accessions, and 19% for both the Navy's and the Air Force's.[15] The corresponding figure for the

Table 5.2 U.S. Military Active Component Enlisted Accessions, Fiscal Year 2019[a]

Race/Ethnicity	Dept of Defense			Civilian Population		
				18–24 Years Old		
	Male (%)	Female (%)	Total (%)	Male (%)	Female (%)	Total (%)
White	72.7	60.7	**70.4**	73.8	72.9	**73.3**
Black	16.9	26.8	**18.9**	14.4	15.3	**14.8**
AmInd/AlskaN	0.7	0.9	**0.8**	1.5	1.5	**1.5**
Asian	4.8	4.9	**4.9**	6.3	6.3	**6.3**
Hawaii/PacIsle	0.6	0.6	**0.6**	0.4	0.4	**0.4**
Two or more	2.2	2.4	**2.3**	3.6	3.8	**3.7**
Hispanic	20.6	24.0	**21.3**	23.1	22.7	**22.9**
Non-Hispanic	79.4	76.0	**78.7**	76.9	77.3	**77.1**
Total	**80.6**	**19.4**		**50.1**	**49.9**	
(# in thousands)	131.8	31.8	163.6	14,587.0	14,508.8	**29,095.7**

a Table constructed from Table B-3, Non-Prior Service Active Component Accessions FY 2019.

Marine Corps was about 10%. These percentages approximate those of Blacks already in these branches of service. Asian accessions were underrepresented in all four branches. American Indian, Alaska Native, Native Hawaiian, and Pacific Islanders have representation in the military near their very small numbers in the recruit-age population.

Finally, Hispanics and Latinos constitute 22.9% of the recruit-age population in the U.S.A. In 2019, they made up 21.3% of all Active-Duty accessions. However, this was distributed very unevenly across the services: they made up 33% of Marine Corps accessions but less than 20% for the other three services. All told, this is a very different U.S. military than that of pre-All Volunteer eras. We will continue this discussion in Chapter 6.

At least some military planners suspected they would need some change in policies regarding women in the military if they were to meet their numbers for the AVF. The AVF quickly rendered the gender-segregated approach untenable. In 1976, women were admitted as cadets and midshipmen at the U.S. military academies, and in the years 1976 through 1978 the gender segregated units were eliminated in all the services (see our discussion in Chapter 7). At first women served in mainly traditionally "female" jobs – administration, medical services, and logistics. Other occupational specialties soon opened to them as well, including, in 2016, combat specialties in infantry and armor.

In the first year of the AVF, 1973, women made up 6% of the U.S. military. That percentage has grown steadily, leveling off a bit in the first decade of 2000 and then rising again through 2019.[16] In 2019, 19.4% of the total enlisted accessions for Active-Duty were women. Women now make up almost 17% of the enlisted ranks, with the highest concentrations in the Air Force and Navy (20%). Their numbers in the Army and Marine Corps are 15 and 9%, respectively. Women service members have shown a similar progression in the officer corps.[17] From 4.3% in 1973, women now comprise 22% of the officers in the Air Force, and 19% in the Army and Navy. Eight percent of Marine officers are women.

We conclude this section with an interesting observation on the intersection of gender and race/ethnicity. The difference between female and male accessions is almost a full 8 percentage points for African American enlistees (26.8 vs. 16.9%) and 4 percentage points for Hispanic enlistees (24 vs. 20.6%). This is not a one-year anomaly. In the enlisted ranks, non-White and Hispanic women service members *in every branch* exceed the proportion of male service members who are Non-White and Hispanic. These proportions come about because non-White and Hispanic women have higher rates of both accession and retention than do non-White and Hispanic men. Likewise, among officers who are non-White, a larger percentage are women in every branch than men. The comparisons in the civilian labor market, those with college degrees who are 21–39 years of age, are the reverse for minorities. This suggests minority women are finding more opportunity and steadier career paths in the military than in the civilian sectors, and therefore ones they are more willing to commit to. We revisit this issue in Chapter 7.

Why Do They Fight?

In Chapter 3, we noted anthropologist Keith Otterbein's definition of war: *armed combat between political communities*.[18] This definition contains the basic core of what we call "war": hostility between whole societies or subgroups within them, the central role of groups equipped with potentially lethal weapons, and actual fighting, leading to injury and death of combatants. All this reminds us that war is not simply *interpersonal violence* but is a *group-orchestrated phenomenon*. Doing war calls for getting seriously organized. It requires intentional reconfiguration of priorities, material resources, and societal arrangements. Like collectively performed non-war activities, the process often bolsters *in-group cohesion*, the *inspiration to do great things for a cause* writ large, and the *willingness to sacrifice* mightily on behalf of the group. For such reasons, sociologists Miguel Centeno and Elaine Enriquez remind us in their primer, *War & Society*, that war often entails depths of suffering and misery *and* heights of honor and heroism.[19]

Military sociologist Nora Kinzer Stewart has developed an illustration of these *whys* from the 1982 Falklands war. The war, lasting two and a half months, was a territorial dispute between the U.K. and Argentina. At stake was sovereignty over a set of islands about 300 miles off the coast of South America, known as the Falklands to the British and Islas Malvinas to Argentines.

Spotlight on Sociological Thinking and Research

Nora Kinzer Stewart, *Mates and Muchachos: Unit Cohesion in the Falklands/ Malvinas War*[20]

Like so many conflicts in the modern era, the one in the Falklands/Malvinas has its origin in colonialization by Europeans. The islands since the 1500s have been contested by Portugal, Spain, and the U.K. Since then, there have been settlements having allegiance to one or more of these competitors. Argentina's claim to jurisdiction is that Islas Malvinas were ceded to them as part of Argentina's independence from Spain in 1816. Britain counters that it has had mostly uninterrupted control of the Islands since 1833. Following World War II, Britain and Argentina held talks through the United Nations in search of a settlement. The sticking point was the 70% of some 3,000 Falklanders, as they now call themselves, who are of British descent and wish to remain part of the U.K. All this came to a head on 2 April 1982.

There are several excellent analyses of the hostilities, mostly from the victorious British point of view.[21] A bare bones account goes something like this. On 2 April, a 5,000-strong Argentine military force landed at Port Stanley and within two days overwhelmed the Islands' small contingent of Royal Marines. Britain dispatched a sizable naval task force to retake the island, an enormous undertaking since the islands are some 8,000 miles away from London. Naval and air skirmishes soon followed, resulting in the spectacular sinkings of an Argentine cruiser, *ARA General Belgrano*, and a British destroyer, *HMS Sheffield*. The Argentine navy did not leave port after these encounters.

Britain established its air and naval superiority and soon landed 4,000 troops on the western edge of East Falkland Island on 21 May. A British commander extoled his Marines: "In the Second World War, we marched from Normandy to Berlin. We can bloody well march eighty miles to Stanley" (capital city of the Falkland Islands).[22] The British fought their way to Port Stanley on the eastern side, arriving on 11 June. After three additional days of fighting in Stanley, the Argentines surrendered on 14 June. Britain lost 258 killed and 777 wounded; casualty figures for Argentina were 649 killed, 1,068 wounded, and 11,313 captured.

During the 1970s, Nora Stewart was one of the chief scientists for the U.S. Army Research Institute for the Behavioral and Social Sciences. She held a Ph.D. in

Sociology and Anthropology from Purdue University and, fluent in Spanish, had studied and worked in Spain, Mexico, and Argentina. Combined with her interest and work in military sociology, she was well-suited to conduct a case study of *unit cohesion* by interviewing participants from both sides of the Falklands/Malvinas conflict. She consulted with the military attachés of both the British and Argentine Embassies to obtain lists of officers and enlisted men who fought in the conflict, and traveled to both Great Britain and Buenos Aires to conduct interviews with samples of them.[23]

Stewart begins her analysis by reviewing military sociology's literature on *unit cohesion*. This body of work yields several levels on which to focus in making sense of *why soldiers fight* and the conditions under which they do so successfully (or not). As we saw in Chapter 1, these studies show that soldiers in *cohesive* units endure and carry on even in the most dire and frightful of conditions, while those in less connected ones break and even surrender more easily. But what is *cohesion*? Summarizing themes in these studies, Stewart advances three kinds of cohesion in military units: relationships among soldiers themselves (*horizontal cohesion*), relationships between soldiers and their immediate superiors (*vertical cohesion*), and relationships between soldiers and their military (*esprit de corps*). She adds to this the relationship between soldiers, and then the military, with society at large (Stewart does not label these, but the terms *patriotism* and *civil–military relations* come to mind).

Turning first to the civil–military dimension, the British had experience and modern hardware on their side – and a long and rich military tradition. Its navy had been the world's premier seafaring power, with its colonial army having to play second fiddle. Nonetheless, its army had a proud history as well.[24] Tracing its origins to the 17th-century reign of Charles II, the army is organized along regimental lines. Each regiment has around 1,000 troops which historically have drawn their numbers from specific locales. The British soldier thus was one who had joined the very military unit in which others from his neighborhood, hometown, or region historically have served. It probably was the same unit of his father, uncles, or grandfathers, and its officers and sergeants (NCOs) likely were the sons and grandsons of those who served with a recruit's relatives of previous generations.

Not surprisingly then, soldiers in these units evince strong *esprit de corps*. A new regimental recruit would be able to recite the long history of the unit he was joining. Stewart observes our recruit would have heard these tales from his older family members, in school and in local pubs, and at remembrances and memorial services at local shrines and churches. Once in his chosen unit, he would be reminded by streamers and banners on unit guidons, special crests attached to his uniform, and unit rituals recounting the valor and sacrifice of days past.

If joining the Second Battalion Scots Guards,[25] for example, he would be aware that his unit's origin goes back to 17th-century service in Ireland. He would know how the unit survived intrigues and many battles between England and Scotland, of expeditions against the French and then American revolutionaries in what is now Pennsylvania, and thereafter of battles against the French again, this time in the Napoleonic wars. And so on, through colonial wars in 19th-century South Africa, to WWI and WWII of the 20th century, followed by more colonial wars in Malaya, the Persian Gulf, and Northern Ireland. In 1982, the 2 Scots Guards were on their way to the Falkland Islands to add to this lore. With adjoining units, 2 Scots Guards would carry out a nighttime assault up Mount Tumbledown held by the unseasoned but tough Argentine *Batallon de Infanteria de Marina* 5, designated BIM-5.

Still, as the classic study of German soldiers in WWII by Edward Shils and Morris Janowitz demonstrates (spotlighted in Chapter 1), *esprit de corps* often is not sufficient once the bullets begin to fly. The size of a combatant's world shrinks to his or her most immediate surroundings, where the quality of emotional connections with fellow squad or section members moves to the fore, something Shils and Janowitz called *primary group cohesion* and Stewart termed *horizontal cohesion*. It is the legacy described above, Stewart argues, developed and reinforced by *horizontal cohesion*, that motivated a soldier to stoically charge up Mount Tumbledown with his 2 Scots Guards' comrades.

The Argentines have their own *civil–military* tradition and *esprit de corps* too but arrive there by very different routes.[26] For starters, our Argentine soldier, unlike his British enlisted counterpart, probably is not a volunteer. BIM-5's enlisted Marines, for instance, though led by highly professional officers and NCOs, were mostly conscripts. In those years,[27] Argentina had a military draft for males, who were required to register at age 18 and receive lottery numbers. Called up in sequence, those selected were required to report for one year of military service. Exactly how many were needed depended upon the size of the military, budgetary considerations, and the number of slots left unfilled by volunteers. Despite the luck-of-the-draw nature of it, military service for draftees in Argentina could be highly regarded or detested, depending upon the soldier's political orientation. Especially in the case of the latter, *esprit de corps* could be, to put it charitably, a bit thin.

In any case, there was a strict divide between the officer and enlisted ranks – not necessarily a threat to *vertical cohesion*. Those in the enlisted ranks came mostly from the working and rural classes. Almost all officers came from middle or upper-middle class families in Buenos Aires. All received their commissions by attending the Colégio Militar (Argentine Military Academy). About 90% of Argentines are at least nominally Roman Catholic, and Catholic services and ceremonies have been incorporated into military routines. Argentine military officers are expected to be practicing Catholics, a dissimilar pattern than in Britain but potentially an adroit unifying element in Argentina.

The military in Argentine society is a powerful force. It traces its history to a revolt by a Buenos Aires citizens' council in 1810, a kind of military junta, who took advantage of the war between Spain and France to replace Spain's colonial government with local strongmen. In the 19th century, the military played a key role as the political structure of Argentine society took shape. The legacy of this process is that the Argentine military sees itself as the guardian and, if necessary, the leader of the national government. In the 50 years leading up to the Falklands/Malvinas conflict, Argentina had 19 different heads of state, 10 of whom were generals who stepped in to assume power. A military junta was in power in 1982 and is the one which made the decision to invade the Falklands/Malvinas, perhaps to distract people's attention from the day's economic crisis.

The Argentine military that entered the Falklands/Malvinas conflict was a rather traditional and insular one. While training programs for officers and enlisted ranks alike were quite rigorous, they lacked real-war experience. The Argentine military had not been in a war in more than a hundred years, normally a good thing. Not surprisingly, several Argentine plans and contingencies went awry. The invasion force collapsed, as miscalculations and inefficiencies drained *esprit de corps* and *vertical cohesion*. As for the British, though they had *snafus*[28] of their own, Stewart's descriptions indicate *esprit de corps* and both types of *unit cohesion* remained intact. After the surrender, one Argentine general commented, "We thought we knew about a war, but [the British] have shown us that we only knew about exercises."[29]

What about Terrorists: Why Do They Fight?

If the question above were put to the average person in the street, the response might be: because *they're crazy people* warped by some bizarre view of the world, religion, or such. A more thoughtful reply might be, *it depends, one person's terrorist is someone else's freedom fighter*. We discuss this issue in historical and present circumstances in Chapters 9 and 10, but note here that *terrorism* is a tactic – *the willingness by nonstate actors or paramilitary groups to kill others to achieve some political goal*. Those doing so usually are part of an *insurgency*, that is, a social movement designed to change something about an existing institutional arena.

There are several important parts to this definition. One, it recognizes that the killing is taking place for political purposes, so it is different than lethal violence in which someone is murdered in everyday life in an armed robbery, bar room brawl, or domestic dispute. The use of violence here is *instrumental*, that is, designed to disrupt the daily functioning of a society and/or to intimidate stakeholders in pursuit of some strategic objective. Still, it identifies the killing as unlawful because it takes place outside the legal confines of a lawfully declared war. And, it identifies the killers as nonstate actors or groups, that is, the killing is not being done by a military authorized to kill by international law under specified conditions. The killing of noncombatants by a military also is not permitted but falls under the rubric of *war crimes*, an issue we discuss in Chapter 10.

It is common these days in some circles to simply attach the adjective "Muslim" to the noun "insurgencies," just as in the mid-20th century it sometimes was implied that all insurgencies were Communist ones. The scope of insurgencies is much more nuanced. Bard O'Neill, professor at the National War College in Washington, D.C., has developed a helpful typology.[30] Though the specific goals of insurgencies differ considerably, common to them is violence. Some seek to overthrow the status quo and replace it with something quite different (*anarchists, apocalyptics, restorationists*), others to reshuffle it (*egalitarians, pluralists, reformists*), while still others attempt to defend it (*preservationists*). Finally, *commercialists* seek power to solidify economic interests, some legitimate, some not, and *secessionists* wish to break away and form a separate entity or join some other existing one. The typology is meant to help pin down the motivation for violence, not to suggest such goals are worthy or that a worthwhile goal merits terrorist tactics.

We now place our Spotlight on a recent analysis of terrorist networks, their demographic characteristics, and the processes through which they attract and retain willing members.

Spotlight on Sociological Thinking and Research

Marc Sageman, *Understanding Terror Networks*[31]

Marc Sageman is uniquely situated to study terrorism. He holds both an M.D. in Forensic Psychiatry and a Ph.D. in Political Sociology from New York University. Plus, he is a former Central Intelligence Agency (CIA) service officer who was based in Islamabad, Pakistan from 1987 to 1989. There he worked with the *mujahideen*, Islamic paramilitary fighters, in the final years of their efforts to kick the Soviet Union out of Afghanistan. Since the attack on the World Trade Center in 2001 by an al-Qaeda cell, Sageman has devoted much of his time to the study of the al-Qaeda network. Currently he is a fellow at the Washington, D.C. think-tank, the Center for Strategic and International Studies.

Islam, Sageman notes, is a worldwide religion, and like other of the world religions – Judaism, Christianity, Hinduism, and Buddhism – it contains many contending and contradictory impulses. Of central concern in Islam is the notion of *jihad*, which can have both personal and public connotations.[32] In general, jihad refers to the struggle on the part of an individual Muslim to live a good life in accordance with the principles and rules of the Qur'an.[33] Historically, it also on occasion has encompassed an obligation to defend Islam from outside attackers when so directed by a *fatwa* – a legal opinion formulated by a religious leader. A fatwa calling for the defense of Islam might take several forms, from praying and fasting to the giving of financial contributions to direct involvement in fighting. For example, in 1979 when the Soviets invaded Afghanistan, several Muslim leaders issued fatwas directing Muslims everywhere to assist in some way to repel them. Among those answering the call in those days was a wealthy Saudi citizen, Osama bin Laden.

Although bin Laden's exact role is somewhat murky, he did travel to Afghanistan and appears to have contributed mainly through relying on his personal wealth to purchase supplies and arms in support of the mujahideen. Following the war, he was one of a coterie of radicalized activists who formed a cell devoted to a continuing defense of Islam against both *near* and *far enemies. Near enemies* included local Muslims who did not share their strict, traditionalist interpretation of the Qur'an in rejecting all modern influences. *Far enemies* comprised foreigners who represented or brought with them modern ideas or ways of life, principally the U.S.A. Though they shared a vision of an Islamic state true to their version of an authentic Muslim way of life, the activists themselves disagreed over exactly how this should be achieved. Bin Laden himself considered the use of violence acceptable and necessary against both near and far enemies.

It was he and fellow thinkers on this issue who took over the organizing group known as *al-Qaeda* (the "base"). They were given free rein to operate independently in Afghanistan by sympathizing religious zealots (the Taliban), who had taken over there after the war with the Soviets.[34] Bin Laden's group, referred to by Sageman as "al-Qaeda Central," set up operations and training camps in that country.[35] They functioned as a clearing house for encouraging and financing violence against Muslims who did not share their views and against non-Muslims with influence in the Middle East. Their key activity was the solicitation of volunteers to be organized into cells who could carry out killings against these targets.

During the 1990s, they were successful in establishing three clusters of cells, one in core Arab states (Saudi Arabia, Yemen, and Kuwait), another in North Africa

(Algeria, Morocco, and Tunisia), and one in Southeast Asia (Indonesia and Malaysia). It was a cell from the core Arab states who carried out the attack on the World Trade Center in 2001.

Who is attracted to al-Qaeda as a recruit and why? Sageman details two kinds of explanations. The first, psychological in nature, argues that those who sign up are mentally ill or unbalanced. The contention here is that one willing to kill others in such missions, especially when they themselves often die in the process, must certainly be sick. Sageman, himself a psychiatrist, admits he initially was drawn to this explanation. However, his perusal of terrorist case histories did not find support for this as a central explanation. A second commonly advanced account contends that poverty and its attendant frustrations make extremist appeals attractive to the downtrodden. As a political sociologist, Sageman thought this explanation also held potential. However, this theory would not explain why only some disadvantaged youths cast their lots with the terrorists while others do not.

To provide better answers to these questions, Sageman put together his own dataset.[36] To do so, he enumerated all violent acts claimed or attributed to al-Qaeda and for each developed a list of all those known to be implicated in the acts. From there, using a variety of sources, he constructed the biographies of all on the list with an eye toward the details of their lives before and after their joining al-Qaeda. If one is expecting the typical terrorist to be a young, crazed, uneducated, religious zealot bent on a date with 72 virgins upon glorious death while killing infidels, one will find little evidence for such in Sageman's data.

For starters, those who join appear to be relatively normal in years leading up to recruitment. The typical recruit did not grow up in poverty but came from a family of at least modest means, if not one very well off. Almost none of them went to *madrasas* (religious schools), where the emphasis is on rote memorization of the Qur'an, but to public schools with largely secular curricula. In fact, very few recruits appear to have been particularly religious at all until just prior to their volunteering for al-Qaeda, and their exposure to and depth of knowledge about Islam seems restricted largely to recently acquired versions of the group's propaganda. Nor do they seem to be impressionable youngsters: their average age is 25.7, most are married with children, and almost all have some years of college, if not post-baccalaureate experience. In college, most took coursework or majored in the technical sciences, with almost none of their work in the humanities or social sciences.

One demographic fact did stand out: almost all the recruits had migrated away from their family's place of residence to go to school or search for work, oftentimes out of their home country. Sageman draws on studies by sociologists John Lofland and Rodney Stark on the conversion experience to account for the relevance of this characteristic. Lofland and Stark studied young people who joined eccentric religious cults in California during the 1960s and 1970s. Their description of who converts, and how it occurs, provides a blueprint which Sageman applies to the cases of recruits to al-Qaeda as well (see Figure 5.2).[37]

For example, the nucleus of the al-Qaeda cell which carried out the attack on the World Trade Center were young Muslim men from Saudi Arabia, Egypt, and Lebanon who met each other in Germany while studying engineering at the Technical University of Hamburg. In Hamburg, they stood out as foreigners – not just as ones holding foreign passports, but ones who were ethnically, religiously, and culturally

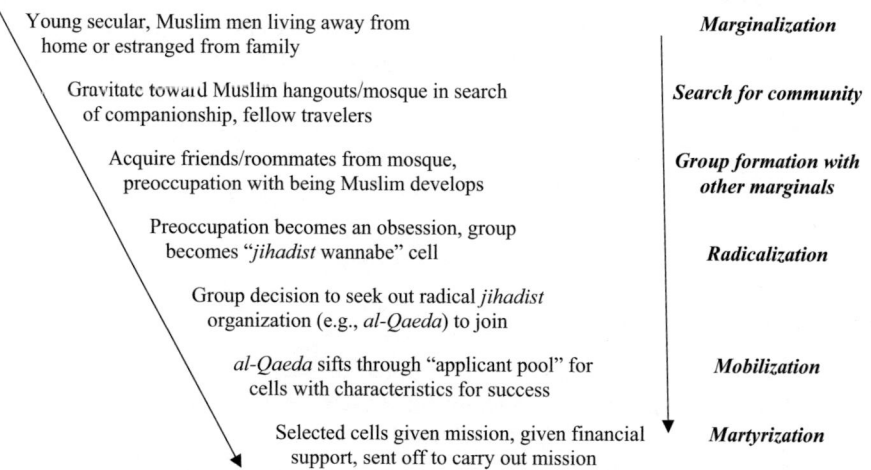

Young secular, Muslim men living away from home or estranged from family	*Marginalization*
Gravitate toward Muslim hangouts/mosque in search of companionship, fellow travelers	*Search for community*
Acquire friends/roommates from mosque, preoccupation with being Muslim develops	*Group formation with other marginals*
Preoccupation becomes an obsession, group becomes "*jihadist* wannabe" cell	*Radicalization*
Group decision to seek out radical *jihadist* organization (e.g., *al-Qaeda*) to join	
al-Qaeda sifts through "applicant pool" for cells with characteristics for success	*Mobilization*
Selected cells given mission, given financial support, sent off to carry out mission	*Martyrization*

Note: The vast majority of the *marginalized* drop out of this process at each step along the sequence, but a very, very small number do not.

Figure 5.2 Sageman's "Conversion" Model: From Ordinary Guy to Jihadist Zealot

Source: Authors' Creation.
Note: Table constructed from Sageman's description of this theoretical model.

distinct. Though only tepidly religious, they gravitated toward a local mosque in search of fellow travelers and companionship. At the mosque, they found friends and eventually roommates with whom they acquired a preoccupation with being Muslim in a strange land. Soon they adopted a more strident Muslim lifestyle in appearance, food, and worship. Then came a perception of, then fixation on, the mistreatment of Muslims in the world. For some who experienced this transformation the preoccupation became an obsession, and a group of such friends began to entertain notions of becoming jihadists in a political rather than a personal sense.

Among those who have reached this point, a few go so far as to actively seek out a radical jihadist organization to join. Recruiters for such groups, including al-Qaeda, often hung out at mosques, coffee shops, and specialized food stores. Al-Qaeda had a reputation for being somewhat picky – for a cell to carry out its mission, it would have to be committed, well-trained, and able to execute a rather detailed plan over a long period of time with minimum supervision. This is not a job description for the crazed and unbalanced. Hence, al-Qaeda recruiters could sift through the applicant pool, so to speak, in search of groups with the appropriate characteristics for success. Once selected, a group would be sent to Afghanistan for rigorous training and the assignment of a mission.

Sageman's account of the conversion process points to the importance of the information and strategic communication dimensions of an effective *war on terror*. Largely this is not a campaign that can be carried out by Western political leaders and media outlets directly. Hence, the importance of emphasizing, as U.S. Presidents George Bush and Barack Obama repeatedly have done, that the war on terror is not a war on Islam, and of cultivating and supporting leaders and outlets within Muslim communities as an essential part of the strategy.

Sageman has followed up his initial book with a more recent one, *Leaderless Jihad*.[38] He argues that the U.S. efforts in Afghanistan have significantly disrupted al-Qaeda Central to the point where it no longer is able to orchestrate missions by jihadist cells. That is the good news. The bad news is that individuals or groups seeking to become jihadists in the violent sense now operate outside organizational control. Here, the internet and chat rooms have become important media in which information and inspiration for bad deeds are conveyed. Hence, impeding and dissuading them has become much more difficult.

Questions for Discussion

1   French sociologist Auguste Comte is quoted as saying, "demography is destiny." Explain how demography (specifically, birth rates, death rates) shapes who serves in the military and how this shapes the security of countries.
2   In 1973, the U.S.A. transitioned from a conscription-based to a volunteer-based military. According to David Segal, how did class and sex shape who served under the draft? How does class and sex influence who serves under a volunteer system? What about the role of race?
3   The willingness to fight in war can be conceptualized through Nora Stewart's terms of *vertical and horizontal cohesion*. Imagine yourself as a British and then an Argentine soldier in the Falklands War. How does cohesion influence your willingness to fight?
4   The average person's image of an al-Qaeda operative is a poor, uneducated, deranged young man rocking back and forth at a madrasa (Muslim religious school) mindlessly chanting verses from the Qur'an. According to Marc Sageman, what is wrong with this picture?

Notes

1   Leo Tolstoy, [1869] 1904, *War and Peace*, translation by Constance Garnett, New York: McClure, Phillips and Company, p. 268, quoted in Samuel A. Stouffer, Edward A. Suchman, Leland C. DeVinney, Shirley A. Star, and Robin M. Williams, Jr., 1949, *The American Soldier: Adjustment During Army Life*, vol. 1, Princeton, N.J.: Princeton University Press, p. 3.
2   Adolphe Landry, [1934] 1982, *La Revolution Demographique: Etudes et Essais sur les Problemes de la Population*, Paris: Insitut National d'Etudes Demographiques, cited and discussed in Kirk, op. cit., pp. 361–363. Landry attributed the decline in fertility to "egotistical" reasons – concerns over the cost of having children, the pain and distress they cause their parents, and to difficulties in childbearing.
3   E.A. Wrigley, 1985, "The Fall of Marital Fertility in Nineteenth Century France: Exemplar or Exception?" *European Journal of Population* 1 (January): 31–60, pp. 50–53.
4   Richard Tomlinson, 1985, "The 'Disappearance' of France, 1896–1940: French Politics and the Birth Rate," *The Historical Journal*, 28 (June): 405–415.
5   Ibid., p. 408.
6   Great Britain fielded an army of 5.4 million and suffered a 44% casualty rate. http://europeanhistory.about.com/cs/worldwar1/a/blww1casualties.htm, downloaded 26 July 2012.
7   David R. Segal, 1989, *Recruiting for Uncle Sam: Citizenship and Military Manpower Policy*, Lawrence, Kan.: University of Kansas Press.
8   Ibid., pp. 18–19. Segal notes, for instance, that "the Massachusetts Militia Act of 1647 exempted officers, fellows, and students of Harvard College; church elders and deacons; schoolmasters; physicians; surgeons; captains of ships more than twenty tons; fishermen who were employed

year round; people who had physical problems; members of the General Court; and people who were excused from service by the General Court or by the Court of Assistants."

9 This provision was discontinued under the Selective Service Act of 1917.

10 At the end of World War II, millions of soldiers, sailors, airmen, and Marines returned from war, married, and began families. The sheer magnitude of this movement created a spike in the birth rate beginning in 1946 that did not fully subside to pre-war levels until 1964 – hence, the term, the baby boom.

11 National Guard units became an attractive way for those with local political connections to avoid service in Vietnam. If one could get into the Guard, one could claim military service during the war without any likelihood of being sent to Vietnam. Two high-level politicians from wealthy families who later took flack for selecting this route are former Vice President Dan Quayle (Indiana National Guard) and former President George W. Bush (Florida Air Guard).

12 Segal, op. cit., pp. 34–38.

13 Both former Vice President Dick Cheney and former President Bill Clinton avoided military service during the war through lengthy college deferments. Former Vice President Al Gore served on active duty during the war and completed a short tour in Vietnam.

14 Under Secretary of Defense, Personnel and Readiness, 2021, *Population Representation of the Military Services, Fiscal Year 2019, Summary Report*. Downloaded from www.cna.org/research/pop-rep, pp. 12–14.

15 Ibid., pp. 25–29.

16 Ibid., pp. 23–25.

17 Ibid., pp. 35–40.

18 Keith Otterbein, 2004, *How War Began*, College Station, Tex.: Texas A&M Press, pp. 9–10.

19 Miguel A. Centeno and Elaine Enriquez, 2016, *War & Society*, Cambridge, U.K. and Malden, Mass., U.S.: Polity Press. See especially Chapter 2, "War of the Warrior."

20 Nora Kinzer Stewart, 1991, *Mates and Muchachos: Unit Cohesion in the Falklands/Malvinas War*, Washington, D.C.: Brassey's, Inc.

21 See, for example, Max Hastings and Simon Jenkins, 1984, *The Battle for the Falklands*, New York: W. W. Norton.

22 Stewart, op. cit., p. 41.

23 Ibid., Appendix A, Methodology, p. 144.

24 Ibid., Chapter 3, "Britain's Military Tradition".

25 See https://www.scotsguards.org/history/.

26 Stewart, op. cit., Chapter 4, "Argentina and San Martin's Legacy".

27 The draft-lottery ended in 2001. Since then, all military service has been voluntary.

28 *Snafu* – military idiom for "situation normal, all fucked up."

29 Stewart, op. cit., p. 56.

30 Bard O'Neill, 2005, *Insurgency and Terrorism: From Revolution to Apocalypse*, second edition, Washington, D.C.: Potomac Books, Chapter 2: "The Nature of Insurgency".

31 Marc Sageman, 2004, *Understanding Terror Networks*, Philadelphia: University of Pennsylvania Press.

32 Ibid., Chapter 1, "The Origins of *Jihad*".

33 The religious book regarded by Muslims as the word of God. It contains the writings of the prophet Muhammad in the 7th century AD which were compiled into a single text after his death.

34 Following the expulsions of the Soviets from Afghanistan, the country fell into factional violence. An ultra-conservative Muslim group, the Taliban, came to power with the promise to end factional fighting. It did so by imposing its own singular set of rules for governing the country.

35 Sageman, op. cit., Chapter 2, "The Evolution of *Jihad*".

36 Ibid., Chapter 3, "The Mujahadeen".

37 Ibid., Chapter 4, "Joining the *Jihad*". John Lofland and Rodney Stark, 1965, "Becoming a World Saver: A Theory of Conversion to a Deviant Perspective," *American Sociological Review* 30 (6): 862–875; John Lofland, 1981, *Doomsday Cult: A Study of Conversion, Proselytization, and Maintenance of Faith*, enlarged edition, New York: Irvington Publishers.

38 Marc Sageman, 2008, *Leaderless Jihad: Terror Networks in the 21st Century*, Philadelphia: University of Pennsylvania Press.

Recommendations for Additional Reading

Scott Atran, 2011, *Talking to the Enemy: Religion, Brotherhood, and the (Un) Making of Terrorists*, New York: Ecco Press.
(Describes research by anthropologist Atran who collects data about terrorism from terrorists themselves)
Ben Connable et al., 2019, *Will to Fight: Returning to the Human Fundamentals of War*, Santa Monica, CA: RAND Corporation, Available at: https://www.rand.org/pubs/research_briefs/RB10040.html.
(Report for U.S. and allied militaries, summarizing literature, suggesting doctrine and training)
Global Terrorism Database, National Consortium for the Study of Terrorism and Responses to Terrorism (START), Available at: https://www.start.umd.edu/gtd/.
(Open-source database, opportunity to view data on terrorist attacks worldwide first-hand)
Office of the Under Secretary for Personnel and Readiness (Department of Defense), published annually, *Population Representation in the Armed Forces*, Available at: https://prhome.defense.gov/M-RA/Inside-M-RA/MPP/Accession-Policy/Pop-Rep/
(Detailed DoD reports on characteristics of All-Volunteer Force vis-à-vis the U.S. population)
Denis Winter, [1978] 1985, *Death's Men: Soldiers of the Great War*, (3rd ed.), London: Penguin Books.
(One of first wave of books by U.K. historians to examine WWI from the point of view of the soldiers themselves)

6   Race and Ethnicity in Society and in the Military

Reader's Guide: Race is both a biological and social phenomenon but draws its relevance for the military and war from the significance a society's members attach to it. We review "the American dilemma" – the gap historically between the U.S.A.'s ideals and the realities for its racial and ethnic minorities – along with how the military has led the way in closing the gap. The first Spotlight focuses on the Navajo Indian Code Talkers and their surprising contribution to communication security for U.S. Marines in the Pacific during World War II. The second Spotlight reveals the little-known story of West African soldiers in France's Colonial Army. African Tirailleurs policed French colonies in West Africa and fought in France during World Wars I and II. In France's post-World War II stab at "compassionate colonialism," they became imperial citizens and fought for France in Indochina and Algeria. Back in the U.S.A., the American military's well-thought-out plan for racial integration has been applied to other of its minorities, including Muslim Americans. The third Spotlight spells out directions their story has taken since 9/11.

In 1937, the president of the Carnegie Corporation, Frederick Keppel, put forth an interesting proposal. He sought a non-American scholar to head up a study of "the American Negro problem." He wanted a non-American because he thought such a person would bring a fresh set of eyes to the long-standing issue, reflected in the country's history of slavery, emancipation, and then rigid segregation based in laws and customs known as the Jim Crow codes. The man selected for the job in 1938 was Swedish social economist, Gunnar Myrdal. Though a World War broke out in 1939, Myrdal commissioned 40 studies by White and Black social scientists in the U.S.A. and wrote a two-volume analysis of what he thought these studies had to say about the "problem."[1]

The Carnegie Corporation was hardly a bastion of racial equality, but Keppel believed passionately in education for all. The Carnegie Corporation had financed libraries across the U.S.A. and supported technical and agricultural studies for African Americans at the Tuskegee Institute in Alabama. The idea behind the proposal Myrdal accepted was not to challenge the country's system of segregation, but to expand education like that provided at Tuskegee. However, Myrdal was left to his own discretion about how to conduct the study and what conclusions to draw.

Myrdal's study, *An American Dilemma*, was published in 1944. In it he spelled out a contradiction between America's ideals about freedom and equality, enshrined in the Constitution and Declaration of Independence on one side, and the realities of legalized

DOI: 10.4324/9781003282549-6

slavery and then Jim Crow on the other. Its history therefore has been one of belated attempts to close the gap or stubbornly keep it in place, sometimes peacefully and lawfully, sometimes through violence and deadly force, sometimes moving forward, sometimes retreating and falling backward. He found it fascinating in a peculiar kind of way that Americans almost universally espoused the principle, *all men are created equal with certain unalienable rights*, but still doggedly supported, or at least tolerated, the blatant and debilitating discrimination against African Americans associated with Jim Crow.

He did however sound a note of cautious optimism. Myrdal felt social forces were aligning for a breakthrough. In particular, he noted rays of hope toward reducing this dilemma as had occurred after the Revolutionary War, Civil War, and World War I. In each of these, he stated, African Americans served in the military and there was substantial progress of some sort, albeit followed afterward by big steps backward. Maybe the participation of Black soldiers in World War II, with other initiatives, would provide a similar but more lasting reprieve.[2] Myrdal's study had little immediate impact. World War II was in its crucial final stages, and few in power felt an urgency to tackle its policy implications.

However, the U.S. military was forced to confront segregation during the war in resolving its manpower needs. As we saw in Chapter 1, this resulted temporarily in assigning Black platoons to some otherwise all-White companies, a move we termed "semi-integration." The military reverted to segregation at war's end, only to confront the problem again in the Korean War during the early 1950s. This time, the move toward integrated military units stayed in effect after the war.

Why Talk About Race and Ethnicity?

The simple answer is that people often attach tremendous significance to these characteristics. *Race*, as a biological concept, refers to people sharing some physical feature in common, usually skin color. Persons having the same or similar skin color may or may not have common ancestors and a shared history. *Ethnicity*, on the other hand, denotes a grouping of people sharing a distinct cultural heritage based upon national origin. Racial and ethnic categories may have some overlap. Interestingly, though race is a biological concept while ethnicity is a cultural one, both are *social constructions*, that is, for either to be consequential in a society, people must think and act as if these characteristics were of critical importance.

Origins of the fascination with race in Western Europe and the New World may be found in the era of colonialization. In the 1500s, buccaneers – first from Portugal, Spain, and the Dutch Republic, then from England and France – set out to chart overseas trade routes. They were looking for precious metals, spices, and other riches. They set up forts and trading posts along the African coasts and trade routes to India, sending merchants and soldiers. Traveling overseas westward, they came across a continent, the Americas, heretofore unknown to them. Everywhere they encountered indigenous people of color in hunting-gathering and horticultural-herding societies, sometimes in farming communities as well. And, in what is now Mexico, Central, and South America, the Aztecs, Mayas, and Incas already had built agrarian empires.

The European explorers and settlers needed land, or at least unimpeded access to it. They reorganized economies in the areas of their trading posts for purposes of extracting riches, and the Portuguese and the English co-opted and diverted the African slave trade to do their bidding. The conquerors thus needed some explanation for who these peoples were and why it was acceptable to enslave them or, at a minimum, usurp their resources.

One early rationale was the "civilizing mission" – *missão civilizadora* in Portuguese and *la mission civilisatrice* in French –that is, the duty of Europeans to bring civilization to the *noble savages* of these continents. Another justification was advanced during the 1850s by the French writer and diplomat Arthur de Gobineau (briefly introduced in Chapter 3).[3] The spectrum of human societies, he claimed, ran from "black" to "yellow" to "white" because of natural biological differences among them, the lower end representing inferior stock, the upper end the superior ones.

Gobineau's book became favored reading after his death in 1882 among the colonial powers of Europe, in the American South, and later in World War II's Nazi Germany. Its guise of *scientific racism* seemed to provide validation for the subjugation that had taken place of non-White peoples by ones of European descent. There are however many historical inaccuracies in his account. If one wants to pay attention to the "color" of who did what, the first civilization was created in the Fertile Crescent – what is now Syria and Iraq – by "brown" people. Other civilizations were developed independently in China ("yellow") and in the Americas ("red"). One explanation developed by UCLA professor Jared Diamond for why civilizations developed independently in these locales points to geography, not skin color.[4] Plants that can sustain human life through large-scale agriculture – wheat, rice, corn, and the potato – are indigenous to these areas, respectively. The cultivation of them spread to other of the world's areas, including Europe, through human migration and cultural diffusion.

Germane to our discussion too is the work of population geneticist, Luigi Cavalli-Sforza, long-time professor at the University of Parma, Italy, and Stanford University in the U.S.A.[5] He is known as the father of genetic geography (the study of human genes from locations around the globe), in Cavalli-Sforza's case, examining the deoxyribonucleic acid (DNA) of fossils from human migrations over the millennia. He also is one of the founders of the Human Genome Diversity Project, started in 1990 to detail the exact genetic content of humans as a species. Findings from his work point to the following: biological differences among peoples of different colors are small and inconsistent, and often do not match the racial categories commonly used in societies; categorizing people based on what they "look like" is messy and frequently arbitrary; people cannot be effectively classified into racial categories using DNA; and, consequently, genetic differences by race are not linked to significant behaviors.

In any case, for almost two-thirds of its history, the original colonies and the U.S.A. had legalized slavery.[6] *Chattel slavery* (one in which enslaved persons are permanently owned by whoever purchases them, as are their children) in the core Southern states was the nation's most divisive issue in the 19th century. This fueled conflict over the issue of whether slaves who fled to Southern border or Northern states had to be returned to their Deep South masters, the requirement that only equal numbers of new "free" and "slave" states could be admitted to the Union, and so on.

At the apex or nadir of that history, depending upon your point of view, the Civil War, 1861–1865, pitted Northern states, loyal to the United States of America, and Southern states, who seceded to form the Confederate States of America. The small professional army of the U.S.A. divided into the two warring militaries, with top officers on each side fellow graduates of West Point, and a military draft and state militias on both sides activated for war. About 680,000 soldiers, Union and Confederate, died in that war – more than the combined total of American military deaths in all other wars in U.S. history. All this was followed by almost 100 years of Jim Crow segregation, a legacy not instantly undone by the passage of the Civil Rights Act in 1964. There has been substantial progress, but the journey is not yet complete.

American Indians in the Military

The Native American experience has been quite different from that of African Americans and all other ethnic groups who immigrated to what became the United States. Archeological evidence indicates that the first immigrants to the Americas crossed the Bering Strait from Eurasia into what is now Alaska around 12,000 BC. Over the millennia, bands of these folk migrated to more hospitable lands, fanning out far and wide to the southeast and southwest, and as far south to what is now South America. However, the continent of the Americas was unknown in the 15th century AD to European explorers, and its land mass was not on their maps as they navigated the seas. Thus, when Christopher Columbus "discovered" America in 1492, an estimated 10 million native settlers already lived across its vast continent. These natives came to be called Indians, probably because Columbus at first mistakenly thought he had landed on the coast of South Asia or India.

The European immigrants who followed usurped Indian lands piece by piece. By the late 19th century, native tribes across the continent had been displaced, sometimes by purchase, other times by hook or by crook.[7] As a result of this history, the U.S. federal government recognizes remaining American Indian tribes as *semi-sovereign nations*, a unique legal status not held by any other racial or ethnic group in the U.S.A. These days, Indians with tribal ties reside on *reservations* (such as the Navajo in southeastern Utah and northeastern Arizona) or in formally designated *traditional tribal areas* (such as the Cherokee in northeastern Oklahoma). Although some tribes gained U.S. citizenship as part of their treaty negotiations, the 14th Amendment to the Constitution in 1868 which guaranteed full rights to all persons born or naturalized in the U.S.A., specifically excluded Indians who were not already citizens. This was corrected in 1924 when Congress passed the Indian Citizen Act, extending U.S. citizenship to all American Indians.

Despite this troubled history, American Indians frequently have been willing participants in the U.S. military and its wars. We focus now on a particularly interesting case, the Navajo "code talkers" of World War II.

Spotlight on Social Science Thinking and Research

Samuel Holiday and Robert S. McPherson, *Under the Eagle: Samuel Holiday, Navajo Code Talker*[8]

Robert McPherson now is Professor Emeritus of History at Utah State University's campus in Blanding. He is known and respected as a historian for his understanding and many books on Native American history and culture, especially in the Four Corners region, the area where the states of Utah, Colorado, Arizona, and New Mexico join. In 2010, he was approached by Samuel Holiday's daughter, Helen Begaii, to help complete an oral history of her father's life and his experiences as a code talker during World War II. The interviews were recorded in 2011 by McPherson and two Navajo research assistants. Holiday recounted his story in Navajo, the language in which he felt more comfortable, fluent, and nuanced. His spoken memories were translated into English, and McPherson added context and commentary for each stage of the text. This is Holiday's story.

Samuel Holiday was born in the summer of 1924 on the Navajo reservation in a hogan, a rounded hut of logs and earth, near Utah's Eagle Mesa in Monument Valley – more than 90,000 acres of stark highland desert with sandstone rock formations so striking it is one of the most photographed places in the U.S.A. His given name in Navajo was *Awéé' zhóni* (Beautiful Baby).[9] He grew up with his mother and sister, tending their flock of about 200 sheep and living a traditional Navajo way of life. Living conditions on the reservation were harsh and economic opportunities sparse.

Holiday's older brother gave him the English name, Samuel, when he turned 12 and was about to depart for the federal government's Bureau of Indian Affairs (BIA) school in Tuba City, Arizona. Like most BIA boarding schools, the one in Tuba City was designed to strip its young students of Indian culture as a way of promoting assimilation. Students there were not allowed to speak their native language, and the school curriculum and regimen of daily living stressed "White ways" at the expense of Indian ones. Holiday struggled but slowly became proficient in English, a skill he felt he needed to succeed as an adult.

When the Japanese bombed Pearl Harbor, Holiday was in the seventh grade.[10] The following spring brought military recruiters to the reservation. Indians were thought to make good scouts and infantrymen, and recruiters found them to be willing volunteers. Young Navajo men joined for a variety of reasons, both economic (they could send monthly pay home to their families) and patriotic (they had pride in their Navajo heritage and newfound U.S. citizenship). Holiday and a couple of buddies from the Indian school signed up together. His father, a medicine man, anointed him and gave him a special eagle feather and sacred amulets containing corn pollen to carry for protection. In the Phoenix, Arizona induction station, Holiday by chance ended up in the line of those receiving physicals for the Marine Corps. So, from there he went on to boot camp at Camp Pendleton outside San Diego in California, all huge steps for a rather sheltered young Indian man.

By this time, a white Marine technical sergeant, Philip Johnston, had convinced Marine Corps brass that Navajos could provide a communications code the Japanese almost certainly would be unable to break.[11] Johnston had lived on the Navajo reservation as a child, where his father was a Christian missionary, and was reasonably fluent in Navajo, an obscure and incredibly difficult language to learn if not acquired as a child. Sgt. Johnston arranged for a demonstration in which a Navajo Marine was given a message in English to transmit in Navajo to another Navajo Marine in an adjacent room, who received the message and translated it back to English. They completed this task in a fraction of the time it would take Marine radiomen with their cumbersome but still vulnerable codes.

An initial group of 29 Navajo Marines was selected to develop a special alphabet and vocabulary of 600 plus words for war in the Pacific theater. Even a Navajo speaker, not part of the program, could not readily follow what they said in code. These 29, and all 460 subsequent Navajo Marines assigned as code talkers, committed this special code completely to memory.

Holiday did well in boot camp. Of the experience, he stated:[12]

> To me, the military was much easier than BIA school .... I was not frightened but ready to fight for my country. When my ancestors went on the Long Walk (in the 1860s when about 10,000 Navajos were forcibly marched several

hundred miles to a reservation), many suffered and died, going far beyond what I was facing in preparation for this war. I believed the Holy People guarded me, and my mother's prayers gave me strength to carry on. These were my real weapons to fight with.

Though the Marines had established a ninth-grade education as a minimum for formally entering the Navajo Code Talker program, Holiday's superior deportment in boot camp and bilingual abilities carried the day. Assigned after boot camp to communications school, he quickly fit in. The code talkers' training area was marked "top secret," and the Navajos were instructed to tell no one about the program. The training included courses in radio communications and repair and in jungle warfare.

Upon finishing the program, Holiday and fellow code talker Dan Akee, a buddy from his Tuba City BIA School days, were assigned to the 4th Marine Division, 25th Regiment.[13] The plan was for one pair of code talkers to move about with the forward units of each of the Regiment's battalions and another pair to remain with each battalion's command post. Only the battalion commanders and White, eight-man radio teams, who would accompany the code talkers to help tote and repair the 80 lb radios, knew who they were, though nothing about the code. Holiday and Akee were designated to accompany their battalion's forward units.

The Pacific battle strategy, begun in Guadalcanal in 1942, consisted of "island hopping," that is, a plan for taking one island after the other, bringing the Americans ever closer to the point where their long-range bombers could reach and devastate mainland Japan.[14] In January of 1943, the 4th Marines boarded ships off Camp Pendleton bound for Hawaii, then the Marshall Islands. In the following 14 months, the Division made four amphibious landings to take the islands of Kwajalein, Saipan, Tinian, and Iwo Jima. In each, though successful, the Marines sustained dreadful casualties.

Holiday and Akee typically were in the second wave ashore, and Holiday recalled the pain of seeing the shores and islands littered with the bodies of fellow Marines. And, in each, the code talkers performed well. Occasionally fighting for their own lives, they maintained a flow of information and communication, updating commanders of the forward situation on the ground, detailing where artillery and other supporting fire were needed, and providing the locations from which casualties could be evacuated. The Japanese suspected some strange American Indian language was being used but could not break the code.

How typical was Holiday's experience? McPherson's review suggests the following.[15] Code talkers comprised a distinctive subset of Navajos who were extensively bilingual and bicultural. Soldiers of Navajo descent raised off the reservation and highly assimilated into Anglo culture were unlikely to have these skills. Neither would most raised solely on the reservation – Holiday was exceptional in that regard. Most code talkers had completed more years of schooling than did Holiday, who, as noted above, was given a waiver. Many had converted to Christianity, even if their Christian practices were intertwined with Navajo religious ones. As one who did not see his first "White man" until the age of 12, Holiday was traditional in his values, especially Navajo religious ones.

The four code talkers who were killed in combat served in forward units, as did Holiday. Finally, one danger for the code talkers was being mistaken by fellow Marines or adjacent Army units as wily "Japs in Marine uniforms." This was enough of a problem that White radiomen in forward units were instructed always to closely accompany their code talkers, even when the talkers were stepping aside to relieve themselves.

Holiday's story does not end with Japan's surrender.[16] Discharged and on the way home by bus, he sought an overnight hotel room in Flagstaff, only to be told, "White men only." Back on the reservation, he found work as a janitor, then left the reservation for several years to work on the railroad. He returned, married, worked as a policeman, then a forest ranger, and immersed himself in native religion and its cleansing post-war ceremonies.

Still, disconnected from his fellow Marines and his status as a code talker – a program that remained top secret after the war – he often felt empty and adrift. In the late 1960s, Holiday heard from another former code talker that the 4th Marine Division had plans, at Sgt. Johnston's behest, to declassify the Navajo Code Talker program and honor former code talkers at their next convention in Chicago. Holiday and other code talkers soon were regulars at 4th Marine reunions, where they found much-needed recognition and meaning for their wartime contributions.

Colonialization and Race

In November of 1884, representatives from 14 nations, including the U.S.A., gathered in Berlin to pore over maps of Africa. In the ensuing three months, they carved up the African continent, reaching agreement among themselves over who – Great Britain, France, Germany, Italy, Spain, Portugal, and, one individual, King Leopold the Second of Belgium – would have unimpeded access to what. It was King Leopold's lustful eye on the resource-rich central area, the Congo, that had prompted the conference.[17] Now, all could go about their business with a minimum of jostling with the others. Significantly, no Africans were represented at the table. Without any input from its rightful inhabitants, all of Africa, with the exception of what is now Ethiopia and Liberia, had been divided up among the completely self-righteous and self-interested colonizers.

Portugal, Britain, and France in the previous centuries already had set up outposts along the coasts of Africa to facilitate trade, including slaves from interior regions. However, none had penetrated very far inland, mainly because of susceptibility to malaria and lack of suitable transportation equipment to do so. Discovery of quinine for the treatment of malaria, along with the production of steamboats to navigate inland rivers, now made access possible. The demand created by early industrialization in Europe for raw materials, especially rubber and certain minerals, made the costly expansion of colonial empires very profitable. We focus on France's holdings in West Africa, *Afrique Occidentale Française* (AOF): Senegal, Mauritania, French Sudan (Mali), French Guinea (Guinea), Ivory Coast, Upper Volta (Burkina Faso), Dahomey (Benin), and Niger.[18]

African tribal groups did not take these invasions laying down, and the colonizers had to pursue their course violently. Most found it necessary to hire African troops to help them accomplish these means, and the French were no exception. They already had established colonial armies in each part of their empire made up of French soldiers and Legionnaires – *La Coloniale Blanche* (the White Colonial Army) – augmented by "conscripts" from the colonized locals, Troupes de Marines. In AOF, this latter group was designated *Tirailleurs Sénégalais* (Senegalese Infantry) who were drawn from all parts of AOF, not just Senegal.

Since these African Tirailleurs were an instrument of the French regime, fought in World Wars I and II, and, later, in the wars of decolonialization, they were regarded alternately as unsung heroes, victims of the colonizer, or despicable villains, depending upon who one was talking to and at what point in time. We turn our spotlight to a study of African troops in AOF.

Spotlight on Social Science Thinking and Research

Ruth Ginio, *The French Army and Its African Soldiers*[19]

Ruth Ginio is Associate Professor of History and Chair of the History Department at Ben-Gurion University in Negev in southern Israel. She obtained her doctorate in African History at the Hebrew University of Jerusalem in 1999. She has been a member of the executive board of the French Colonial Historical Society since 2002 and served as its president from 2010 to 2012. In this book, Ginio breaks new ground by focusing on the French Army's use of its African soldiers and veterans to link France and the African peoples of AOF – a last-ditch effort to maintain its colonial empire there. Her research draws on archival data from the Tamar Golan African Center at Ben-Gurion University, the *Centre des Archives d'Outre Mer* in Aix-en-Provence, France, the *Museé des Troupes de Marine* in Fréjus, France, and the *Archives Nationales* in Dakar, Senegal. She also reviews interviews with African veterans at Cheick Anta Diop University in Dakar.

Ginio presents a nuanced picture of colonialization. As we have seen, several European countries and King Leopold set up colonies in Africa, but the circumstances and methods varied, even by country of the colonizer. In the case of France, its efforts in north Africa took on a hard edge (see our discussion of Algeria in Chapter 9). The scene in AOF was more forgiving. Two local groups, the *"métis"* – the offspring of unions between French colonizers and local African women – and the *"originales"* – Africans in coastal Senegal who stood with the French during an inland rebellion in 1848, and their descendants – formed the buffer group in AOF, a kind of middle-man minority. The métis played intermediary roles, particularly in trade arrangements, and the originales could vote, serve as members of local political assemblies, and even were given a representative in the French Assembly in Paris. The originales also had an obligation to serve in the regular French Army, while the métis (and other Africans) could be enlisted to serve in the Tirailleurs Sénégalais. In the last half of the 19th century, the Tirailleurs served as an instrument of coercion and control in AOF.

At the beginning of World War I, the French faced a demographic dilemma: their birth rates had declined since the 1880s more rapidly than those of Germany and Britain, putting them at a disadvantage in the number of available military-age males (see our discussion in Chapter 5). After fractious debate, France became the only European nation to use its African troops on European soil. The justification by French officials was that Africans owed a blood debt to their French colonizers for having extended *la mission civilisatrice* to AOF. More than 170,000 African soldiers from the AOF were mobilized and sent to France to fight in the trenches alongside the regular French Army. Following the war, the French Army, to the chagrin of French colonial administrators, admitted the blood debt now flowed the other way. These administrators worried about uppity African veterans no longer satisfied to be merely colonial subjects.[20]

A similar pattern played out in World War II, where again France was the only colonizer to use its African troops in Europe. This time there was no debate. After its monstrous casualties in WWI, France was in even worse shape demographically to take on WWII. The timing was bad since other groups in AOF – African leaders, students, and workers – had begun staging protests and making demands of their own. French officials thus promised to improve the salaries and benefits of Tirailleurs. About 100,000 African soldiers from AOF already were in France when it surrendered to Germany in 1940, and more than 15,000 subsequently spent the rest of the war in German Prisoner of War camps. Another 100,000 subsequently were enlisted to fight in the Free French Army against the German occupation of France and were part of the Allied forces that liberated it in 1944. However, African soldiers were not permitted to enter Paris – General Charles de Gaulle, leader of the Free French Army, felt it necessary to "whiten" the parade of heroic forces on display in the capital city.[21]

African veterans of WWII quickly became angry and disillusioned. Post-war France was enfeebled economically and in turmoil politically. Thus, prompt return home and financial compensation due to AOF's African veterans were far down the list of priorities. African soldiers waiting in camps on the French mainland under trying conditions became belligerent and organized protests. And, at a holding area in Camp Thiaroye near Dakar, veterans revolted when they did not receive their promised pay. Thirty to forty of them were shot and killed. Realizing they were losing control of their African soldiers and veterans, and worried that the discontent might spread to the colonial population, Army officials sought to alleviate their grievances. They delivered however only pittances such as biscuits and cigarettes.

Meanwhile, serious agitation for independence was underway in other French colonies (see our analyses of Algeria and Vietnam in Chapter 9). The French government, increasingly aware that their colonial empire was in jeopardy, announced the formation of the French Union (*Union pour la démocratie française*) in October of 1946. There would be no more colonies but "one France" consisting of the French Mainland (and Algeria) and all other former colonies – now called *territoires d'outre-mer*, overseas territories – under one governing body in Paris. Former colonial subjects now would be imperial citizens. Imperial citizens gained limited local voting rights, trade unions were authorized to protect the rights of colonial workers, and an African political party, *Rassemblement Démocratique Africain* (RDA), was created to represent the overseas territories in the French Assembly – the first step toward what might be termed *compassionate colonialism*.[22]

French colonial administrators were quite perturbed by this recalibration. However, by now French Army officials had concluded their African veterans were not clamoring for independence but for the full payment of salaries and benefits.

And, significantly, they saw a distinct role the military and its African veterans could play in demonstrating the viability of this new colonial structure, if only they could regain their veterans' goodwill. Emphasizing their shared bond as brothers-in-arms, the Army in 1947 instituted a series of startling reforms, of which the most momentous were equal pay for equal work for French and African Army soldiers alike, and identical benefits for French and African veterans. It also extended financial support to children in the AOF who were war orphans and to those whose fathers were war invalids or former POWs, benefits the children of French veterans had been receiving since 1917. Finally, the Army set up preparatory schools in the AOF to groom young Africans to become sergeants and officers in the Tirailleurs.[23]

These changes by the Army, sharply criticized by French colonial administrators as lavish overkill sure to inflate the troublesome aspirations of AOF's newly empowered imperial citizens, had their intended effect. Voluntary enlistments in the Tirailleurs increased substantially, and African veterans became less cynical in their support of the French Union. This allowed the Army to showcase them as a model minority and to deploy them in the Indochina and Algerian insurrections against French rule. The Algerian case is especially interesting since Algerians and Africans of AOF are both Muslim. The Army thus worried that Tirailleurs sent to Algeria might be inclined to side with the insurrectionists there. To counter this, Army officials launched *action psychologique* (psychological operations campaigns) in which they made a clear distinction between Black Islam and Arab Islam, the former "good" or at least benign, the latter not. They went so far as to set up sponsored trips to Mecca, Saudi Arabia, for those Tirailleurs serving in Algeria who wished to perform *hajj* (the essential once-in-a-lifetime pilgrimage to Islam's holy city).[24]

The French Union became the first of a series of reforms in which the French sought to maintain a grip over its colonial empire. By 1960 or so these would lead, contrary to French intentions, to independence for all French colonies. In the case of Vietnam and Algeria, the wars of independence created complete breaks with France. In AOF, the parting was much more peaceful, and France left with guarantees as the sole source of military hardware and the first-choice beneficiary of certain mineral exports, principally uranium.

Race/Ethnicity and the All-Volunteer Force

As noted in Chapter 1, the U.S. military in WWII went from racially *segregated units* to a few *semi-integrated ones*, back to *segregation* after the war, and during the Korean war to *incrementally integrated units*. Still, the road to *full integration* was a bumpy one. Vestiges of segregation in the military lingered, and many U.S. military installations were in Southern states where racial segregation remained in full force outside their perimeters. The situation remained so until, about ten years after the U.S. military implemented racial integration internally, the U.S.A. initiated the landmark Civil Rights Act in 1964. This Act rendered *racial segregation* unconstitutional as a matter of law and custom throughout the country.

As if the 1960s were not already contentious enough, the first American ground troops in 1965 hit the shores of what would become the war in Vietnam, fanning a controversy that soon would engulf that effort. As we saw in Chapter 5, the draft in that war at first oversampled African Americans because a larger percentage of White eligibles qualified for college deferments. Though steps were taken to correct this, the war continued to pose special dilemmas for Black communities across the country. Journalist Wallace Terry recorded the oral histories of 20 African Americans who served in Vietnam.[25] His presentation of their stories in the book, *Bloods* (short for "*blood brothers*"), reveals themes common in the theater of combat regardless of race and ones unique to Black servicemen. Their recollections capture the incoherence and gritty nastiness of that war, something found in the stories of most Vietnam veterans. But, while racial tensions often took back seat in forward combat areas, racial strife in rear areas and stateside was more common where Confederate flags in the barracks were likely to elicit angry Black Power salutes.

The final withdrawal of American ground troops in 1972 and the shift to an AVF in 1973 altered this landscape. With the AVF's emphasis on *equality of opportunity*, the numbers of properly qualified African Americans soon made up one-fifth of the Army and Navy (see Chapter 5), and there is a larger percentage of Hispanics in the Marine Corps than in the population. Likewise, Native Americans and Pacific Islanders, though small in number, are well-represented. A main reason is that racial and ethnic minorities in the military routinely report having more opportunities there than in corresponding civilian sectors.

This shift has not occurred by chance. Charles Moskos, whom we already have met in Chapters 1 and 4, and military sociologist and Vietnam veteran John Sibley Butler, have documented the Army's efforts to successfully attract and retain minorities.[26] It was evident in 1996 when they did their review that highly qualified African Americans were volunteering for military service in surprising numbers. Moskos and Butler's data show this was no accident, but the result of a "race savvy" plan. The Army, they show, focused on enhancing the skills of its volunteers through training programs, educational initiatives, and mentoring – a process of *compensatory action* rather than *preferential treatment*. The formula called for recruiting enough Black volunteers to attain a *critical mass*, stressing *performance standards*, and emphasizing a *command policy* that discriminatory behavior will not be tolerated. Their study does not deny problems remain but point to the Army's commitment to a workable plan.

We will now see a version of this dynamic in our spotlight on Muslims serving in the U.S. military. For them, *intersectionality* more resembles that of Japanese Americans in WWII: they do not fall somewhere on a *Black–White* continuum, but on an *American–alien* one.

Spotlight on Sociological Thinking and Research

Michelle Sandhoff, *Service in a Time of Suspicion: Experiences of Muslims Serving in the U.S. Military Post-9/11*[27]

Michelle Sandhoff received her Ph.D. in Sociology at the University of Maryland, where she concentrated her focus on military sociology. She recalls that she began her graduate studies a week after the 11 September 2001 attacks on the World Trade Center in New York and the Pentagon in Washington, D.C. Her interest in writing about Muslim Americans came from discussions with a variety of friends about their experiences in a post-9/11 world. This led more narrowly to her study of Muslims who now are in the U.S. military or who were in the military after 9/11. This is an intriguing subject because Muslim Americans recently have received considerable scrutiny, if not acrimony, in large part because the perpetrators of 9/11 were al-Qaeda operatives. Still, Muslims in the U.S. military after 9/11 were likely to deploy to Afghanistan, Iraq, or some other predominantly Muslim country where the enemy was "Islamic." Thus, those in her study are ones who have volunteered to serve, as the book's title states, during "a time of suspicion."

Christianity and Islam are the two largest world religions, the former having just over 2 billion members and the latter just under 2 billion. Within each there is substantial diversity because of denominational, regional, and national variations. Currently, the three countries with the largest Muslim populations are Indonesia, India, and Pakistan, and the three with the largest Christian populations are the U.S.A., Brazil, and Mexico.[28] Conflicts between the two religions go back to the Middle Ages and, Sandhoff notes, some of the legacy of that history carried over into 18th-century America. The Founding Fathers debated whether the common practice of restricting eligibility for holding office only to those of specified Protestant denominations should be added to the Constitution. It was not, but a concern over hypothetical Muslim office holders framed the debate.[29]

In any case, the immigration of Muslims to the U.S.A. remained a trickle until 1965 when Congress's Hart-Celler Act set standards for immigration in terms of needed skills and educational specialties. Since then, highly qualified Muslim immigrants have come from the Arab world, Iran, and South Asia, that is, India and Pakistan. Muslim Americans currently make up about 1% of the U.S. population. In the 1990s, political scientist Samuel Huntington, whom we introduced in another context in Chapter 4, published his book, *The Clash of Civilizations and the Remaking of the World Order*.[30] In it, he argued that post-Cold War conflicts would revolve around civilizational divides, with the West and Islamic split being a salient one. However, it is the attack on 9/11 that has spiked anti-Muslim sentiments and stereotypes, though both Presidents George W. Bush and Barack Obama have emphasized that the U.S.A.'s fight is with terrorism, not the religion of Islam.

Service in a Time of Suspicion features in-depth oral histories with 15 Muslim Americans who are service members or veterans. Sandhoff sought subjects who "self-identify as Muslim" through personal contacts, electronic discussion boards, and strategically placed flyers. They constitute an available sample, so how representative they are of all Muslims who serve in the AVF cannot be readily established. Eleven of them immigrated to the U.S.A. with their families from South Asia or were born in the U.S.A. of parents from South Asia. Muslims from South Asia tend to be multi-lingual: they speak or at least understand Urdu, though it may not be their primary language, and most also can speak Arabic or at least recite verses in Arabic from Islam's holy book, the Qur'an. Of the four remaining interviewees, one immigrated from the Middle East and three are native-born U.S. citizens who converted to Islam after joining the AVF. Three of these four speak Arabic fluently. And, of course, all 15 are completely fluent in English.[31]

The interviewees had been members of the AVF for an average of 10.7 years. Nine of the fifteen were enlisted, and five had deployed to Iraq and/or Afghanistan, three in combat positions. Why did they join the military? Eight of the fifteen gave some version of patriotism as the primary reason for signing up.[32] They typically explained it this way: "giving back to the U.S. as a grateful Muslim," "purely to serve my country," and "a tradition of military service in my family." Three of these also stated a desire to use their language and cultural skills to assist in accomplishing military missions. Six mentioned economic opportunities and a chance to "broaden horizons" as their primary motivations. And, one said, "did it on a whim." Several admitted they were inspired by movies they had seen, for example, *Top Gun*, or books they had read, such as Tom Clancy's military-themed thrillers. These responses would be typical of most any group volunteering for the AVF.

A common stereotype of Muslims is that they are fanatically devout. However, like people who identify as Christian, there is much variation in type and intensity of religious practices among them. For Sandhoff's interviewees, nine described themselves as "devout" or "regularly practicing" Muslims, that is, ones who pray daily, avoid pork and alcohol, and fast during Ramadan. The remaining six said their observance of Muslim religious practices was "so-so," praying and following dietary restrictions only erratically. One said he was "basically an atheist, a Muslim by heritage only."

The U.S. military provides guidelines under which Christians, Muslims, or others can request "religious accommodation" if this does not interfere with the mission. Only four said they had requested religious accommodations, mainly for day-long fasting from all food and drink during the month of Ramadan. Most said that they worked out a way of praying or fasting without accommodation. Military regulations allow Muslim men to wear the *tufi* (and Jews a *yarmulke*), a small skullcap, but only one of the thirteen male interviewees did so. The two women interviewees do not wear a scarf while on duty but do so during off-duty hours.

So, what has been their experience? Ten of the fifteen reported very affirmatively about their time in the AVF.[33] One of them said, "Very positive [experience], full of opportunities ... the military is diverse ... and must follow EO (Equal Opportunity) policies, so a very good place." An infantryman, stated, "Enjoyed all my time in the military, the camaraderie ... my being a Muslim contributed to diversity [and] the mission in Afghanistan ... Being an Urdu speaker, [I] was able to form relationships with Afghan soldiers and village elders." A pilot described his experience as "fantastic"; "diversity in the military is real," he continued, "I felt accepted and I accepted others, including a former member of the Ku Klux Klan." One of the women reported: "Leadership was very supportive, both in my career and religiously ... I considered myself a 'bridge-builder' because others could learn about my faith and I could teach [Muslims not in the service] about the military." Six interviewees reported similar bridge-building experiences. Those who developed ways to speak up constructively when confronted with negative stereotypes felt they were able to have a positive impact.

But not all was roses and some noted concerns.[34] One said, "once some people find out you're Muslim, they act like you're on speed dial with Osama bin Laden." A second infantryman in the study, who deployed multiple times, felt the mission there was "confusing and senseless, ... and [me and] my buddies [just] hung on to make it home alive." Three assigned to language school stated that Arabic classes included no readings or coursework on Islam, leaving those in the class with many of the stereotypes with which they had begun their training. Others reported less acceptance than those who had very positive experiences. They said, for example, that often prayers in unit gatherings were not nondenominational and referenced "only Jesus," as if those present were only Christians. Those for whom things went totally awry pointed to bad leadership. One interviewee, for instance, felt his commander had shafted him because he, the commander, did not like Muslims and often expressed disdain for them. And, the person who joined "on a whim" admitted the military was "a poor fit."

There are several take-aways from this research. The military, reacting to pressures to sustain its manpower needs, has had to develop strategies for integrating diverse groups into its ranks. Because of its hierarchical structure and formal policies of equal opportunity, it is able, potentially at least, to do so more straightforwardly than many civilian organizations. This is facilitated by the necessity of forming cohesive teams for carrying out its often-dangerous missions. Several times, Sandhoff's interviewees mentioned feeling "more vulnerable" outside the military, whether working in the civilian labor force or just walking down the street off-base. Finally, as always, success in developing cohesive teams is very leadership driven. Toxic leaders are anathema.

Questions for Discussion

1 We have presented evidence of the gap between the biological and the social significance of race. Despite this gap, race gained social significance in the United States and other Western countries, especially during the era of colonization. Why? How? Explain.

2 It is worth revisiting Chapter 1, especially the Spotlights on the semi-integration of Blacks and the service of Japanese Americans in the U.S.A. military during WWII, and then to compare their experiences with those of the Navajo Code Talkers. What similarities do you note among the three groups? What differences?

3 A major theme of this chapter has been how military needs shaped the framing of race and how militaries across the world have led the way in racial integration because of military necessity. Apply this theme to the experience of France and its colonial force, the Tirailleurs Sénégalais.

4 One could argue that our final spotlighted group – Muslim Americans serving post-9/11 – experience an intersectionality that does not fall somewhere on a Black–White continuum, but on one of American–Alien. How convinced are you on this distinction, and why?

Notes

1 Gunnar Myrdal, with the assistance of Richard Sterner and Arnold Rose, 1962 (1944), *An American Dilemma: The Negro Problem and Modern Democracy*, Twentieth Anniversary Edition, New York, Evanston, and London: Harper & Row, Introduction. Myrdal did return home during the war for a short period when Sweden was threatened by German invasion. American sociologist Samuel Stouffer, who was doing the American Soldier studies at the time (see Chapter 1), took over Myrdal's responsibilities during this absence.

2 Ibid., pp. 997–1010.

3 Joseph Arthur de Gobineau, [1853–1855], 1915, *An Essay on the Inequality of the Human Races*, 4 vols., translation by Adrian Collins, New York: G.P. Putnam's Sons.

4 Jared Diamond, 1999, *Guns, Germs, and Steel: The Fates of Human Societies*. New York and London: W.W. Norton & Company.

5 Luigi Luca Cavalli-Sforza, Paolo Menozzi, and Alberto Piazzia, 1994, *The History and Geography of Human Genes*, Princeton, N.J.: Princeton University Press.

6 The first slaves were brought to what is now Virginia around 1620 and the arrangement still was in effect when the 13 northern and southern English colonies formed the United States in 1776. Many of the framers of the U.S. Constitution, and Thomas Jefferson himself who penned the Declaration of Independence (which contains the phrase, "*all men are created equal*"), were slave owners. A year later, Vermont became the first state to outlaw slavery within its boundaries and by 1840 all northern states had followed suit. Slavery remained legal in Southern states until 1865. Jim Crow laws began to be put into place in the 1870s and were ruled unconstitutional in 1964.

7 Robert Blauner, 1982, "Colonized and Immigrant Minorities," pp. 501–519 in Anthony Giddens and David Held (eds.), *Classes, Power and Conflict*, Berkeley: University of California Press.

8 Samuel Holiday and Robert S. McPherson, 2013, *Under the Eagle: Samuel Holiday, Code Talker*, Norman, Okla.: University of Oklahoma Press.

9 Ibid., pp. 26–44, 52–60.

10 Ibid., pp. 61–68.

11 Ibid., pp. 85–88, 14–18.

12 Ibid., p. 83.

13 Ibid., pp. 105–110.

14 Ibid., pp. 125–142, 157–167.

15 Ibid., pp. 142–152.

16 Ibid., pp. 183–198, 209–215.

17 For an account of the conference and the role of King Leopold, see Adam Hochschild, 1999, *King Leopold's Ghost: A Story of Greed, Terror, and Heroism in Colonial Africa*, Boston, New York: Houghton Mifflin Harcourt.

18 Ruth Ginio, 2017, *The French Army and Its African Soldiers*, Lincoln, Neb.: University of Nebraska Press, Kindle version, Chapter 1.

19 Ibid..

20 Ibid., location 425–470.

21 Ibid., location 605–660.

22 *Compassionate colonialism* is our characterization of what Ginio describes as the French Union and subsequent reforms.

23 Ibid., Chapter 3.

24 Ibid., Chapter 5.

25 Wallace Terry, 2006, *Bloods, Black Veterans and the War in Vietnam: An Oral History*, New York: Ballantine Books.

26 Charles C. Moskos and John Sibley Butler, 1996, *All that We Can Be: Black Leadership and Racial Integration the Army Way*, New York: Basic Books.

27 Michelle Sandhoff, 2017, *Service in a Time of Suspicion: Experiences of Muslims Serving in the U.S. Military Post-9/11*, Iowa City: University of Iowa Press.

28 Indians are primarily Hindu, but Islam is the second largest religion of the country, about 15%. The countries with the largest Muslim populations are projected in 2060 to be India, Pakistan, and Nigeria, and the U.S.A., Brazil, and Nigeria for Christianity (Jeff Diamant, 2019, "The Countries with the 10 Largest Christian Populations and the 10 Largest Muslim Populations," Pew Research Center, April 1. Downloaded from https://www.pewresearch.org/fact-tank/2019/04/01/the-countries-with-the-10-largest-christian-populations-and-the-10-largest-muslim-populations/.

29 Sandhoff, op. cit., pp. 13–14.

30 Samuel P. Huntington, 1996, *The Clash of Civilizations and the Remaking of the World Order*, New York: Simon & Schuster.

31 Sandhoff, op. cit., pp. 6–9.

32 Sandhoff presents each of the oral histories sequentially by topic (Range of Experience, Importance of Leadership, Role of Diversity). The names she uses are pseudonyms. For our

summary, we charted the characteristics of each on a spreadsheet, so we will provide the pseudonym for the interview from where a quote is taken. For example, Yusef is "give back as grateful Muslim," Jamal "purely to serve," and Hakim is "family tradition." Sadia did it "on a whim."

33 Kareem had a "positive experience" and equal opportunity policies, Pervez is the Urdu-speaking infantryman, and Ahmed is the pilot. Rahma is the woman interviewee with supportive commanders.

34 Yousef is the second infantryman.

Recommendations for Additional Reading/Viewing

Da 5 Bloods (film), directed by Spike Lee, released on Netflix, June 12, 2021.
(Four Black veterans return to Vietnam to search for body of fallen comrade and gold bars they buried during war. Directed by Spike Lee, so you know there is more to it than that. Highly acclaimed film.)

Yasmin Khan, 2015, *India at War: The Subcontinent and the Second World War*, Oxford, U.K. and New York: Oxford University Press.
(An Oxford historian recounts India's role in the war and the tremendous social changes that occurred as a result)

J. Todd Moye, 2010, *Freedom Flyers: The Tuskegee Airmen of World War II*, Oxford, U.K. and New York: Oxford University Press.
(Account of the U.S.A.'s first African American pilots based on 800 interviews collected by the Tuskegee Airmen Project)

Bernard Nalty, 2003, *Long Passage to Korea: Black Sailors and the Integration of the U.S. Navy*, Washington, D.C.: Department of the Navy Naval Historical Center.
(History of race relations and ultimately racial integration in the U.S.A. by a well-respected U.S. military historian)

David E. Rohall, Morten G. Ender, and Michael D. Matthews (eds.), 2017, *Inclusion in the American Military: A Force for Diversity*, Lanham, MD: Lexington Books.
(This edited book reviews the ways the military has addressed race, ethnicity, gender, religion, and sexuality)

7   Gender and Sexuality in Society and in the Military

Reader's Guide: The U.S.A. instituted a gender-segregated Women's Army Corps (WAC) in World War II with the goal that it would constitute 2% of the force. We review their contributions during that war and the Korean and Vietnam wars. The transition to an All-Volunteer Force (AVF) in 1973 led to gender integration. Fifteen percent of the U.S. military currently are women, but this has not transpired without some backlash. The first Spotlight appraises the Congressionally mandated gender integration of the U.S. military service academies in 1976, and details what went right and what did not work so well. In the second Spotlight, the first study analyzes gender harassment as a form of protest among some Army men, and the second looks at responses of women in the Portuguese and Dutch militaries to integration dilemmas. The final two studies, again presented in tandem, review the response to lifting the ban against gays and lesbians in the British, Canadian, and Australian militaries, and the work of the Department of Defense's Comprehensive Review Working Group on the implementation plan for replacing "Don't Ask, Don't Tell."

In human history, the use of weapons has been as uniquely male as childbearing has been female.[1] We broached this vital topic in Chapter 3's first Spotlight, *War and Gender*.[2] It is not coincidental the two phenomena appeared in tandem. Until the advent of industrialization, exceedingly high birth rates were required to offset similarly high death rates. This formula necessitated a notable amount of *gender polarization*, that is, societally defined gender roles that accentuated childbearing and domestic chores for women, and for men hunting and hence the possession of weapons. We noted two caveats. The amount of gender polarization almost always exceeded the amount required to meet the *demographic imperative*, and the consequent concentration of power in the hands of men lingered long into the 20th century pretty much intact.

Thus, the history of women's rights is one of a haltingly unfolding contentious process. We already have reviewed historical exceptions to the exclusive warriorhood of men (again, see Chapter 3), so continue here with U.S. women in World War II. Early in that war, military planners sought to alleviate its burgeoning "manpower" demands with women volunteers. In May of 1942, Congress passed controversial Public Law 77-554 to establish the Women's Army Auxiliary Corps (WAAC). WAACs were not part of the regular Army but filled clerical and support positions, thereby relieving male soldiers for reassignment to combat slots.[3]

DOI: 10.4324/9781003282549-7

Despite hostilities toward the very idea of women in the military – and slanderous gossip about any woman who might sign up for such duty – their military usefulness led Congress in July 1943 to pass Public Law 78-110. It replaced the WAAC with the WAC. WACs were incorporated into the regular Army and accorded official military status, albeit in gender segregated units, with the goal of 240,000 volunteers, that is, 2% of the total force. About 140,000 WACs served during World War II[4] and they carried out dozens of military jobs not stereotypically associated with women.[5] More than a thousand African American women served in the WAC, as did a smaller number of Japanese American women.

Military sociologist Brenda Moore scoured official archives and conducted in-person interviews with veterans of both these groups,[6] of special interest because of the issue of *intersectionality*, in this case of how race and gender interact to *enhance privilege* or *heighten disadvantage*. Economic opportunities for Black women were few and far between. The desire to step up and to "serve my country, serve my race" prompted those who responded. African American WAC officers and enlisted women lived and worked in racially segregated facilities. Black enlisted WACs mostly were assigned to low-level clerical positions and menial jobs cleaning quarters and doing laundry, the latter assignments not permitted for White WACs.

Although some all-White WAC units in 1943 already had been sent overseas in support of combat operations, there were no plans to deploy any Black WAC units. It was controversial enough to rely upon *women* soldiers in anything besides the Army Nurse Corps, where about 59,000 women nurses served in World War II.[7] However, in late 1944 a massive backlog of undelivered letters and Christmas packages sat in the U.S. military depot in Birmingham, England. In response to pressure to utilize African American units more fully, a call went out for Black WACs to volunteer for the newly formed 6888th Postal Directory Battalion and deployment to England. Some 800 enlisted women and 30 officers responded, and, in February of 1945, the "6-triple-8" became the only African American WAC unit during World War II to serve overseas.[8]

Like other WAC units, women of the 6-triple-8 gained respect for how well they performed their jobs but, at the same time, took the usual defamatory flak for being women soldiers. Being both women and Black, they encountered gossip they had been deployed to "service" American soldiers, an expectation sometimes expressed by soldiers themselves. Their commander, Major Charity Adams Earley, countered sexual and racial innuendo adamantly but with style and dignity, and the women of 6-triple-8 followed her example with pride.[9]

Moore shows that *intersectionality* was defined differently for Japanese American WACs. As we saw in Chapter 1, Japanese Americans were banned from living in the west coast region after Lieutenant General John L. DeWitt designated areas along the Pacific Ocean as security zones, and some 120,000 Japanese Americans were forcibly removed from their residences and businesses into ten inland "relocation" camps. In 1942, Japanese Americans therefore were not on a racial continuum somewhere between "White" and "Black," but toward the end of a separate continuum ranging from "American" to "Foreign Alien."

Pressing manpower needs produced cooler heads. The situation had become one of "all hands on deck," and in 1943 Secretary of War Henry Stimson declared, "It is the inherent right of every faithful citizen, *regardless of ancestry*, to bear arms in the nation's battle" (emphasis added).[10] Also, the War Relocation Authority (WRA), which oversaw the ten relocation centers, was looking for ways to reduce expenses by releasing "loyal" Japanese Americans, so the military plunged ahead with a plan for a segregated unit of

Japanese American women. However, interviews with women at the camp found them opposed to being in a racially segregated unit. They sought to serve, in terminology later supplied by law professor Frank Wu, as patriotic "honorary Whites."[11]

The WAC set a goal of attracting 500 Japanese American women.[12] Those who volunteered were anxious for the opportunity both to get out of the camps and to demonstrate their patriotism. Contrary to the initial plan, there were no "Japanese" WAC units, so all were dispersed throughout the WAC with other women. All held enlisted ranks and were assigned mainly to clerical positions. A small number became Japanese language specialists attached to intelligence units, and some served overseas in Tokyo starting in January of 1946 after the cessation of hostilities.

Women and the All-Volunteer Force, Round 1

Post-World War II demobilization reduced the WAC to about 10,000 officers and enlisted women by the end of 1946. However, post-war assessments recognized the notable contributions of women soldiers and nurses and, hence, the desirability of solidifying their status in the military. In 1947, the Army Nurse Corps (ANC) became a Regular Army unit in the Army's Medical Department and, after several contentious rounds of Congressional hearings, the WAC in 1948 was incorporated into the Regular Army as well.[13] These changes afforded women soldiers more official standing and benefits, but still maintained gender-segregated units. The first to change this feature was the ANC. In 1955, Public Law 294 authorized the commissioning of male nurses.

Meanwhile, the Korean War, June 1950 to July 1953, brought the usual wartime remobilization of the Army, and the same occurred during the undeclared war in Vietnam, where American ground troops were involved from 1965 to 1972. WACs served in both conflicts, primarily in clerical and logistical positions. WAC recruitment goals proved challenging, but during the Korean War about 24,000 women served in the WAC and the Air Force's new unit, Women in the Air Force (WAF), and another 8,000 in the Navy's WAVES (Women Accepted for Volunteer Emergency Service).[14] In Vietnam, around 24,000 women served in the WAC, WAF, and WAVES, and two small administrative WAC detachments were stationed at the U.S. Army Headquarters in Saigon and Long Binh at the request of the U.S. Forces commander, General William Westmoreland. About 5,000 Army nurses served in-country with evacuation hospitals, about 20% of whom were men.[15]

By 1970, full-scale discussions for doing away with the military draft were underway at the direction of President Richard Nixon. It was evident to a few analysts that a *draftless peace- and war-time military* would necessitate a substantially larger percentage of women volunteers. General Westmoreland, in his role now as Chief of Staff, requested a steering committee to update the image of women soldiers and their career prospects. Changes included newer, more stylish uniforms, an increase in the military occupations open to women, and a provision allowing women officers temporarily assigned to non-WAC support units to command men. The committee also reviewed the WACs long-standing policy of automatic separation upon pregnancy.

Fortuitously, the implementation of these changes occurred in concert with a burgeoning women's movement and Congressional approval of the Equal Rights Amendment in March of 1972 prohibiting discrimination based on sex.[16] Still, the resultant surge in women volunteers caught many off guard. The WAC and WAF had a difficult time finding enough uniforms, equipment, and facilities to handle all its new, incoming personnel. To provide enough female officers, the Army and Navy began experimental Reserve

Officer Training Corps programs at a few colleges and universities (the Air Force had begun so in 1969). The ROTC programs too attracted more women applicants than they could handle. Notably, these swells in women volunteers took place as male volunteerism slowed to a dribble in the waning of America's involvement in Vietnam.

On 1 July 1973, Secretary of Defense Melvin Laird announced the end of the draft and the beginning of an AVF. The Department of Defense set a goal of 60,000+ women in the military, about 6% of the AVF.[17] Another question soon arose: Should women be admitted to the nation's service academies? Chiefs of Staff of the Army and Navy responded negatively, while the Air Force's Chief indicated the Air Force Academy would do so if directed by Congress. We turn our Spotlight to the integration of women into the service academies.

Spotlight on Social Science Thinking and Research

Judith Hicks Stiehm, ***Bring Me Men and Women: Mandated Change at the U.S. Air Force Academy***[18]

Now retired, Judith Hicks Stiehm is a long-time, distinguished Professor of Political Science at Florida International University. She completed her B.A. in Asian Studies at the University of Wisconsin in 1957 and her doctorate in Political Theory at Columbia University in 1969. Stiehm's doctoral research resulted in her first book, *Active and Passive Resistance in America*. Published in 1972, it focused on the history and use of *nonviolent resistance* in promoting social change from 17th-century Quakerism to the anti-war protests in the 1960s. One reviewer described it as "a fair, tough-minded introduction to nonviolence as a mode of political action ..., well designed to puncture the illusions of both the romantic enthusiast and the skeptical critic."[19]

Soon, an initiative to allow women admission to the military's service academies – the U.S. Military Academy, West Point, N.Y., the U.S. Naval Academy Annapolis, Maryland, and the U.S. Air Force Academy (USAFA), Colorado Springs – was winding its way through Congress. The debate surrounding the initiative occasionally was heated. On the one hand, those favoring the change emphasized that most noncombat military occupational specialties were open to women, although mostly in gender-segregated units, and that women already were enrolled in Reserve Officer Training Corps programs across the country. Allowing admission to West Point, Annapolis, and USAFA was thus a matter of fairness in preparing highly talented women for officer careers.[20]

Those opposed lamented the creeping reliance on women in the military and emphasized the service academies' mission should be to prepare officer-leaders for combat. For example, Jacqueline Cochran, a pilot with more than 15,000 flight hours who had directed the training of the Women Airforce Service Pilots (WASPs) during World War II, testified against the initiative. She argued, "a woman's primary function in life is to get married, maintain a home and raise a family," and pointed to the long-standing WAC policy (since rescinded) that a woman exit the military upon pregnancy. General Westmoreland, by then retired, stated in a magazine interview, "Maybe you could find one woman in 10,000 who could lead in combat, but she would be a freak, and the Military Academy is not being run for freaks."[21] However, Congress had just recently passed the Equal Rights Amendment, and it was difficult to make an airtight *noncombat-only argument* since a large number of male cadets were commissioned each year in noncombat specialties.

Congress passed Public Law 94-106 directing the service academies to admit women and to do so by the incoming 1976 cohort, and President Gerald Ford signed it into law on 7 October 1975. Stiehm by then was head of the University of Southern California's program for the Study of Women and Men in Society. As the fall of 1976 approached, she was given free rein to roam USAFA for the academic year. She reviewed documents and "repeatedly interviewed" those who planned and implemented policies for the first gender-integrated cohort.[22] Her research focused on the views of USAFA's mostly male administration and staff, and those of male cadet leaders who applied many Academy policies at the squadron level. Studies of the women cadets themselves were conducted separately by Washington State University sociologist Lois DeFleur.

Decisions about whom and how many to admit, Stiehm notes, is "one of *demand and use*, not one of *supply and qualification*."[23] The Superintendents – general officers at the three-star level – of the Naval Academy and West Point were the least enthusiastic. In the case of the Navy, service in combat positions meant being repeatedly at sea, either to man warships or fly planes from the decks of carriers. Hence, regular rotation to seaside assignments was important in maintaining morale. Since Congress prohibited the assignment of women to ships (except the *USS Sanctuary* hospital ship), an infusion of women officers could disrupt this balance. As women entered Annapolis, the Navy requested they be allowed to serve on noncombat vessels and on combat ships while not doing combat missions. Meanwhile, Annapolis sought to make the arrival of women a "nonevent." [24]

West Point's Superintendent declared he would resign rather than oversee gender integration at his institution, a position he quietly reversed in respect for the mandate signed by the President as Commander-in-Chief. West Point geared its training toward producing infantry officers and had a well-deserved reputation for being the most physically demanding of the three academies. West Point proceeded grudgingly, emphasizing that all existing standards must be strictly maintained. One could argue much of this training was not directly relevant for women cadets, but if so, this also would be true for the many male cadets who received assignments other than infantry. Even top-ranked cadets, who were given first preference, often listed branches other than infantry atop their wish lists.

USAFA had appointed a panel two years earlier to develop a plan and assumed an optimistic posture. In the Air Force there were three combat slots for officers: pilots, navigators, and missileers, the first two of these the more prestigious and the last of these a distant third. Most Air Force officers carried out technical, logistical, and administrative duties in support of its core of pilots. Congress mandated that at least 60% of each USAFA graduating class must be physically qualified to become pilots, but the number sent to flight school varied in response to needs of the Air Force. Hence, many male cadets were commissioned each year in noncombat specialties also open to women in WAF units.

The Air Force had experience preparing women to become officers at Lackland Air Force Base's Officer Training School. In July of 1976, the Air Force canceled its remaining gender-segregated WAF units and integrated their enlisted women and officers into existing Air Force units. (The Army and Navy discontinued the WAC and the WAVES in April of 1978.) Meanwhile, USAFA's initial plan for putting women in a separate squadron commanded by a female Air Force officer was scrapped. Cadet women would be billeted in a separate dormitory with female Air Force junior officers for off-duty supervision and counseling. This followed the Air Force model at Lackland Air Force Base of "work together but live and be administratively controlled separately."[25]

For their incoming 1976 cohorts, the Naval Academy admitted 81 women, West Point, 119, and the Air Force Academy, 157. Hence, women at the three academies made up about 5, 8, and 10% of the new admissions, but less than 1, 2, and 3% of the total number of midshipmen and cadets, respectively.[26] At Annapolis and West Point, women were scattered two or three per company. At USAFA, clusters of at least ten women cadets were assigned to only some of the squadrons to give them more opportunity for interaction with each other. Most squadrons did not have women in them.

USAFA's planning had anticipated that adjustments in training might be necessary because of physiological differences between men and women. National studies had suggested an overall physical differential in maximum performance of about 15% due to greater musculature among men. The Physical Fitness Tests (PFTs) at the military academies emphasized *endurance* and *upper body strength*. These traditional PFTs were developed to differentiate fitness so defined among men, with untested assumptions about how they might equate with military performance. Monitoring by USAFA's Department of Physical Education revealed that cadet women, most of whom had been high school athletes, could narrow the gap to 10%. Thus, portions of the USAFA's PFT were adjusted to this norm.[27]

West Point had entered the summer of 1976 with the notion that women giving "serious effort" would not be disenrolled for physical reasons. By year's end, its physical training regimens were changed to "challenge but not exhaust" cadets. Women and men cadets were divided into three categories based on physical capabilities and each given demanding routines to complete. Even the lowest group met an exacting standard, and those in the two higher groups still were put "in a taxing situation."[28]

Finally, USAFA sought to improve the acceptance of cadet women. Male cadets largely were outright hostile to the idea. They resented the fanfare surrounding the admission of cadet women, felt standards were being watered down, and resented any suggestion they should give up cherished traditions of *male culture*, such as pin-up pictures of seductive sweethearts and scantily clad playmates in their Academy calendars and publications. More concerning was their uneasiness, if not unwillingness, to accept women cadets and officers in positions of authority, including an occasional instance in which some male cadets refused to salute a female Air Force officer.

Planners hoped that "contact" would reduce male resistance, but as we saw in Chapter 1's discussion of the *contact hypothesis*, contact alone is not sufficient. Working *together as co-equals* on a *team* to accomplish *demanding tasks* is required. For the most part, that was not the cadet experience. Although attitudes among male cadets did improve over the first year in those squadrons having women in them, for many *upper body strength* remained a code for "weakness of women."

Success in that first year had modest goals: find the right number of capable women admittees and maintain a retention rate in keeping with that of men in the 1976 cohort. At the end of basic training leading into the fall semester, West Point had the highest dropout rate (15.9%). Navy and USAFA fared better at 7.2 and 2.5%, respectively. However, over the course of the year, the USAFA women's dropout rate increased to about 16%, a matter of concern but still lower than the 23% for the first-year male cadets. Alarmingly, in year two the dropout rate for USAFA women cadets rose to 32%, embarrassing because this exceeded the rates for West Point and Annapolis. This left many questions about the implications of differences in approach by the three academies.[29]

Women and the All-Volunteer Force, Round 2 and Beyond

Difficulties in achieving gender integration at the military academies were not unique to them. Initiatives during the 1960s to admit women into previously all-male colleges and universities at Georgetown, Princeton, Johns Hopkins, and Yale Universities encountered blowback from male students and, at Sarah Lawrence and Vassar Colleges, from women following the admittance of men. Years later, "Save the Males!" was the rallying cry of male cadets protesting the first female cadet in 1994 at The Citadel, South Carolina's state-supported military academy.[30] In all these cases, grim *resistance* against "the interlopers" muted the first new admittees' chances of success.

In the AVF during the 1970s, the problem of attracting – and affording – volunteers in competition with the civilian labor market, plus a weak pulse of volunteerism among eligible males, kept a focus on recruiting women. In May 1977, the Department of Defense under President Jimmy Carter issued its *Use of Women in the Military*.[31] The report stated that the demonstrated effectiveness of women gave reason to further expand their role. It set a goal for a total of 208,000 women for the three services by 1984, a projected 11.6% of the AVF. It reiterated the position of removing "all unnecessary restrictions on women."

An account of these and subsequent events by Bernard Rostker, former Director of the Selective Service System and Under-Secretary for the Department of Defense, confirms that all was not well.[32] Some military leaders chafed at the inroads being made by women or believed that the changes were occurring too quickly. With the election of Ronald Reagan as President in 1980, the Army took it upon itself to declare a *"woman-pause"* and roll back its previously set recruitment goals for women soldiers. Secretary of Defense Casper Weinberger's staff immediately called for an assessment of current and projected numbers for women, as well as estimates of the costs and likelihood of recruiting additional male volunteers. The Army meanwhile was allowed to pursue its reduced 1984 end-strength proposal for women of 65,000 rather than the originally planned 80,000.

As the assessments were being completed, it became clear the Army had miscalculated. Badly. In July of 1981, a news story broke in the *Washington Post* revealing the Army had suddenly realized it would need 100,000 additional soldiers for the missions outlined by the Reagan administration. Worse yet, it hinted the draft might have to be temporarily revived to make this happen. Major General Jeanne Holm, the first female general officer in the Air Force, later recalled that the story "landed like a live grenade on the third floor of the Pentagon."[33] In response, a spokesman for a "livid" Secretary Weinberger sternly stated that a return to the draft was completely off the table, and "the Pentagon saw the increased, rather than the decreased, use of women as important to the Reagan buildup."

It was time to shut up and color. Army leaders recommitted themselves to how to make the numbers and participation of women work in the best way possible.

To this point, the AFV had been largely a peacetime military. That changed in 1989 with Operation Just Cause in Panama and, on a bigger stage with larger stakes, in 1990–1991 with Operation Desert Storm in the Middle East. Reports to Congress emphasized not only the superior military response by the AVF itself, but also of its women soldiers, sailors, Airmen, and Marines who "served in hundreds of different [skilled] positions … [and] as commanders and key staff officers."

Still, these advances were not without problems. Military sociologist Mady Wechsler Segal, whom we met in Chapter 1, reviewed factors affecting women's participation in the military across nations in the 1990s.[34] As we have seen, increases in the participation of women almost always have been initiated by a male-based but "needy" military. If the *need* was dire enough, *necessity* won out. Still, culture remained a player: the *combat warrior as an ideal type* had been defined in terms uniquely male, said to justify *masculine hegemony*. So, Segal noted, the more *gender differences* were viewed as *categorical* (e.g., "men are stronger than women") without recognizing *individual differences* ("some women are in fact stronger than are many of the men"), the *more resistance there is and the more limited women's military roles* are.

Sometimes that resistance took the form of sexual *harassment*, with several high-profile incidences in all three services. However, a less noticed problem was *gender harassment* carried out by a small number of disgruntled male servicemen. We now turn our Spotlight to this issue and the responses of women to the two forms of harassment.

Spotlight on Sociological Thinking and Research

Laura L. Miller, "Not Just Weapons of the Weak: Gender Harassment as a Form of Protest for Army Men"[35]; and Helena Carreiras, "From Loyalty to Dissent: How Military Women Respond to Integration Dilemmas"[36]

Laura Miller is a senior scientist at the RAND Corporation. Her specialty is the study of military personnel in their "natural habitats," that is, where they live and work. As a result, she has conducted research over the last 20 years at more than 30 military installations in the U.S.A. and with troops on operations in Europe and the Middle East. She completed her doctoral work in Sociology at Northwestern University under the tutelage of Charles Moskos. Conversing with the "common soldiers" and seeking out their opinions on topics of concern to them was Moskos's *modus operandi*, and clearly has been Miller's as well. In 2021, Miller was named President of the Inter-University Seminar on Armed Forces and Society, military sociology's top professional organization.[37]

"Not Just Weapons" examines the resistance of some military men to expanded roles for women in the military. Data for Miller's study are drawn from her conversations, open-ended interviews, focus groups, and surveys carried out with Army troops at multiple continental U.S. and overseas military sites between 1992 and 1994. She derived the theoretical theme from political anthropologist James C. Scott's ethnography of a small rice-growing village in Malaysia. Scott detailed the tensions within the *class structure* of the village and shows how those at the bottom used the *ordinary weapons of the powerless* – foot dragging, feigned ignorance, slander, pilferage, and the like – to mitigate the domination others had over them. The novel application in Miller's study is this: Why do some Army men, when men are in most positions of power, employ these *weapons of the weak?*[38]

The first step in solving the riddle is to note that power is not always binary, as in *the powerful* and *the powerless*. There are multiple levels of power in the military based upon rank, occupational specialty, mission experience, age, race, gender, and the like. *Intersectionality* may be an issue: some combinations of these may afford more power and privilege, while others may sap the amount of power and privilege one rightfully should have. So, do *male officer/female officer* have more status and authority (legitimate power) than *male enlisted/female* enlisted? Or, are *officer/ enlisted women* categorically thought less of by some and given less authority by them than their due?

Women made up about 14% of the Active-Duty Army at the time of Miller's study. Her anonymous survey data indicated that 44% of White Army men of all ranks opposed the roles in effect for military women and any expansion of them, compared to 34% of Black and other soldiers of color of all ranks. While plenty of officers, White and otherwise, were very supportive of women was the military, among all male subgroups defined by the *intersection of rank and race*, the one group most opposed to expanded or current roles for women was White officers.

The above *intersections* are defined *objectively*, but individuals' *subjective percep-tions* of their relative power are of critical importance. In her focus groups and open-ended interviews, Miller noticed that some male officers and enlisted men saw themselves as a *subordinate minority* and women as the military's *privileged and powerful group*. One sergeant, speaking for himself and "many others," identified himself as an *oppressed minority*, stating, "I feel that the *white enlisted male* has *more prejudice against him* than any other [group] in the military." Some officers felt they were in unfair competition for promotion, made "easier" for women because they are not in combat roles. Others felt women in their units created problems they, as leaders, were responsible for resolving. Some had no direct experience with women but based their projections on rumors from other units.[39]

Miller estimates the percentage of Army men who feel embattled, who think there are too many women in the Army or that women have no place in the military, at 15% or so. These represent, Miller states, "a category of [military men] who have assumed and enjoyed *gender privilege*, and who rebel because their *public voice* has been deemed sexist and has been silenced."[40] Here, "the resisting group is not entirely powerless [but] does not wish to relinquish one of the realms in which it holds power." These soldiers have limited options in challenging the status quo. So, they harass.

"Sexual harassment" refers to unwanted sexual comments or overtures. Here, however, Miller is talking about *gender harassment*, that is, harassment used to enforce traditional gender roles or perceived violations of those roles. Some exam-ples include: *resistance to authority* (failing to comply with directives of a female superior, e.g., by feigning ignorance of what is meant or expected), *constant scru-tiny* (closely monitoring actions of female superiors to catch mistakes of individual women or criticizing women's abilities in general), *rumors and gossip* (passing on innuendo or stories to impugn the integrity of military women), and *indirect threats* (ominous warnings of "what might happen" to women), such as this comment on an anonymous survey: "I know, I'm a man ... [Women will] be highly harassed, if not molested, if they join combat arms. Trust me."[41]

Many women soldiers mentioned having experienced these slights. Like the tactics of the disadvantaged Malaysian villagers, these actions are *passive aggressive* attempts to re-exert control. Here, they are used by a small subgroup of those dominant in the institution. On the plus side, this represents progress in making Army men aware it is inappropriate to make openly sexist statements. But, it also suggests a problem – these soldiers feel they cannot speak openly in what they see as a *zero-sum game*: women advance at the expense of male soldiers, so, to regain what males have lost, women must be pushed down.

Helena Carreiras currently is the Portuguese Minister of Defense and Associate Professor of Sociology and Public Policy at the Instituto Universitário de Lisboa, Portugal. Her many writings over the last 20 years focus on gender integration in military institutions and gender dimensions of international security. Her research includes studies of Portuguese officers, cadets at all three Portuguese service academies, and women officers in European militaries and peacekeeping missions. Carreiras received her Ph.D. in Social and Political Science from the Istituto Universitario de Europeo Firenza, Italy, in 2004, and served as the President of the European Research Group on Military and Society (ERGOMAS, introduced in Chapter 1) from 2017 to 2019.

"From Loyalty to Dissent" begins with the work of Harvard Professor of Business, Rosabeth Kanter (who holds a doctorate in Sociology).[42] Carreiras reviews Kanter's three central bases of organizational success in an integrated work environment: sufficient access to *opportunity* and *power*, and, vitally important but often overlooked, *relative numbers*. There are good reasons for beginning cautiously with small numbers when integrating women into stereotypically masculine occupations, but there are some downsides.

When the numbers of women are small, the tendency is for them to become *tokens*, that is, representatives of a suspect category. The result *is heightened visibility* with consequent pressure to perform, *exaggeration of gender differences* by some in the organization, and the *expectation women will assume a stereotypical role type*, such as the *pet*, *mother*, *battle-axe*, or *seductress*. (There may be some predefined roles for men as well, such as *great man*, *hot-shot*, *buddy*, or *yes-man*.) Kantor observes from her research that the *token-effect* is most likely when women make up less than 15% of an organization's members. The optimum point, when organizational work roles become less *gender-based* and open to both men and women, is thought by her to be around 30%.

Social psychologist Janice Yoder, also known for her long-time study of leadership and tokenism, has criticized Kantor's thesis. While scarce numbers certainly contribute to the problem, so do *prejudice against women outside the organization*, the extent to which an *occupation is seen as a departure from traditional gender norms*, and the amount of *intrusiveness expressed by the organization's male members*. These exist along with practical concerns that motivated organizational leaders to bring women into the fold.

Carreiras conducted in-depth interviews in 2000 with 29 women officers in the Portuguese and Dutch armed forces. The Portuguese military provides for the country's defense and joins in peace-keeping missions for NATO and the United Nations. The opening of its armed forces to women began without fanfare or legislation. In 1988, two women applied and were admitted to the Portuguese Air Force Academy and combat slots in the Army opened in 1996 when a woman applied to the Army's Artillery school, an "evolving policy" Carreiras characterizes as *ex-post facto pragmatism*.[43] In 1991 the Portuguese military began an incremental conversion to an AVF. The Air Force and Navy had attracted enough volunteers to complete the transition; however, in 2000, about 25% of the Army still were conscripts. Of the 42,000 active-duty members, about 3,000 were women making up 3 and 6% of the Navy and Army, and 13% of the Air Force. Women served mainly in support positions, but in the Air Force, more were in operational slots as helicopter pilots, parachutists, and air police.

The Dutch armed forces are chartered to defend the Kingdom of the Netherlands and to protect the international rule of law through peacekeeping missions. However, the *merchant class* has been the dominant segment of the Netherlands' *power elite*, with the prestige of the military – except for its *peacekeeping* and *peace-enforcing missions* – ranking well below that.[44] This has implications for military recruitment and for stressing deployment capacities over military hardware. The admission of women without restrictions to military institutes and training centers began in 1978 and gender-segregated units were suspended in 1982. The Dutch military began moving to an AVF in 1989 and ended the draft in 1996. In 2000, 9% of the 50,000-member Dutch armed forces were female, with about 13% of women in the Army, compared to 19 and 33% for the Air Force and Navy, respectively. For much of the AVF, the assumption has been that *increases in the numbers* of women would change the *military's masculine culture*.

Carreiras assessed the experiences of women officers in these two militaries. The interviewees reminisced about their times in the military, good and bad. When asked early on if they themselves had personally encountered barriers or hostile reactions, none admitted having experienced these directly, but many said they knew this happened to others. However, as the interviews unfolded, most of the subjects revealed coming up against such problems. A Portuguese 1st Lieutenant, a parachutist, after stating early in the interview, "I don't have any reason to complain," later disclosed: "I am a deputy company commander, and they won't give me the position of company commander because I am a woman. There is even a company where the commander is missing. ... I have shown I am as capable as a man but I think ... at the leadership level a woman is not welcomed."[45]

As the women spoke, Carreiras also recognized consequences of *tokenism*. For example, a 39-year-old Dutch Air Force Captain, a helicopter pilot, remarked: "You are always in the picture and they are not even aware of it, but as soon as you make a mistake everyone notices. If a man makes a mistake nobody cares, we do not even remember his name anymore" (*heightened visibility*). Likewise, several women reported feeling excluded from male colleagues in their unit, either because they were not invited or because they were not interested in sharing "boys' talk," "drinking and smoking in the bar," or "watching porn movies" (*exaggeration of gender differences*). Finally, a 24-year-old Portuguese Air Force Lieutenant, a navigator, stated: "When a woman joins the forces they are immediately labeled: looking for men. Since I came [into the Air Force] it's been a hard fight to escape that label" (*stereotypical role-typing*).[46]

How did women deal with these encounters? Carreiras identifies four responses: if accepting of the status quo, women can opt for becoming more masculine (*assimilation*) or can emphasize their femininity in accordance with male expectations (*complicity*); or, if dissenting from the status quo, they may dilute their femininity to get along (*conformity*), or adopt a more decisive, even militant, feminine posture (*assertiveness*).

The modal response, the choice of two-thirds of the Portuguese officers and about one-half of the Dutch, was *conformity*. Conforming women try to de-emphasize their femininity as a way of reducing their visibility. A 31-year-old Portuguese Navy psychologist explained: "I always wanted to be ... more neutral ..., not being an excessively feminine stereotype nor the opposite." And a 24-year-old Dutch Army Lieutenant in the medical services put it this way: "The main thing is not complaining. When you complain ... you are not accepted. ... So if I have a tough time I walk or something, they cannot see me crying and complaining." About 20% though chose *complicity*. Said one interviewee:

> I love to dress [in] my skirt and do everything to wear it because I am a woman. ... My skirt, my shoes, my earrings ... they see me and they say 'she is a woman and an officer, but she is careful to show she is a woman'.
>
> (27-year-old Portuguese Army Lieutenant, administration)[47]

The Military as a Change Agent

Research over the past 20 years has shown that the notion of a *single masculine military culture* is an inadequate conception. As Miller noted above, the *intersection of mission, occupation, and gender* have created new spaces where military women and men carry out their work. Further, the same military unit may have more than one culture. As we saw in Chapter 3's Spotlight on Joseph Soeters and associates' analysis, a unit's culture changes dramatically on a continuum from *cold* to *hot* in response to moving from garrison to combat zone. In the early years of the AVF, the more staid, *cold cultures* of peacetime units likely diluted the positive effects of *contact between military women and men*.

Still, the prediction was that actual warfare would increase the participation of women only to a point short of combat roles. However, the nature of the *new wars* in Iraq and Afghanistan (see Chapter 10) meant that military women in noncombat positions often accompanied combat units on *hot* military operations, providing support and expertise but sharing risk and at times being drawn into the fight. Consequently, British military sociologist Anthony King, whom we met in Chapter 4, observes that American and British forces "ironically have more *female soldiers who are genuine combat veterans* than any other nation."[48] The Israelis might want to debate that.

King reports that new gender designations have begun to develop. For example, Connie Brownson, who conducted oral histories with women who served in the Marine Corps, suggests a concept of "equivalence": women Marines were not accepted as "absolute equals" but as *valued equivalents*, that is, *sisters* or *female comrades* who added functional capacity to the unit. King continues: "[A]mong Western, Anglophone forces, it has become increasingly common for female soldiers to describe how on recent operations they have been accepted as '*one of the boys*,' '*one of the lads*'." He concludes: "[H]ow good they are at their job ... and professionalism ... are the crucial determinants of [this] cohesion ..., a form of *solidarity based upon competence*."

We turn our attention now to the integration of another group into the military: gays and lesbians.

Spotlight on Social Scientific Thinking and Research

Aaron Belkin, "Don't Ask, Don't Tell: Is the Gay Ban Based on Military Necessity?"[49]; and Jonathan Lee, "The Comprehensive Review Working Group and Don't Ask, Don't Tell Repeal at the Department of Defense"[50]

Aaron Belkin is Professor of Political Science at San Francisco State University. His primary area of specialization is the study of military masculinity and sexuality in the armed forces. He wrote this article while at the University of California at Santa Barbara and Director of the Center for the Study of Sexual Minorities in the Military (later called the Palm Center for Research on Gender, Sexuality, and the Military). He obtained his doctorate in Political Science at the University of California at Berkeley. Belkin's professional efforts have been devoted to ways of using social science research to engage non-academic audiences, an approach reflected in this article.

"Don't Ask, Don't Tell" (DADT) is the label for a policy developed in the first years President Bill Clinton was in office. At the time, policy put into effect by President Ronald Reagan in 1982 prohibited gays and lesbians from entering the military and subjected them to separation if so discovered once in the service. During his presidential campaign in 1992, Clinton often stated his intention to overturn this policy. Days after he took office, Clinton suspended the "incompatibility" policy. After six months of contentious hearings, Congress passed the National Defense Authorization Act. The bill contained a new policy: gays and lesbians could serve in the military if they kept their sexual orientation secret (*Don't Tell*); however, military authorities could not inquire about a servicemember's sexual preference (*Don't Ask*). Still, if discovered somehow, *known homosexuals* would be discharged from the military under the incompatibility principle.

DADT thus represented a stilted compromise with which neither those favoring or opposing the incompatibility policy were completely comfortable. Belkin reviews the support and opposition to DADT in 2003 and adds a twist: he contacted military leaders and social scientists in Canada, Great Britain, and Australia (Anglo cultures most like that of the U.S.A.) and in Israel (whose military is highly respected) to study the issue in those militaries. Each of these militaries had recently *confronted the same question* with many of the same objections and reasons for change found in the U.S.A., but *had lifted the ban*. What had happened after their policy changes and what could the U.S.A. learn from their experiences? To answer these questions, Belkin interviewed "every identifiable *expert*, pro-gay and anti-gay alike," who had studied and testified for or against the changes. This produced a sample of 104 in-depth interviews and 622 written documents and articles.

The most common reason given for opposition to gays in the military in these four countries prior to reversing their policy, and in the U.S.A. in 2003, is that homosexuals in the military would infringe on *military effectiveness*. Each country's culture had some homophobic elements, and most service members surveyed beforehand did not want gays to serve. For example, an earlier survey of military members in Great Britain had revealed that about two-thirds of male respondents said they would no longer wish to serve if gays were admitted, and about half of a similar survey of Canadian military men stated they would refuse to work with a gay soldier. Professor Hugh Smith, an expert on the Australian military, recalled some officers agreed with one who claimed, "Over my dead body, I'll resign if this occurs."[51] So the very presence of gays, the argument went, would undermine *unit cohesion*. Most of the vitriol, incidentally or not, was directed by males at gays rather than lesbians.

In each of the four cases, the change in policy had come about for different reasons. Canadian federal courts had ruled against the Canadian military's ban in 1992. That same year, the liberal government of Prime Minister Paul Keating voted to lift the gay ban in Australia. Israel had removed its ban in 1993 in response to public opposition to the exclusion of gays and lesbians. Israeli public opinion endorsed the position they should be allowed to serve like everyone else. Finally, the European Convention on Human Rights in 2000 provided the basis for eliminating Great Britain's ban.

So, what happened when the bans were removed? The short answer is, "not much." Australia's Admiral-rank Commodore R.W. Gates termed it "an absolute non-event." Steven Leveque of the Canadian Department of National Defence stated that this is "not that big a deal for us ..., there has not been much of a change." Stuart Cohen of Israel's Center for Strategic Studies told Belkin, "[T]he entire subject is very marginal indeed as far as this military is concerned." And, as for the British, an internal government report, tasked with tracking the response, reported the changeover as "a solid achievement ... with fewer problems than might have been expected." And, Professor Smith reported that the officers he had spoken with were very much alive and still in the Australian military.[52]

What accounted for these expectedly smooth transitions? In all four countries, there were no consciousness-raising sessions, no briefings on queer theory. Rather, military leaders made clear they expected their members to approach and carry out the change as professionals. Of equal importance, they emphasized *standards of conduct were the same for both straight and gay soldiers*. For example, the Australian military's directives on fraternization do not mention sexual orientation at all; they simply state what constitutes fraternization and what behaviors are unacceptable. An Australian officer explained: "Our focus is on the work people do, and the way they do work, and that applies to heterosexuals, bisexuals, and homosexuals [alike]." These emphases seemed critical in providing soldiers, irrespective of how they felt about homosexuality itself, sufficient motivation to *work together as a team* – that is, to *change behavior now, let attitudes follow whenever*.

Finally, there was "no mass coming out of the closet." Some military peers of gays already in the military knew their sexual orientation, and a few more discreetly revealed their orientation after the lifting of the ban, but the majority did not. A lesbian soldier serving in the Canadian military explained it this way: "Gay people [in the military] have never screamed to be really, really out. They just want to be really safe from not being fired."[53] None the less, it seems that at least some military peers serving in the same unit as gays and lesbians know these days what their sexual orientation is, but that does not seem to matter a great deal.

Jonathan Lee served as General Counsel for the Deputy Secretary of Defense during a critical time in the administration of President Barack Obama. President Obama had repeatedly on the campaign trail prior to his election in 2008 indicated his intention to seek the repeal of DADT. Obama stated his approach in his 2010 State of the Union address: he would work with the military to propose legislation in Congress to overturn DADT – a law "that denies gay Americans the right to serve the country they love because of who they are." Secretary of Defense Robert Gates and Chairman of the Joint Chiefs Admiral Michael Mullen formed a Comprehensive Review Working Group (CRWG) to conduct a thorough investigation. Headed by the Defense Department's General Counsel Jeh Johnson and the Commander of U.S. Army Europe, General Carter Ham, it was to provide a report within nine months. Lee was assigned to CRWG as a Special Assistant to General Counsel Johnson.

The importance of Lee's summary of events lies in *laying bare the diligence of CRWG's inquiry* and the extent to which it *opened a conversation within the military* about DADT. Admiral Mullen testified before Congress that he was in favor of the review. His personal opinion, he stated, was that DADT should be repealed but, at that time, he did not have sufficient data to render an adequate professional opinion. CRWG would walk a tightrope: it was to provide the basis for determining if repeal *could* be done without "unacceptable impacts" to the military, while at the same time leaving the decision about whether repeal *should* take place in the hands of Congress.[54]

General Counsel Johnson and General Ham set up four subgroups within CRWG. A Survey team would collect information from the components of the military; a Policy team would review all relevant rules and regulations; a Legal team would analyze laws relevant to the issue; and an Education and Training team would assess leadership responsibilities and "best practices." In the interest of full disclosure, one of this book's authors, Wilbur Scott, served on this latter committee.

CRWG collected an enormous amount of information. CRWG representatives visited 51 military installations in the U.S.A. and overseas where they held 95 town halls (*information exchange forums*) and held 140 follow-on *focus group sessions* with interested military members and their spouses. To allow anonymous feedback, they set up an online webpage and received 72,284 entries. They hired a professional research firm, Westat, to conduct a survey of military personnel to assess their opinions and actions if DADT were to be repealed. A stratified random sampling technique sent out 400,000 online surveys to military members, and 150,000 paper versions to military spouses. A secure and confidential online portal was open for a five-week period to allow self-identified gay and lesbian servicemembers to share their experiences and opinions, and 296 did so. Finally, CRWG compiled other sources of research, including Aaron Belkin's studies of repeal in other militaries and the RAND Corporation's prior assessments, as well as the views of chaplains and religious groups.

Working against the clock, CRWG compiled a 151-page report of all these findings and an 87-page implementation plan and released it for public consumption on 10 November 2010, arguably the most extensive study of a personnel issue in the history of the U.S. military. General Counsel Johnson's and General Ham's bottom line assessment was straightforward[55]:

> Based on all we saw and heard, our assessment is that, when coupled with the prompt implementation of the recommendations we offer below, the risk of repeal of Don't Ask, Don't Tell to overall military effectiveness is low. … [W]ith a continued and sustained commitment to core values of leadership, professionalism, and respect for all, we are convinced the U.S. military can adjust and accommodate this change, just as it has others in history.

On 15 December 2010, Congress's House of Representatives approved the repeal of DADT by a vote of 250 to 175, and the Senate followed suit three days later, 65 in favor, 31 opposed. President Obama signed the DADT Repeal Act into law on 22 December 2010.

Questions for Discussion

1   The concept of intersectionality was framed by legal scholar Kimberlé Crenshaw to capture the way characteristics like race, class, and gender intersect to shape one's lived experience. Why is this concept so critical to understanding the experiences of women in the military?

2   Judith Stiehm reviews the differing ways the three U.S. service academies responded to the mandate to admit women. Review the differences and comment upon how these differences came about and how they might have affected the success of the first integrated cohorts. Would you like to have been there then?

3   Laura Miller's research shows that, though women are fewer than men and occupy lower positions in rank, some military men consider themselves the persecuted minority. State why and explain. Review the responses of Portuguese and Dutch military women. In their shoes, how would you respond?

4   The integration of openly gay and lesbian servicemembers across Western countries turned out to be an "absolute non-event." From your reading of it, what take-aways and leadership strategies accounted for these smooth transitions?

Notes

1 Judith Hicks Stiehm, 1981, *Bring Me Men & Women: Mandated Change at the U.S. Air Force Academy*, Berkeley, Los Angeles, and London: University of California Press, p. 2.
2 Joshua Goldstein, 2001, *War and Gender: How Gender Shapes the War System and Vice Versa*, New York: Cambridge University Press.
3 Bettie J. Morden, 2000, *The Women's Army Corps, 1945–1978*, Washington, DC: Center of Military History, United States Army, pp. 7–8. Six months later, the Navy, Coast Guard, and Marine Corps set up their own units for women volunteers.
4 Ibid., pp. 24–35. This total does not include the much smaller numbers of women volunteers in the Navy, Coast Guard, and Marine equivalents of the WAC.
5 See, Women's Army Corps recruitment video, *It's Your War Too*, produced by the War Activities Committee of the Motion Picture Industry, https://youtu.be/TnjKPy0-vEY.
6 Brenda L. Moore, 1996, *To Serve My Country, To Serve My Race: The Story of the Only African American WACs Stationed Overseas during World War II*, New York and London: New York University Press, and 2003, *Serving our Country: Japanese American Women in the Military during World War II*, Piscataway, N.J.: Rutgers University Press.

7 About 200 African American nurses served in an all-Black Army medical unit.

8 Moore, 1996, op. cit., Chapter 4.

9 Ibid., pp. 133–138.

10 Moore, 2003, op. cit., p. 15.

11 Frank Wu, 1995, "Neither Black nor White: Asian Americans and Affirmative Action," *Boston College Third Law Journal*, Summer, pp. 225–226; Referred to by Moore, 2003, op. cit., p. 20.

12 Moore, op. cit., pp. 28–30 and Chapter 5, "Service in the Women's Army Corps".

13 Morden, op. cit., Chapter II.

14 Ibid., pp. 93–101.

15 Ibid., pp. 241–254.

16 The Equal Rights Amendment had to be ratified by 38 of the 50 states within three years to stand as an Amendment to the U.S. Constitution. The three-year requirement was extended several times, but by 1977 only it had been ratified by only 35 states, failing to gain approval in the core Southern States (Alabama, Florida, Georgia, Louisiana, Mississippi, North Carolina, South Carolina, and Virginia) plus Arizona, Arkansas, Illinois, Missouri, Nevada, Oklahoma, and Utah. In 2017 the Nevada State Legislature reversed its position, followed by Illinois in 2018, and, in 2020, Virginia became the 38th state to ratify the Amendment.

17 Bernard Rostker, 2006, *I Want You! The Evolution of the All-Volunteer Force*, Santa Monica, Calif.: RAND Corporation, Chapter 15, "The Role of Women in the All-Volunteer Force", p. 563.

18 Stiehm, op. cit.

19 Lane Davis, 1974, review of *Active and Passive Resistance in America* by Judith Stiehm (1972, Lexington, Mass., Toronto, London: D.C. Heath), *The Western Political Quarterly*, 27 (June): 338–340, p. 339.

20 Stiehm, op. cit., p. 32.

21 Ibid., p. 1.

22 Ibid., pp. 3–4.

23 Ibid., pp. 94–95.

24 Ibid., pp. 131–132.

25 Ibid., pp. 110–113.

26 Ibid., pp. 84–85.

27 Ibid., pp. 148–152.

28 Ibid., pp. 121, 129–130.

29 Ibid., pp. 282–287.

30 See https://www.nytimes.com/1994/05/23/us/save-the-males-becomes-battle-cry-in-citadel-s-defense-against-woman.html

31 Rostker, op. cit., pp. 563–564.

32 Ibid., pp. 564–567.

33 Major General Jeanne Holm, USAF, 1992, *Women in the Military: An Unfinished Revolution*, revised edition, Novato, Calif.: Presidio Press, p. 395, quoted in Rostker, op. cit., p. 566.

34 Mady Wechsler Segal, 1995, "Women's Military Roles Cross-Nationally: Past, Present, and Future," *Gender & Society*, 9 (December): 757–775.

35 Laura L. Miller, 1997, "Not Just Weapons of the Weak: Gender Harassment as a Form of Protest by Army Men," *Social Psychology Quarterly*, March (60): 32–51.

36 Helena Carreiras, 2008, "From Loyalty to Dissent: How Military Women Respond to Integration Dilemmas," pp. 161–182 in Helena Carreiras and Gerhard Kümmel (eds.), *Women in the Military and Armed Conflict*, Wiesbaden, Germany: VS Verlag für Sozialwissenschaften.

37 https://www.iusafs.org/about-us/the-ius-president.

38 Miller, op. cit., pp. 34–36.

39 Ibid., pp. 44–49, quote appears on p. 48.

40 Ibid., p. 48.

41 Ibid., pp. 36–39, quote appears on p. 39.

42 Carreiras, op. cit., pp. 62–65.

43 Helena Carreiras, 2002, "Women in the Portuguese Armed Forces: From Visibility to 'Eclipse,'" *Current Sociology*, 50: 687–714, pp. 687–690.

44 René Moelker and Jolanda Bosch, 2008, "The Visibility of Women in the Netherlands Armed Forces," pp. 80–127 in Helena Carreiras and Gerhard Kümmel (eds.), *Women in the Military and Armed Conflict*, Wiesbaden, Germany: VS Verlag für Sozialwissenschaften, pp. 82–85, 92–98.

45 Carreiras, 2008, op. cit., pp. 166–168, quote appears on p. 167.
46 Ibid., pp. 168–170.
47 Ibid., pp. 174–178.
48 Anthony King, 2017, "Gender and Close Combat Roles," pp. 305–318 in Rachel Woodward and Claire Duncanson (eds.), *The Palgrave International Handbook of Gender and the Military*, London: Macmillan.
49 Aaron Belkin, 2003, "Don't Ask, Don't Tell: Is the Gay Ban Based on Military Necessity?," *The U.S. Army War College Quarterly: Parameters*, 33 (Summer): 108–119. Belkin presented this and similar research at least half a dozen times at the invitation of then Col. Gary Packard, Chair, Dept. of Behavioral Sciences & Leadership, U.S. Air Force Academy. Col. Packard also served as one of the two authors of the final report of the Comprehensive Review Working Group.
50 Jonathan Lee, 2013, "The Comprehensive Review Working Group and Don't Ask, Don't Tell Repeal at the Department of Defense," *Journal of Homosexuality*, 60 (2–3): 282–311.
51 Belkin, op. cit., p. 110.
52 Ibid., pp. 110–111.
53 Ibid., p. 117.
54 Lee, op. cit., pp. 283–284, 287.
55 Ibid., p. 301.

Recommendations for Additional Reading/Viewing

Allan Bérubé, 1990, *Coming out under Fire: The History of Gay Men and Women in World War II*, Chapel Hill, N.C.: University of North Carolina Press; Randy Schilts, 1993, *Conduct Unbecoming: Lesbians and Gays in the U.S. Military, Vietnam to the Persian Gulf*, New York: St. Martin's Press.
(Two books that changed the debate, the first by an historian, the second by a journalist)

Ruth Margolies Bietler and Sarah M. Gerstein, 2021, *Women and the Military: Global Lives in Focus*, Santa Barbara, Calif.: ABC-Clio.
(Two West Point professors explore the roles and challenges in militaries around the globe)

Connie Brownson, 2015, *Lady Leathernecks: The Enigma of Women in the United States Marine Corps*, Stillwater, Okla.: New Forums Press
(A former Marine-turned-sociologist interviews women Marines and reveals their experiences, opinions, and suggestions)

Invisible War (film), 2012, Directed by Kirby Dick, Produced by Amy Ziering and Tanner King Barklow.
(The film investigates the incidence and mishandling of sexual assault in the military)

Gayle Tzemach Lemmon, 2015, *Ashley's War: The Untold Story of a Team of Women Soldiers on the Special Ops Battlefield*, New York: Harper Perennial.
(An account of women soldiers attached to Special Ops in Afghanistan before the legality of women in combat)

Kayla Williams, 2005, *Love My Rifle More Than You: Young and Female in the U.S. Army*, New York: W.W. Norton.
(Popular "funny, frank, gritty" memoir by former Arabic linguist in the U.S. Army)

8 Spectrum of Conflict
"Big" Wars

Reader's Guide: When people think of wars, they usually have in mind "Big" ones: state-supported, near-peer militaries clashing with each other until one is pounded into submission. Though these occur less frequently than "Small" wars, they are the kind for which the U.S. and other modern militaries are configured, equipped, and trained to fight. For the U.S.A., World War II is the ultimate and the best, the Big One. Wars like these have become incredibly destructive because of modern weaponry and the extent to which civilians perish in them. We advance the concept of "multicides" to capture the full range of casualties. The first Spotlight addresses why the Allies prevailed over the Axis Powers in World War II, by no means a given in 1939. About two-thirds of the casualties in WWII were civilians. Among the most horrific is the systematic rounding up and killing of Jews by Nazi Germany. The second Spotlight investigates the pattern of Jewish victimization in the Holocaust. In the final Spotlight, we turn to the issue of nuclear weapons and why, despite their ominous destructive capacity, nations continue to acquire them.

"War is Cruelty … War is Hell!" This phrase was penned late in the American Civil War by Union Army General William Tecumseh Sherman as part of his explanation to the mayor and city council of Atlanta for why he intended to burn their city to the ground. He stated:

> You cannot qualify war in harsher terms than I will. War is cruelty, and you cannot refine it … [T]he only way the people of Atlanta can hope once more to live in peace is to stop the war … We don't want your Negroes, or your horses, or your lands, or anything else you have, but we do want and will have a just obedience to the laws of the United States. That we will have, and if it involves the destruction of your improvements, we cannot help it.[1]

General Sherman uttered a second statement, his address to the Michigan Military Academy's graduating class on 19 June 1879. He chided them because he had entertained similar cravings at his own graduation from West Point in 1840. So, he admonished:

> It's … natural that there should beat in the breast of every one of you a hope and desire that some day you can use the skill you have acquired here. Suppress it! You don't know the horrible aspects of war. I've been through two wars and I know. I've seen cities and homes in ashes. I've seen thousands of men lying on the ground, their dead faces looking up at the skies. I tell you, War is Hell![2]

DOI: 10.4324/9781003282549-8

In the two wars for which he is known – the Civil War and the subsequent Indian Wars in the American West – General Sherman had a reputation for making sure war indeed was hell for those he fought against. The term *total war* connotes his approach. It extends the conduct of war beyond the killing of opponents to the destruction of infrastructure and economy, spreading the violence and mayhem to civilians and their livelihoods. In the words of Civil War historian Mark Grimsley, Sherman's was the "hard hand of war."[3]

It is well to start with the U.S. Civil War, for it remains the most lethal one in American history. As noted in Chapter 6, more American soldiers died in that war than in all our other wars combined. And that includes the other "Big" one, World War II, fondly referred to by Americans as "the good war." If there is an *American way of war*, to borrow a term from Russell Weigley,[4] it is found in the lore of WWII and, more recently, in the Persian Gulf War of 1991–1992.

What makes these *good wars*? First, involvement was *just*, originating in response to the attack on Pearl Harbor and the invasion of Kuwait, respectively. The combatants were *near-peer militaries* – the Allied militaries vs. those of Germany, Italy, and Japan in WWII, and the U.S.-led coalition against the Iraqi Army in the Gulf War (regarded at the time the fourth strongest in the world). Though complicated by the presence of urban populations, the action took place on battlefields with identifiable "fronts" and "rears." Finally, *victory was absolute* and attained by *annihilating the adversary's military* and then some. These are the marks of *Big Wars*, how to go about them, and how to end them successfully.

How Lethal Is Human Violence?

There is some controversy about exactly what ought to count as *casualties of war*. Without getting bogged down in the details, we draw on recent work by statistician Matthew White who maintains a website documenting the scope of "horrible things."[5] White focuses on death tolls resulting from societally sponsored violence directed either against its own people or those of other societies. He terms these death tolls *multicides* because they include *all deaths whose proximate cause is traceable to such violence.*[6] His estimates thus include the deaths of soldiers, civilians who are killed by them or other civilians, and those who die due to causes such as starvation or disease occurring because of such violence.

Table 8.1 contains a listing of the top 12 multicides in history gleaned from White's accounting. The listing has several interesting characteristics. First, all occur within the past few hundred years and five of them took place in whole or in part during the 20th century. This does not necessarily mean that ancient peoples were not killers of each other, but that recent societies have larger critical masses of people and more highly destructive means to kill them.

For example, among White's top 100 multicides of all time, only 11 took place in the years BC. When hunting-gathering bands (in which most humans lived before recorded history) attacked and killed each other, the numbers killed might have been at worst in the dozens or hundreds. The first large-scale agrarian societies appear around 5,000 BC, and the oldest multicide on White's list occurred around 500 BC (number 96, the Second Persian War with about 300,000 deaths).

Secondly, the carnage is extensive. The numbers who perished range from 15 to 66 million – most of these, except for the two World Wars, accomplished with relatively primitive weapons. This observation though lacks appreciation for the role of *starvation and disease* wrought by large-scale violence. In the conquest of India and the Americas,

Table 8.1 History's Top 12 Multicides[a]

Multicide Event	Type	Estimated Number of Deaths
Second World War (1939–1945)	World war	66,000,000
Reign of Chinggis Khan, Mongolia (1206–1227)	Regional war	40,000,000
Reign of Mao Zedong, China (1949–1976)	Institutional oppression	40,000,000
British Empire India famines (18th–20th centuries)	Commercial exploitation	27,000,000
Fall of Ming Dynasty, China (1635–1662)	Failed state	25,000,000
Taiping rebellion, China (1850–1864)	Messianic uprising	20,000,000
Reign of Josef Stalin, Russia (1928–1953)	Institutional oppression	20,000,000
Slave trade, Mideast (7th–9th centuries)	Commercial exploitation	18,500,000
Reign of Timur, Central Asia (1370–1405)	Regional war	17,000,000
Slave trade, Americas (1452–1807)	Commercial exploitation	16,000,000
Conquering of Americas (1492–1890)	Colonial conquest	15,000,000
First World War (1914–1918)	World war	15,000,000
		319,500,000

a Matthew White, 2012, *The Great Big Book of Horrible Things: The Definitive Chronicle of History's 100 Worst Atrocities*, New York: W.W. Norton, constructed from information on pp. 529, 543–554.

for instance, most deaths are attributable to famine and infectious disease instigated by disruptive colonialization of native peoples. Still, the outright killing of natives contributed substantially to the problem.

Only four of the twelve are identified specifically as "wars" (World Wars I and II, and the reigns of Chinggis Khan and Timur), though several – the British-colonial India famines, the reign of Joseph Stalin, the slave trades in the Middle East and the Americas, and the conquering of Native America – contain brutal elements of inter-societal conquest. This leaves three internal (civil) wars: the reign of Mao Zedong, the collapse of the Ming dynasty, and the Taipei Rebellion.

World War II: The Ultimate Big War

WWII deserves extended comment. Before it was over, all the world's major nations and multiple minor ones had participated, and the rules had devolved to a particularly gruesome version of *total war*, the bellicose equivalent of "almost anything goes." There were several milestones. Of the 66 million deaths attributed to it by White, more than twice as many civilians died as did soldiers (46 million vs. 20 million), most of them targeted by one or another of the militaries involved.[7] Officially the war spanned 2,174 days (from 1 September 1939 to 2 September 1945) and averaged just over 30,000 multicides, that is, war-related deaths, *per day*.

The war saw full use of industrial machinery and weapons, attacks on unarmed cargo and troop ships by submarines, the carpet-bombing of cities, the operation of bureaucratically efficient extermination camps, and the deployment of atomic bombs. It had the two bloodiest battles in human history – the sieges at Leningrad and Stalingrad. Both transpired in Russia as the Red Army and Russian people desperately – and, ultimately, successfully – sought to stop the advance of the German Army. Also on the Eastern Front were the largest Nazi extermination camps: Auschwitz, Belzec, Majdanek, Chelmno, and Sobibor, all located in Poland. There the Germans executed about 2.6 million civilians – Jews, Bolsheviks, gypsies, and others designated as enemies of the state. A similar number died in other concentration camps or holding facilities and at the hands of *Einsatzgruppen*, special death squads who systematically lined up their victims and shot them.

The concept of *total war* came to be equated with bombing cities and, hence, the wholesale killing of civilians. In response to the bombing of the Dutch port city of Rotterdam by the German Luftwaffe in May of 1940, the British Royal Air Force (RAF) retaliated. In tit-for-tat fashion, the Germans conducted aerial attacks on Coventry, London, Southampton, and Liverpool, and the RAF bombed Frankfurt, Hamburg, Cologne, Wurzburg, and Berlin, just to name a few English and German cities. At first, the U.S. Army Air Corps (USAAC) restricted itself to "military targets," but eventually watered down the definition. The napalming of Dresden in February of 1945 – a city of negligible military significance – by the RAF and USAAC remains one of the most controversial bombing raids of the war. Allied planes dropped 3,900 tons of high explosives and napalm in three nights of bombardment on Dresden, killing 25,000 civilians.

Meanwhile in the Pacific theater USAAC Gen. Curtis LeMay was reassigned in August of 1944 from the European theater with an order to increase the lethality of bombing runs over Japan.[8] Relying on the B-29 *Superfortress*, he increased payloads by reducing the number of defensive armaments the planes carried and instructed his pilots to deliver incendiary bombs at extremely low altitudes (4,500 to 8,000 feet). The 10 March 1945 bombing runs over Tokyo ignited the city's largely wooden structures, destroying 25% of the metropolitan area and killing an estimated 100,000 people. The use of atomic bombs five months later in the destruction of Hiroshima and Nagasaki, and their essentially civilian populations, led to Japan's surrender in August of 1945.

Images of World War II conjure up for most Americans all the warm fuzzies of a good war, or, to borrow a phrase from historian Michael C. C. Adams, "the best war ever!"[9] Like anyone else, Americans think of this war ethnocentrically from their own experience or, these days, the prevailing folklore about it. Except for the bombing of Pearl Harbor, the war's deadly violence occurred outside the U.S.A. Because the war took place else-where, the U.S.A. incurred "only" about 450,000 multicides, the vast majority of whom were combatants. The U.S. military heavily censored news stories and photos to keep reporting of the war to the American public cleansed and upbeat, and the dominant narrative since has been largely self-congratulatory.

There is some justification for this celebratory mood. However, consider how the same war, but in a different theater with different circumstances, was etched in the Russian psyche. The German Army invaded Soviet territories in 1941 and a good portion of World War II thereafter was fought on the Eastern Front. The Soviets – also fighting a good war, in this case, a valiant, desperate, all-out defense of the homeland – incurred around *27 million multicides*, the vast majority of whom were civilians. Here, there is not much nostalgia for "let's do a war like that again."

To lend perspective on all this, we focus our Spotlights on one of the premier social science analyses of what transpired and a sociological analysis of genocide in WWII.

Spotlight on Social Science Thinking and Research

Richard Overy, *Why the Allies Won*[10]

Well-known British military historian, Richard James Overy, is one of the foremost experts in the study of WWII. Born in London in 1947, he received his baccalaureate degree from Cambridge University's Gonville and Caius College in 1969, and a Ph.D. in History from the University's Churchill College in 1979. He served as Fellow and Lecturer in History at Cambridge's Queen's College and London's King's College, and as Professor of History at King's College and the University of Exeter. He has published more than 20 books on WWII, many of which provide baseline data for *Why the Allies Won*.

Why the Allies Won is not a blow-by-blow historical account of the war. Rather, as the title states, it addresses the *whys* of the Allied victory over the Axis powers – the *Allies* narrowly defined as Britain and Canada, the U.S.A., and the U.S.S.R. (Soviet Union), and the *Axis powers* as Germany, Japan, and Italy. Overy focuses on the whys because the Allied *total victory* by war's end appears to have left many with the misleading impression that this was the war's natural course or preordained outcome, a "given" as it were. Such is hardly the case. Hence, this book.

Any prediction in the first months of 1939, Overy contends, would have favored the Axis powers (see Figure 8.1). Germany had the world's state-of-the-art army and air force. Its army featured highly mobile tank (*Panzer*) divisions and, in defiance of the Treaty of Versailles at WWI's end – which limited German forces to 100,000 – had almost 8 million troops. Japan's military prowess was much less, but it had a modern navy and supporting air force. Italy's military was uneven but capable of making trouble. On the other side, Britain had Europe's most formidable navy but a smallish colonial army, and the U.S.S.R.'s military was sizable but ponderous and outmoded. France, still reeling from World War I, would quickly prove to be a military no-show in WWII (see Chapter 5). As for the U.S.A., it was the world's foremost naval power but its skeleton of an army, in Overy's estimation, ranked 18th internationally in size and fighting capacity in 1941.[11]

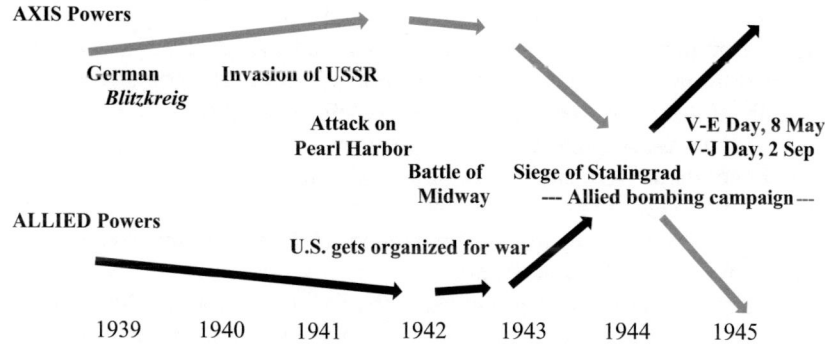

Figure 8.1 Overy: Military Strength/Capacity of Axis/Allied Powers, 1939–1945

Source: Authors' Creation.
Note: Constructed from Overy's summary description of the course of the war, Richard Overy, 1997, *Why the Allies Won*, New York and London: W. W. Norton & Company., Chapter 1.

Hence, many battles and events which could have gone either way had to end fortuitously for the Allies, none more so than the 1942 naval Battle of Midway in the Pacific, and the 1942 through 1943 massive, all-out sieges at Leningrad and Stalingrad on the European theater's Eastern Front. And, the Allies would have to rapidly accelerate their military capacity and efficiency throughout the war to contain, then crush, the Axis powers.

Germany's military invasion of Poland on 1 September 1939 denoted the official start of WWII. However, Axis powers already had conducted offensive operations before then. In 1931–1932, Japan had invaded and conquered China's northeast region, Manchuria, and in 1935 Italy had done the same in Abyssinia (now Ethiopia). Japan seized control of French Indochina and Italy grabbed Egypt, both in 1940. Germany, without a shot, had forcibly annexed Austria and Czechoslovakia before its all-out attack on Poland. By 1941, Germany had stormed through the Netherlands, Belgium, France, Denmark, Yugoslavia, Greece, and Norway. And, on 22 June of that year, Germany embarked on its largest military operation yet – it invaded the Soviet Union. That would eventually prove to be a monstrous mistake.

Despite these land invasions, WWII to this point still was very much a naval war. Britain was heavily dependent upon worldwide shipping lanes for importing food-stuffs and industrial supplies, and confronted the Axis powers with its Royal Navy. At first its superior naval capacities held sway, but this advantage, rooted in a traditional view of sea power, evaporated as the Germans began attacking ships from the air. Not even the biggest, most well-armed ships were invincible to attacks by gnat-like bombers. In 1940, the *Luftwaffe* sank over a half a million tons of British shipping and over a million tons in 1941. By the end of 1941, Britain had lost 1,299 ships to German and Italian bombers.

The U.S.A. officially had remained aloof from these goings-on. However, stubborn *isolationism* was off the table after the bombing of Pearl Harbor, Hawaii, on 7 December 1941. At 7:48 that morning, dive- and torpedo-bombers launched from six nearby Imperial Japanese Navy aircraft carriers attacked battleships, destroyers, and other U.S. vessels in the harbor. Eight battleships were damaged and four sank, the *Arizona*, *Oklahoma*, *West Virginia*, and *California*; 2,335 sailors, soldiers, and airmen and 49 civilians died in the attack.

If the German invasion of Russia would prove to be the Axis's first monumental miscalculation, Japan's attack on Pearl Harbor would be its second. An infuriated U.S.A. fully hurled itself militarily into the war effort. While its *war-fighting capacities* were enormous, the U.S.A.'s immediate *war-fighting capabilities* (except for its navy) were comparatively anemic. It would take time for its capacities to mature into capabilities. Meanwhile, the Japanese navy held sway over most of the Pacific. The basic American strategy in 1942 was to retain a foothold there while preparing to be relevant in the European theater.

The westernmost U.S. military outpost in the Pacific was the tiny Midway Atoll – islands, as the name suggests, roughly halfway between the continental U.S.A. and Japan. In June 1942, Japanese Marshal Admiral Isoroku Yamamoto sought to take these islands to gain fuller control of the Pacific. The formidable flotilla he assembled for the task had four aircraft carriers (the *Akagi*, *Kaga*, *Soryu*, and *Hiryu*), seven battleships, twelve cruisers, forty-four destroyers, as well as fuel and supply ships. His plan of attack depended upon surprise, but U.S. naval intelligence at Pearl Harbor had broken the Japanese communication security code. Admiral Chester Nimitz, commander of the Pacific fleet, thus was privy to Japanese intentions. Nimitz's plan was to confront the attack with U.S. Army Air Corps (USAAC) and Navy aircraft land-based at Midway itself, and Navy bombers from three aircraft carriers to the northeast of Midway.

On 4 June, amidst dense fog and unaware of the American naval formation northeast of Midway, Japanese dive bombers armed with explosives for land targets departed their carriers at daybreak. The dive bombers damaged Midway's runways and other facilities. However, Japanese pilots radioed back another bombing run would be necessary to disable the base, so Yamamoto ordered dive bombers held in reserve on his carriers to be armed with land-bombs for a second run. Thirty minutes later, a Japanese reconnaissance aircraft spotted the American aircraft carriers. The stunned Japanese commander could not ignore the American naval force. He therefore gave the order for the reserve dive bombers to be refitted with bombs for attacking the American ships while his initial wave of dive bombers continued their return to the carriers and landed.

The decks of the Japanese carriers thus were cluttered with planes, bombs of various sorts, petrol tankers, and fuel hoses when 41 American *Devastator* torpedo bombers arrived on the scene. Mitsubishi *Zero* fighter planes surgically picked apart the unescorted *Devastators*. However, circling nearby were 54 American *Dauntless* dive bombers. They had had difficulty locating the Japanese carriers and were dangerously low on petrol, but a squadron commander ordered one bomb run before returning to refuel. Obscured by the morning sun, the *Dauntlesses* had an unimpeded approach to the Japanese carriers.

The results were devastating. The *Kaga*, hit by three bombs, and *Akagi*, hit by four, burst into roaring flames as planes and fuel lines on the flight deck ignited and stacks of bombs cooked off in secondary explosions. Three of the *Dauntlesses'* bombs inflicted similar damage to the *Soryu*. The *Kaga* and *Soryu* would both sink that day after burning furiously and by morning the *Akagi* too was a total loss. Overy summarized the carnage: "Within ten minutes the heart of the Japanese navy's strike force was destroyed by a mere ten bombs on target."[12]

The skirmish at Midway thus proved a first decisive turning point for the Allies. The Japanese navy, which to this point had done pretty much what it wanted, had been dealt a semi-crippling blow and Japanese aspirations for a firm hold in the Pacific were thwarted. Though a grueling Pacific war would continue island by island for another three years, the U.S.A. now could divert war materiel, and later troops, to the European theater where the Germans were on a rampage.

Meanwhile, the Germans had launched its massive land invasion to take Soviet Russia. Adolph Hitler ordered the attack over the advice of his generals, citing a hatred of Bolshevism, dislike for "Asiatic" races, and the German master race's need for more *Lebensraum*, "living space." He also craved their resources, especially oil. Three million German troops, spearheaded by fast-moving *Panzer* (tank) divisions and backed by another nearly 700,000 soldiers from other Axis powers, quickly conquered Soviet Ukraine and turned their attack toward Moscow. The Red Army was utterly overwhelmed. As it retreated, it employed a *scorched-earth* policy, destroying bridges, crops, and anything else that might be of use to their attackers. The overextended German advance stalled short of Moscow during the frigid winter months. In the spring of 1942, Hitler revised his plan, again over the counsel of his generals.[13]

Hitler redirected the assault force toward the eastern and western cities of Stalingrad and Leningrad, respectively. Overy focuses on the siege of Stalingrad. Sprawling for 40 miles along the Volga River, Stalingrad – Tsaritsyn before the 1917 Bolshevik Revolution and its renaming after Soviet Premier Josef Stalin – was a key transportation hub and industrial city with a population of 200,000. Hitler loved the idea of taking the city because of its name. Stalin would defend the city at all costs for the same reason.

The attack, which began on 23 August caught the Red Army in the region short-handed. The Germans thus advanced rapidly toward the city with the *Luftwaffe* and *Panzers* leading the way, but the advance stalled as it reached Stalingrad's outer suburbs short of the Volga River. The river's transportation capacity and cargo already had been reduced to uselessness by the *Luftwaffe* and the Soviet's scorched-earth policy. The Luftwaffe now indiscriminately bombed the city to clear the way for a final thrust by the army.

As panic swept the city, Stalin issued his famous Order #227: *Not one step back!* To capitalize on the revulsion against German brutality, Stalin implored all citizens able to shoulder a weapon to stand in defiant defense of Mother Russia.[14] Men of all ages responded, and thousands of women took up the call as well. Desperate, Stalin also reassigned one of his most trusted and experienced military advisors, Marshal Georgi Zhukov, from Moscow to Stalingrad. Zhukov recognized immediately what Hitler's dissenting military staff had feared: the rapid German advance on the city had created a vastly overextended supply chain stretching more than 300 miles. This he would exploit, but first he had to buy time.

Zhukov placed General Vasily Chuikov in charge of Stalingrad's defense. The Germans' carpet-bombing of the city had leveled it into piles of rubble that could be exploited to slow the *Panzers* to a crawl. Chuikov reorganized defenses into small clusters and reduced the fighting to block-by-block warfare. He introduced the tactic of "hugging" the enemy (engaging only when within "the throw of a hand grenade") to neutralize the Germans' superior firepower, made extensive use of snipers, and developed "storm groups" for attacking German positions at night. His soldiers and civilian defenders relished this style and as they became more and more adept, the fighting became house-by-house and then, legend had it, room-by-room. As the tempo shifted, German morale slumped.[15]

Now it was Marshall Zhukov's turn. On 19 November, a huge Soviet force reinforced by tanks and a revived Soviet air force struck the flanks of German Army supply lines. Within days, the attacking Soviets completely encircled the estimated 270,000 German soldiers still in Stalingrad. Hitler forbade them to retreat and promised to resupply them by air. Soviet gunners and fighter jets took on the resupply planes, and as bitterly cold winter weather set in, only a fraction of the needed supplies got in. Trapped, undernourished, short on ammunition and fuel, ill-prepared for the –20 degrees Fahrenheit temperatures, and hounded now day and night by Soviet storm groups and snipers, the Germans were caught in a deadly vise.

In late December, Zhukov positioned fresh Soviet divisions, guns, tanks, and aircraft for a final push. The German resistance imploded. Tens of thousands of panicked German soldiers surrendered,[16] and those that did not were flushed out of their bunkers and tanks with flamethrowers and grenades.[17] The siege of Stalingrad ended on 2 February 1944. The total number of Axis deaths is estimated at 800,000.

On the Soviet side, more than 1 million troops and civilians perished. The German Army had been dealt a stunning defeat and, by 1945, it would be pushed out of the Soviet Union. Altogether, the German military would suffer more than 90% of its WWII casualties in the Soviet campaign.

The battles of Midway and Stalingrad thus were turning points that gave the U.S.A. time to develop its *war capabilities*. As we showed in Chapter 4, America converted its world-class mass production system in 1942 and 1943 from making automobiles and consumer products to manufacturing war materiel. Though Germany and Japan too were top-notch industrial nations, their production systems, though awesome, could not match the sheer immensity of America's output.[18]

Still, these developments, Overy argues, do not fully explain the Allied success. The missing pieces are the *moral contest* and the *role of Axis and Allied leaders*. For starters, there would have been no World War II without Adolf Hitler. Thoroughly racist and xenophobic, thin-skinned, and quite willing to put his ideology brutally into practice, he was a charismatic leader fully in touch with a simmering current in German society: *the desire to restore Germany to its rightful place* after its humiliating defeat in World War I. While only a minority of Germans bought into his agenda, many stood by and tolerated his antics and then his plunging of the nation into world war.

On the Allied side, Hitlerism provided a clear ideological threat to democratic ways of life.[19] The unprovoked invasions by Germany and Japan compounded this threat and gave the Allies an undisputed moral high ground. It is ironic then that the three major Allied powers were Britain, the U.S.A., and the Soviet Union – the latter a totalitarian, communist state headed by Josef Stalin, a rather ruthless dictator in his own right. The marriage between Britain and the U.S.A. was based genuinely in shared values, but their relationship with the Soviets was purely an arrangement of necessity. That the relationship worked so well and endured the course of the war is, Overy states, a testament to the extent Hitlerism – and Germany's invasion of the Soviet Union – motivated a united opposition.

Differences in how the leaders executed their roles also had significance.[20] A corporal in WWI embittered by a belief in *Dolchstosslegende* – the myth that the German army did not lose WWI but was "stabbed in the back" by a military and government enmeshed in a secret Jewish conspiracy – Hitler was not given to sharing power or trusting advice from others. Germany's military had as talented a pool of military minds as anyone, but Hitler *disdained dissenting advice*, instead *favoring lockstep loyalty*. An amateur at best, he fancied himself a more astute strategist and tactician than his generals and had no appreciation for the organizational underpinnings and operational contingencies incurred by his orders. Against weaker militaries, German military prowess might have overcome these shortcomings.

Though each had quirks of their own and were not above making bone-headed decisions, the Allied leaders – Roosevelt, Churchill, and Stalin – were in an important respect the opposites of Hitler. While they took seriously their responsibility to define war strategy, they wisely conspired with their military experts. Roosevelt and Stalin especially were known to demand unvarnished situation and battle reports and to expect advisors to challenge their thinking. Of the three, Stalin perhaps was the most surprisingly adaptive during the war. Accustomed before the war to being an unquestioned tyrant (a role he quickly reassumed after the war), he sensed that the German invasion could not be repelled without a united and Herculean effort by the Russian people – something that could not simply be ordered and coerced.

The Holocaust

We already have noted the staggering numbers of civilian deaths related to combat on the Eastern Front, for example, during the sieges of Leningrad and Stalingrad. However, Nazi Germany also orchestrated the willful extermination of certain noncombatants as a matter of policy. About two-thirds of the Jewish citizens of Germany and surrounding European countries were gassed or shot to death or otherwise died of starvation and disease between 1939 and 1945, a *crime against humanity* known as the Holocaust, or in Hebrew, *Shoah*.[21] The ostensible justification for these actions lay in the Nazi Party's racist ideology proclaiming a belief in the biological superiority of Germanic peoples and the inferiority of Jewish ones. The policy's legal basis was set in a series of laws passed by the Party between 1933 and 1939 which first marginalized Jewish citizens, then declared them substandard human beings and enemies of the state.[22]

These laws in themselves did not call for the extermination of Jews. Rather, they laid the groundwork for the goal of a *Judenfrei* ("clear of Jews") Germany, and then Europe, without specifying how this was to be achieved. At first, the answer in Germany and Austria was to deport as many Jews as possible. However, as the number of countries invaded and dominated by Germany began to swell – the Netherlands, Demark, Poland, Czechoslovakia, Hungary, Rumania, and so on – deportation was no longer feasible, and a more ominous *Endlosung* ("final solution") came into play: annihilation. Hence the extermination camps of Eastern Europe we mentioned earlier and the dirty work of the special death squads (*Einsatzgruppen*). In 1944, the Polish jurist, Raphael Lemkin, coined a term for this policy: *genocide*, the planned and purposeful destruction of people who are members of a religious, racial, or ethnic group.

We turn our Spotlight now to one of the pioneers of the study of genocide, Helen Fein.

Spotlight on Sociological Thinking and Research

Helen Fein, *Accounting for Genocide: National Responses and Jewish Victimization during the Holocaust*[23]

The sociologist who has systematically researched genocide is Helen Fein. She completed her Ph.D. in Sociology at Columbia University in 1971 and since has written about the Holocaust, genocide, and other forms of collective violence for more than 40 years. She is the founder and first president of the International Association of Genocide Scholars, and has served as the Executive Director of the Institute for the Study of Genocide at the City University of New York into her 80s. Born in 1934, Fein passed away in 2022.[24]

Accounting for Genocide, first published in 1979, still stands as a classic sociological treatise on the subject. The proportion of Jewish citizens who died was not the same in all regions dominated by the Nazis. Rather, the percentage of fatalities ranged from a low of less than 5% of Jews in some areas to a high of more than 90% in others. *Accounting for Genocide* sought to explain this variation. Her data were derived from post-war documentation of the numbers of Jews who were displaced and/or died during WWII, memoires and diaries of Jews who experienced this period, transcripts of war trials, and numerous secondary sources.

Fein's theory is depicted in Figure 8.2. In broad strokes, Fein begins with the preconditions that led to the genocide in Turkey of Armenians in World War I and

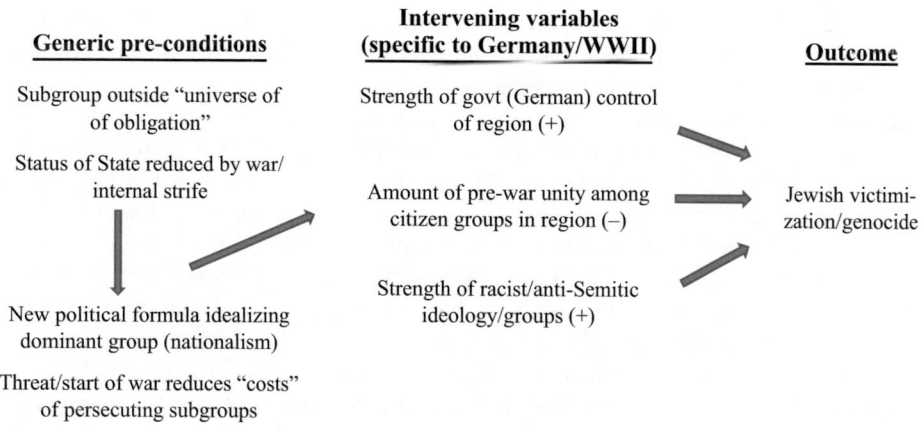

Figure 8.2 Fein's Theory of Jewish Victimization/Genocide

Source: Authors' Creation.
Note: Constructed from Fein's description of her theory.

of Jews in Germany during World War II.[25] Some multi-religious/multi-ethnic states, she notes, especially democratic ones, specify that *all its citizens are within the universe of obligation*, that is, enjoy protection under the law. However, more autocratic ones often *reserve that benefit for its dominant groups* and limit legal protections for some groups defined as lesser. Such was the case in pre-WWI Turkey and pre-WWII Germany.

Also, nations sometimes lose status or rank in the world community because of a setback in war or because of decay in what previously made them strong. A steep *loss in standing* creates an especially volatile situation in states which have historically relegated one of its religious, ethnic, or racial groups *outside the universe of obligation*. Turkey's loss of its Ottoman Empire and Germany's setback in WWI were jolting demotions of this sort. States in this predicament are quite susceptible to political formulas that *equate nationalism with special privileging of dominant groups* and, should war or the threat of war exist, with an *identification of less franchised groups as enemies of the state*.

The next phase of her study provides a careful data analysis of the specifics for European Jews in WWII. The dependent variable is the *rate of Jewish victimization*, that is, the "sum of all persons killed, subjected to fatal trauma, or dying by their own hand ... because they were socially labeled or recognized as Jews."[26] Victimization was a long, often complicated process: Jews had to be identified and stripped of their rights, segregated into specific areas, isolated from the population, and finally moved out for the *final solution*. The chain potentially could be broken or interrupted at any step along the way, and each phase required concerted effort on the part of German and local authorities.

The nations and regions conquered by Nazi Germany between 1939 and 1945 varied by the *extent of control the Nazis* were able to exert locally, the *strength of interconnections among local citizen groups* during crises before the war, and of the pre-war *presence of racist and anti-Semitic groups*. The first and third of these variables were expected to increase victimization, the second one to mitigate it.

Fein's multivariate analyses confirm these hypotheses.[27] In areas where there was a more forceful German presence, already strong anti-Semitic organizations and sentiments, and few connections between non-Jewish and Jewish community groups, Jewish victimization rates were catastrophic. For instance, in parts of Austria, Croatia, Germany, and Serbia, more than 85% of Jews who lived there died under German occupation, and the rate was above 94% in parts of Hungary, Lithuania, and Poland. On the other hand, significant evasion of the final solution occurred in France, Italy, Belgium, Norway, and in some parts of Rumania and Hungary, where less than 30% of Jews were victimized. In Denmark and Finland, fewer than 2% were. Here, the Germans had less of an occupational presence and there were sparser histories of anti-Semitism.

Importantly, Fein also was able to show the mechanisms through which these two variables work.[28] For example, where Germans had more control, they were able to co-opt local Jewish councils (*Judenraten*) by cajoling, bullying, and using other forms of coercion. This made it easier to deceive the local population about the true end of "deportation" programs. And, where there were historically strong anti-Semitic organizations, it was easier to obtain state cooperation in discriminating against Jews, segregating them, and isolating them from the larger population. Both these conditions produced very high rates of Jewish victimization.

The last part of the book contains three case studies, each presented from the victims' point of view: the ghetto in Warsaw, Poland, the case of the Netherlands, and that of Hungary.[29] All three were especially high victimization areas. The Warsaw Ghetto was especially notorious – more than 97% of the Jews who lived there at one time or another during the war became victims. It also was the site of fierce resistance. Since fighting back meant almost sure death, survival in any of the German-occupied lands meant knowing when to muddle through by complying with local authorities and when to risk everything by striking back. Conditions were bad enough in Warsaw to make the latter a frequent option, though usually to no avail.

The Netherlands represents an outlier, that is, a state where the model would have predicted low victimization. However, an identification-card system there developed early on by a faceless bureaucrat made the documentation of Jews easy, and the strategy by Dutch Jews of protesting peacefully backfired. German occupiers went to great lengths to make tragic examples of those who protested. Hungary was a country where strong anti-Semitism existed long before the war, local authorities cooperated enthusiastically with the Nazis, and the Roman Catholic church, unlike in some other countries, remained especially silent. All these conditions facilitated a quick deportation of more than half a million Jews to Auschwitz, Poland, for extermination. Their story is told through the war-time diary of Eva Heyman and the post-war recollections of one of the best-known Holocaust survivors, Eli Wiesel.

Fein's work, along with that of historian Raul Hilberg,[30] stands among the important social scientific treatments of the Holocaust. *Accounting for Genocide* challenges the oft-stated notion that the Holocaust was carried out just by maniacal Nazis while others had no idea what was going on. The study makes clear that the nature and scope of the Holocaust would not have been possible without many willing facilitators: lukewarm

Nazis who went along with the flow, compliant local administrators, non-Nazi anti-Semitic groups, bureaucrats, and others just doing their jobs, silent churches and civic institutions, and even non-Jewish friends and neighbors. All participated directly or indirectly or else stood by without raising a hand to intercede.

Fein's work inspired much research since then devoted to the *role of active and complicit bystanders*. One of the most comprehensive of these studies is that of social psychologist Ervin Staub.[31] Himself a young child in Hungary during World War II, he has devoted his professional life to the study of genocide. Like Fein, he begins with preconditions for intergroup violence: *difficult life conditions* and *cultural heritages favoring "outrage against others."* These structural conditions often translate into *psychological traits that encourage intergroup violence*. The model then highlights the role that *bystanders* play in converting these structural and psychological characteristics into actual incidents of intergroup violence. Bystanders facilitate the violence in many ways even when they themselves may personally oppose what is taking place: *passive bystanders* give tacit approval and hence permission to proceed, as it were, while *active bystanders* enable violence more directly by performing specific tasks. In either case, *complicity* is a key factor in making the violence possible.

Thinking the Unthinkable: Nuclear War

In the latter part of WWII, American scientists worked feverishly in Los Alamos, New Mexico, to develop an atomic bomb[32] to use against Germany. However, the war in the European theater ended in April of 1945 just as preparation was in the final stages. That left only the war against Japan in the Pacific theater. Here fierce resistance by the Japanese in island battles such as Iwo Jima, and the desire to revenge Pearl Harbor, already had legitimated the U.S. bombing of Japanese cities with high explosives and napalm, including Tokyo.

The tenuous relationship of necessity between the U.S.A. and the Soviet Union soured quickly after the surrender of Nazi Germany in May of 1945. The two powers broke ranks to gain advantage in dictating the terms of post-World War II settlements. The issue of nuclear weapons came to the fore almost immediately and would prove to be the most dangerous. Without informing the Soviet Union, President Harry S. Truman in the summer of 1945 authorized the U.S. Army Air Corps to proceed with plans to bomb Japan with two atomic warheads.

Code-named *Little Boy* and *Fat Man* to denote differences in their size and shape, components of the two bombs were transported by ship in July of 1945 to the island of Tinian, where final reassembly took place. Colonel Paul Tibbetts's B-29 *Superfortress* crew dropped *Little Boy* with the explosive equivalent of 13,000 tons of TNT over Hiroshima on 6 August 1945. Three days later, Captain Frederick Bock's B-29 crew delivered *Fat Man* over Nagasaki with a similar payload. The original target for *Fat Man* was the city of Kokura, but the target area was covered by clouds and smoke from the previous day's fire-bombing of nearby Yahata. The official death toll from the immediate blasts were set at 66,000 and 30,000 for Hiroshima and Nagasaki, respectively, with the numbers adjusted to 140,000 and 90,000 for subsequent deaths from burns, radiation, and the like. Virtually all these casualties were civilians.

In his public statements, President Truman indicated that he never lost any sleep over his decision to employ the atomic bomb. Yale University Cold War historian, John Lewis Gaddis, has observed that Truman's actions suggest otherwise. Like almost all Americans, Truman had never heard of Hiroshima and Nagasaki prior to August of

1945. He had left the decision as to where and when to bomb to the Army Air Corps. However, after the bombing of these cities, Truman took the unusual step of placing America's atomic arsenal under tight civilian, rather than military, jurisdiction. He explained to his advisors[33]:

> It is a terrible thing to order the use of something that … is so terribly destructive, destructive beyond anything we have ever had …. So we have got to treat this differently from rifles and cannon[s] and ordinary things like that.

The U.S.A. continued to make atomic bombs after WWII. However, Truman's control over them was so thorough that the Pentagon was excluded from even basic information about their numbers and capabilities. It got to the point where, Gaddis observed, "Soviet intelligence knew more about American atomic bombs than the United States Joint Chiefs of Staff did."[34] In fact, the U.S.S.R. had begun spying on the U.S.A.'s secret nuclear program during WWII when the two were uneasy allies. Stalin thus knew the U.S.A. had the atomic bomb before President Truman announced it to the world and, following the war he, Stalin, ordered an all-out effort to develop such bombs of their own. The U.S.S.R. became the second nation with nuclear weapons in 1949. By the time of the Korean War the U.S.A. had 369 second-generation hydrogen bombs – each thousands of times more potent than an atomic bomb – and the Soviets had five.

The division of Korea into North and South at the 38th parallel by the Soviets and the Americans was part of the post-WWII territorial agreements among the victors in Potsdam, Germany. The sum of these agreements gave rise between 1946 and 1991 to an edgy game of chess between the U.S.A. and the U.S.S.R. short of all-out-war, the so-called "Cold War." The "hot" wars that took place did so through intermediaries. In Korea, the North, with Stalin's approval and military support from China, invaded the South on 25 June 1950. The North's surprise attack within days drove South Korean and U.S. forces from Seoul (located near the 38th parallel) to the southern tip of the country near the city of Pusan. Sensing the North Koreans had badly overextended their reach, U.S. General Douglas MacArthur executed a daring but dangerous amphibious landing with 10,000 Marines at the port city of Inchon to cut the North Korean advance in half.

The counter-attack worked and, with the shift in momentum, MacArthur pushed into North Korea. The Chinese Army counter-attacked with 300,000 troops and routed U.S. and South Korean forces on the dead run back to the 38th parallel. U.S. military planners requested nuclear weapons from Truman to stem the tide, but the President refused. The only suitable targets were Chinese cities and military facilities, and Truman did not want to risk Soviet retaliation. As for Stalin, he had little invested in Korea and did not want to chance reprisal by the U.S.A. So, the war ground on in a stalemate. In Gaddis's words: "The only decisive outcome of the war was the precedent it set: there could be a bloody and protracted conflict involving nations armed with nuclear weapons – and that they could choose not to use them."[35]

General Dwight D. Eisenhower won the 1952 U.S. presidential election and soon confronted the same nuclear reality. At first, he proclaimed that nuclear weapons would be treated as any other entry in the arsenal. A keen student of military history and one who had read von Clausewitz's *On War* very carefully, Eisenhower soon realized the constraints of this position. If war is an instrument of policy, then its scope must be *limited* to actions destined to attain policy goals. However, the destructive force of nuclear weapons by this time was so great that nuclear war threatened every user's very existence.

Eisenhower remarked to his staff, "[we] literally would be in the business of digging our-selves out of the ashes, starting [all over] again."[36]

To his advisors' chagrin, Eisenhower thus rejected the notion of *limited nuclear war*, opting instead for *total nuclear war*. In short, his administration's only plan for use of the U.S.A.'s nuclear weapons was to launch all of them at once. This stance forced the Soviet Union to adopt a like policy. If the U.S.A. would only do *total nuclear war*, a *limited nuclear war* by the U.S.S.R. against America was a losing proposition, so it too could only do *total nuclear war*. Oddly, this situation thus created an inhibition against using nuclear weapons at all, a principle eventually and ironically called *mutually assured destruction* (MAD).

In the decades since the 1950s there has been a proliferation of nuclear-armed nations. Great Britain, France, and China developed the capability in response to the Soviet Union, India in response to China, Pakistan in response to India, and so on. The implication is that proliferation has occurred essentially for security reasons. We turn our Spotlight now to political scientist Scott Sagan, who offers a more nuanced perspective.

Spotlight on Social Science Thinking and Research

Scott Sagan, 'Why Do States Build Nuclear Weapons?'[37]

Scott Sagan is Professor of Political Science at Stanford University and Senior Fellow of the University's Center for International Security and Cooperation. He is a recognized expert on nuclear strategy and deterrence. At the time of Sagan's analysis, Iran, Iraq, Libya, and North Korea were in the market for nuclear arms. A key question was, given their enormous destructive capacity, costliness, and, ironically, limited usefulness: Why are states nonetheless motivated to build or acquire nuclear weapons? Note this wording directs the inquiry to investigate the problem from the *demand* point of view, that is, why do nations want nuclear weap-ons? Sagan offers three lines of explanation for why they do: the *security, domestic politics,* and *norms* models.

The *security* model is a *realist explanation* for a state's motivation: nations seek a nuclear capacity when faced with a threat they cannot resolve with conventional arms. In face of such a threat, states do what they can. Strong states seek to develop their own nuclear capacities, and weak ones try to join alliances with those capable of providing deterrence. In a phrase, "proliferation begets proliferation." At work is what political scientists call a *security-dilemma paradox*: state A does something to increase its security, in so doing weakening the security of states B, C, and D, who then take corrective action, which in turn diminishes the security of state A. The circularity creates both an *arms race* and an *erosion of trust and cooperation.* The description above of nuclear proliferation after WWII seems to fit this model.

Under what circumstances then would a nation give up its nuclear weapons? The *security* model predicts such a possibility with the removal of an external threat. An example ostensibly can be found in the case of South Africa. The country's rationale for initiating its nuclear program during the 1970s cast it as a response to Soviet interventions in Africa and the introduction of Cuban forces in nearby Angola. South Africa dissolved its small nuclear arsenal in the early 1990s after the decline of the Soviet Union arguably reduced the country's need for having nuclear weapons.

The United Nations' 1970 Nuclear Non-Proliferation Treaty (NPT) – which became a permanent agreement in 1995 endorsed by 190 nations – represents a *supply*-side approach to the problem. Essentially it seeks to limit the availability of nuclear weapons. Its goals are the disarmament of nuclear weapons among those who already have them, a ban on the acquisition of new nuclear weapons among those who do not, and the redirection of nuclear capabilities to peaceful endeavors.[38] This creates at least one *demand*-side implication. A state might prefer to be the only one in its region with nuclear weapons but, since that is unlikely to be so, it might be willing to refrain if its neighboring states agree to abstain also.

How well does the *security* model explain things? Generally, pretty well, in the sense it matches what state leaders typically say in defending their decisions regarding nuclear weapons. However, the *domestic politics* model offers another dimension. Sagan notes three categories of actors often in play: the *state's nuclear industry establishment* (scientists, interested businesses, and civilian contractors), elements of the state's *professional military*, and *politicians and parties* favoring or opposing nuclear weapons. Where these three groups form strong coalitions, nuclear weapons programs are very likely.

This line of explanation has been influenced by the literature on the *weapons procurement process* in the U.S.A. and the Soviet Union. It emphasizes the role of individuals and organizations as active participants in constructing narratives justifying or opposing military expenditures and programs. To the extent this is so, this activity obscures the "pure" security issues. Realists preferring the *security* model admit that parochial interests often are in play but assign them a marginal role. Still there are empirical examples where *domestic politics* appears paramount.

In the U.S.A., for instance, the base-closure program – the effort to reduce the number of military installations in the U.S.A. – has been beset with political complications. Essentially, the military has not been able to close some installations it says it no longer needs because it would cause job and revenue loss in the districts of certain powerful politicians. It therefore is forced, as it were, to continue the operation of some bases rather pointlessly from a security point of view. For similar reasons, this model contends, it has been difficult to preclude the production of more and more nuclear weapons. Consequently, the U.S.A. (and the Soviet Union) during the Cold War each built many more nuclear weapons than could possibly be used. The resultant stockpiles had the capacity to destroy literally the entire planet several times over, obviously a theoretical number of times – one time, it seems, would do it.

Above we stated that India got the bomb in response to China developing that capability. Sagan argues that such simple characterizations often are not quite accurate. Here, China's joining of the nuclear club set off a debate within India between those who wanted the country to develop nuclear weapons and those who wished to remain a good-standing member of the NPT. This debate, Sagan states, delayed the initiation of nuclear programs for ten years until Indian scientists sold the idea of making a "peaceful nuclear device." Then Prime Minister Indira Gandhi, facing a tough re-election, supported their re-imaged proposal as a resolution to the domestic squabble.

Finally, Sagan's *norms* model underscores the symbolic importance the acquisition of nuclear weapons may hold for a nation. This explanation originates from sociological work showing a convergence among modern states resulting from an inclination to mimic each other. The model thus emphasizes the appeal nuclear weapons may hold for denoting a country's status as a truly modern state and recognized world power. Here, Sagan explains, nuclear weapons may serve a similar *symbolizing status* for a nation that having its own national airline or Olympic team does. These markers make an otherwise poor or marginal country look modern and important – think Philippine Airlines or Botswana Air. In this sense, being a signatory member of the NPT, even when a country has no capability of producing nuclear weapons, carries with it important symbolic weight. Having one's own nuclear weapons takes it to another level.

Sagan's analysis makes clear there may be several rationales at work motivating the acquisition of nuclear weapons. With proliferation, the fear is there are many more opportunities for miscommunication and misuse. Worse yet, nuclear devices might fall into the hands of actors or groups who relish the thought of triggering mass destruction.

Questions for Discussion

1   The Civil War, U.S. Indian wars, and WWII are synonymous with *total war*, and hence civilian death and destruction as a significant part of *multicides*. Does it seem possible for near-peer militaries to fight without total war? What conditions would need to be present for a total war not to happen?
2   Overy argues that it was not a given that the Allies would win WWII. What lessons can be learned that can be applied for future conflicts regarding: (a) logistics (i.e., getting supplies and food to troops), (b) willingness to fight, and (c) leadership, particularly at the highest levels?
3   Fein's work is an example of theory-building and then using data to test validity. At what level of analysis (i.e., micro-, meso-, and/or macro-level) is her work? Based on her model, is it possible to initiate micro- or meso-level interventions to prevent genocide?
4   Even with their enormous destructive capacity, nation-states continue to build and maintain nuclear arsenals or aspire to do so. Which of Sagan's explanations do you think creates the most risk for a nation (and by extension, the world)?

Notes

1   Quoted in William Tecumseh Sherman, Charles Royster (editor), 1990, *Sherman: Memoirs of General W.T. Sherman*, New York: Literary Classics of the United States.
2   Ibid.
3   Mark Grimsley, 1997, *The Hard Hand of War: Union Military Policy against Southern Civilians, 1861–1865*, Cambridge: Cambridge University Press.
4   Russell F. Weigley, [1960] 1977, *The American Way of War: A History of United States Military Strategy and Policy*, Bloomington, Ind.: Indiana University Press.
5   Matthew White, 2012, *The Great Big Book of Horrible Things: The Definitive Chronicle of History's 100 Worst Atrocities*, New York: W.W. Norton.

6 Ibid., p. 555. White explains his methodology as such: "I count all of the deaths of living, breathing individuals that result from a specific outbreak of coordinated human violence and coercion, both directly (war, murder, execution) and indirectly (aggravated disease, avoidable famine), as long as they are the obvious result of the event. I count all deaths the same, whether military or civilian, malicious or accidental, negligent or authorized. I count only deaths that occur immediately or follow closely – no cancer deaths, no long-term complications from wounds, no suicides among haunted victims, no unexploded ordnance that blows up farmers fifty years later."

7 Ibid., pp. 401–405.

8 Max Boot, 2007, *War Made New: Weapons, Warriors, and the Making of the Modern World, 1500 to Today*, New York: Gotham Books, pp. 268–294.

9 Michael C.C. Adams, 1993, *The Best War Ever*, Baltimore, Md.: Johns Hopkins University Press.

10 Richard Overy, 1997, *Why the Allies Won*, New York and London: W. W. Norton & Company.

11 Ibid., pp 27–29.

12 Overy, op. cit., pp. 40–43, quote appears on p. 42.

13 Ibid., pp. 64–66.

14 Ibid., pp. 66–70, quote appears on p. 68

15 Ibid., pp. 73–75.

16 About 91,000 German soldiers surrendered, about 5,000 or so of whom eventually returned home. The rest died in Soviet prisoner-of-war camps.

17 Overy, op. cit., pp. 82–84.

18 Ibid., Chapter 7.

19 Ibid., Chapter 9.

20 Ibid., Chapter 8.

21 A number of high-ranking leaders of Nazi Germany were prosecuted (and found guilty) after the war under the auspices of an International Tribunal at Nuremberg, Germany. The charge, *crimes against humanity*, had been encoded in the Fourth Hague Convention of 1907, which explicitly stated that "the murder of civilians in time of war" is an international crime (Helen Fine, 1990, "Genocide: A Sociological Perspective," *Current Sociology*, 38 (Spring): 1–125, cited material, pp. 1–2).

22 William F. Meinecke, Jr. and Alexandra Zapruder, 2011, *Law, Justice and the Holocaust*, Washington, D.C.: United States Holocaust Memorial Museum.

23 Helen Fein, 1979, *Accounting for Genocide: National Responses and Jewish Victimization during the Holocaust*, New York: The Free Press.

24 https://zoryaninstitute.org/the-zoryan-institute-remembers-dr-helen-fein/.

25 Fein, 1979, op. cit., pp. 19–26.

26 Fein, 1979, op. cit., p. 38.

27 Ibid., pp. 61–75.

28 Ibid., pp. 137–147, 354–355.

29 Ibid., see Part Two: "The Victims' View".

30 Raul Hilberg, [1961] 1985, *The Destruction of the European Jews*, one volume edition, New York: Holmes and Meier. Other important sociological analyses are: Anna Pawelczynska, 1979, *Values and Violence in Auschwitz: A Sociological Analysis*, translated by Catherine S. Leach, Berkeley, Calif.: University of California Press; and Frank Chalk and Kurt Jonassohn, 1990, *The History and Sociology of Genocide: Analyses and Case Studies*, New Haven, Conn.: Yale University Press.

31 Ervin Staub, 2011, *Overcoming Evil: Genocide, Violent Conflict, and Terrorism*, New York: Oxford University Press, Part One, "The Origins of Mass Violence".

32 An "atomic bomb" is a device that acquires its tremendous explosive capacity through the process of nuclear fission, i.e., by the splitting of atoms in its core of uranium or plutonium. The resultant chain reaction increases the explosive capacity of a small amount of matter exponentially. The first operational atomic bomb was developed by American scientists working in the so-called Manhattan Project.

33 John Lewis Gaddis, 2005, *The Cold War: A New History*, New York: The Penguin Press, p. 53.

34 Ibid., p. 54.

35 Ibid., p. 50.

36 Ibid., p. 66.

37 Scott Sagan, 1996–1997, "Why Do States Build Nuclear Weapons? Three Models in Search of a Bomb," *International Security*, 21 (Winter): 54–86; and 2017, "The Korean Missile Crisis: Why Deterrence Is Still the Best Option," *Foreign Affairs*, (November/December).

38 A notable loophole is the "supreme national interest" clause: a nation is permitted to opt out of the agreement by declaring the perception of an external threat requiring a nuclear deterrence.

Recommendations for Additional Reading/Viewing

Dr. Strangelove, or, How I Learned to Stop Worrying and Love the Bomb (film), 1964, written and directed by Stanley Kubrick.

(Academy Award winning film, this satire of the Cold War is regarded as one of the era's most culturally and historically important films)

John Keegan, 1976, *The Face of Battle*, New York: Penguin Group.

(In-depth presentation of three battles – Agincourt, Waterloo, and the Somme – that has become a classic in military history)

Mahmood Mamdani, 2001, *When Victims Become Killers: Colonialism, Nativism, and the Genocide in Rwanda*, Princeton, N.J.: Princeton University Press.

(Riveting account, and resolution, of genocide in Rwanda by Ugandan academic and writer)

James M. McPherson, 1988, *Battle Cry of Freedom*, Oxford, U.K. and New York: Oxford University Press.

(Highly respected, balanced account of the U.S. Civil War – the title reflects both the Northern and Southern view of the war – that details the political, social, and military dimensions of this deadly war)

Richard Rhodes, 2010, *The Twilight of the Bombs: Recent Challenges, New Dangers, and the Prospects for a World without Nuclear Weapons*, New York: Vintage Books.

(Author well-known for incisive series of books on nuclear weapons describes the nuclear policies of Russia, China, France, Great Britain, and the U.S.A.)

9 Spectrum of Conflict

"Small" Wars

> *Reader's Guide*: "Small" Wars also go by names such as "irregular warfare" for a reason. They are ones between mismatched modern militaries or between modern militaries and paramilitary groups revolting against an established political order. Success here often is not attained militarily but politically, especially if the lesser group has the option of when and where to engage. Further, the application of awesome lethal force, an advantage in Big Wars, can be a liability here. We spotlight three illustrative cases, each with a pair of studies. The first is the Algerian war of independence from France, with accounts of the conflict by a historian and by a military-officer known as a specialist in counter-insurgency warfare. We turn then to the war in Vietnam, with commentary by one of the founders of the National Liberation Front, and a critical analysis by a military-officer/organizational theorist of the U.S. military's failures as a learning organization. We conclude with an examination of "the Troubles" in Northern Ireland through organizational analyses of the Irish Republican Army and of the Ulster Volunteer Force.

The division of Chapters 8, 9, and 10 into "Big" Wars, "Small" Wars, and "New" Wars is done with some trepidation on our part. The process of getting to the core of these concepts can be a bit messy, not unlike eating soup with a knife. Recall from Chapter 8 that we cast *Big Wars* as ones in which the contestants are near-peer militaries, usually locked in a series of set-piece battles. In *Small Wars* the contestants are not peers, that is, there is great asymmetry in capabilities. This definition would include mismatches between formal militaries, but we are especially interested in situations in which militaries confront paramilitaries, that is, take on groups doing organized violence for political purposes who are not part of any state's formal armed forces.

How to prevail in Small Wars calls for a different calculus: while wars such as these have a distinct military dimension, they usually do not have a purely military solution. This line of thinking is reflected in the 1940 *U.S. Marine Corps Small Wars Manual*, which notes: "The application of purely military measures may not, by itself, restore peace and orderly government because the fundamental causes ... of unrest may be economic, political or social."[1]

The "eating soup with a knife" phrase does not originate with us but comes from British Army Colonel T. E. Lawrence. He took part in one of the 20th century's first Small Wars: the 1916–1918 Arab Revolt against the Ottoman Empire, in which small

DOI: 10.4324/9781003282549-9

Bedouin bands took on the Ottoman Turkish Army.[2] Lawrence was fluent in Arabic and thoroughly knowledgeable in Bedouin ways. Donning a Bedouin headdress and white robe, he avidly became one of the revolt's leaders and articulated a strategic mindset compatible with Bedouin culture. He later described in his memoir[3]:

> Taking practical account of the area … I began idly to calculate how many square miles: sixty, eighty, one hundred, perhaps one hundred and forty thousand square miles. And how would the Turks defend all that? No doubt by a trench line across the bottom, if we came like an army with banners; but suppose we were … intangible, invulnerable, without front or back, drifting about like a gas? Armies are like plants, immobile, firm-rooted … We might be a vapour, blowing where we listed.

Lawrence assisted the Arab Bedouins, desert nomads who moved about on camels, in effectively applying their traditional hit-and-run fighting style to take on the Turks. Within a year, the Lawrence-led Arabs reclaimed Medina and the port city of Aqaba, and they assisted British forces in the taking of Damascus. Of course, all was not that simple. Lawrence lamented that doing warfare like this was messy and slow, not unlike eating soup with a knife. Colonel Lawrence's exploits eventually made him quite famous. The publication in 1926 of his memoir, *The Seven Pillars of Wisdom*, enjoyed wide popular acclaim. A 1962 American-made movie about his life, *Lawrence of Arabia*, won seven Academy awards.

Insurgency/Counterinsurgency Theory

At the end of World War II, European powers ruled colonies they had seized throughout Africa and Asia without consent of their inhabitants (see our discussion of this in Chapter 6). The decade after World War II was marked by dozens of Small Wars as the Europeans' colonies in Africa and Asia, one by one, demanded and fought for their independence. Military theorists of the day advanced the term *insurgency* to designate a violent attempt by non-ruling groups to destroy, overthrow, or otherwise reformulate an existing political system, and *counterinsurgency* to reflect the range of a government's responses.[4]

It was difficult for colonized peoples to visualize how to overcome the vast military machinery of the colonizers. In World War II, Japan invaded and assumed control of French Indochina. That an Asian nation did so illustrated that modern European militaries were not invincible. More dramatically, a Communist revolutionary, Mao Zedong led a peasant-based revolt against strongman Chang Kai-shek's Republic of China, culminating in the defeat of the Chinese Army in 1949. Mao's 1930s writings on warfare became textbooks on how to succeed as an insurgency.[5]

Mao did not start out as a violent revolutionary. He came to realize after participating in uprisings ruthlessly repressed by the Chinese military that "power grows out of the barrel of a gun." Those early days also pointed to the wisdom of pushing the struggle into a long, drawn-out quagmire and of choosing to fight only when having a momentary advantage.[6] Mao famously underscored this point in a four-line limerick:

> When the enemy advances, we withdraw;
> When the enemy rests, we harass;
> When the enemy tires, we attack;
> When the enemy withdraws, we pursue.

But the most important lesson concerned the vital role of the population. Small-war military thinking on how to quell insurgencies was to levy unforgiving military force on the malcontents and others in their midst. This represented what came to be identified as an *insurgent-centered* response. Mao's work suggested an alternative *population-centered* approach. The population, Mao wrote, is the sea within which the insurgents, like fish, must swim. The military too needs the support and cooperation of the people to sort through all of this and conduct effective operations. It is in this sense that insurgency–counterinsurgency becomes a tug-of-war for "hearts and minds."

To help visualize what it takes to prevail we draw upon Robert K. Merton's structural-strain theory[7] (see Figure 9.1). Written in the 1930s, Merton meant the theory as an explanation for organized crime in many large U.S. cities. He begins with the observation that every society defines standards of material well-being and rightful means for achieving them. Where these *legitimate goals and means* provide realistic, genuine opportunities, there is a high degree of *conformity*, that is, most people willingly follow the rules to get what they want.

When there is a big gap between goals and means, people devise other strategies. Among those who accept the societally defined goal of material well-being, some concoct illegal means for getting there, a form of deviance Merton labels *innovation*.[8] Another adaptation to hopelessly bad conditions is apathy. Some people just go through the motions (*ritualism*), while others drop out of the larger culture (*retreatism*), centering their lives, for example, around drugs or alcohol. Finally, some people are anything but apathetic: the disparity between goals and means motivates them to seek to overthrow the system altogether (*rebellion*).

Insurgencies usually occur where economies offer little genuine opportunity, polities are rigged to favor a privileged few, and police and military forces protect the elites but oppress the masses. In such cases there is an extensive amount of *innovation*: people form gangs and militias to provide their own security, develop subterranean economies and black markets to meet material needs, and devise informal proceedings for meting out justice. In this context, *rebels* do not stand out as exceptionally strange. While people may not agree with the rebels' agenda, there is a tendency to tolerate their presence. So, rebels often take things a step further. Lacking legitimate means to retool the system, they use violence and intimidation to harass government entities and to bully the population into active acquiescence.

The presence of a military to assist or prop up the government adds another layer of complication. Addressing the broader dimensions of the problem is not the military's

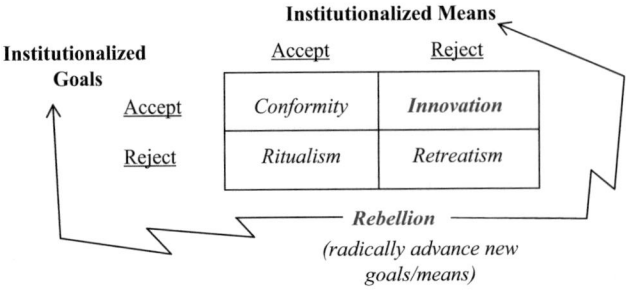

Figure 9.1 Merton's Structural-Strain Model

Source: Authors' Creation.
Note: Figure reconstructed from Merton's description of his theory and commonly known depictions of it.

preferred skill set. Yet, in Merton's terminology, a significant portion of the counterinsurgency effort must be devoted to converting *innovators*, and, yes, recoverable rebels, into *conformists*. To convince these folks they have more to gain by "playing the game" than by resorting to violence, it really helps if there actually is a legitimate game in town worth playing.

The military usually does what it is configured and equipped to do: it employs lethal force. Further, if using only its insurgent-centered lens, the military is prone to view all who oppose them as committed rebels to be killed or imprisoned. Unless dedicated to do otherwise, the military is destined unwittingly to convert large segments of people into rebels. Similarly, reckless violence by not-so-shrewd rebels can drive the population into the government camp. Sometimes it comes down to who, insurgents or counterinsurgents, are more judicious in their use of violence.

Illustrative Case: The French in Algeria

We begin with the French war in Algeria – a complicated and particularly nasty Small War. The official French position characterized Algeria as a region of the French homeland. If so, native Algerians were its second-class citizens. France invaded and assumed control of what is now Morocco, Algeria, and Tunisia in 1830. Indigenous to the area were Berber and Arab tribesmen and villagers. The two groups differ culturally, have different native languages (Berber and Arabic), and Berbers often were referred to as "white Mediterraneans" because of their blue eyes and lighter skin tone. Both though are Sunni Muslims.

When France took over, it ran the colony directly from Paris. Settlers from France, but also nearby Spain and Italy, migrated to *Algérie française* and commandeered the most desirable farmlands. Called *Pieds-Noirs*,[9] they also occupied virtually all local positions of governance inserted by the French. This made for a resolute *three-tiered, political-economic system*: the French on top, then the *Pieds-Noirs*, and, at the bottom, the vast number of Berbers and Arabs, about 85% of the population. Thus entrenched, *Pied-Noirs* became Algeria's third ethnic group and an intractable roadblock to any facile solution once the Algerian struggle for independence began.

To explore how this cozy relationship escalated into all-out war, we shine our Spotlight on two classic analyses of the Algerian insurgency and the French counterinsurgency.

Spotlight on Social Science Thinking and Research

Alf Andrew Heggoy, *Insurgency and Counterinsurgency in Algeria*[10]; and Lt. Colonel David Galula, *Pacification in Algeria, 1956–1958*[11]

Alf Heggoy was born in Algeria in 1938 where his parents, originally from Norway, were Methodist missionaries. His family resided in a fortified town, Ft. National, located 60 miles to the east-southeast of Algiers in the rugged hills of the Berber Kabylie region – interestingly, the very same area in which a young Captain Galula later would serve during the war. By then, however, the Heggoy family had moved to Virginia in the U.S.A. Heggoy completed his doctorate at Duke University and became a long-time professor of History at the University of Georgia.

Insurgency and Counterinsurgency draws upon archival data and Heggoy's interviews with Algerians and former *Pieds-Noirs*. It examines the pre-revolutionary period and phases of the war: 1954 to 1956 when the rebels held the initiative; 1956 to 1958, the period in which the French Forces of Order seized control of the military course of the war; and 1958 to 1962, the years of a painful, circuitous search by France to mediate a transition to *Algérie algérienne*.

Early in the 20th century, small groups of Algerians began challenging the status quo. These early agitators criticized the three-tiered system and sought a modicum of political autonomy. WWI intervened, and Algerians were subject to France's military draft while others worked in factories as replacements for drafted Frenchmen. Both groups had their eyes opened by Allied propaganda espousing the principles of freedom and self-determination. The mainland French politicians did extend some incremental reforms, but, outnumbered in Algeria by nine-to-one, the *Pieds-Noirs* adamantly opposed any hint of self-determination for Muslim Algerians.

In 1931, a Muslim cleric, Cheikh Abd-el-Hamid Ben Badis, founded the Association of Reformist Ulama. The Ulama ran small schools, teaching Arabic and religious studies calling for devotion to Islam. Though Ben Badis at first was avowedly nonpolitical, his religious program carried strong nationalist sentiments, and the organization slowly became recognized, both by itself and by the French authorities, as a political vanguard.

The issue exploded at the end of WWII, when demonstrations broke out in several Algerian towns, notably Sétif. There, Muslims carrying nationalist placards scuffled with police and then attacked *Pieds-Noirs* bystanders, killing 21 of them. French military units responded with ferocious beat-downs that lasted several days, killing 8,000 Muslims. A long, uneasy truce followed, and in 1954 the more radical Algerian leaders formed the *Front de libération nationale* (FLN). Now there was no question: the goal was total independence, and the means would be violent.

FLN kicked off its revolution in November of 1954. The main tactic consisted of guerrilla raids against French forces – about 50,000 at the time – and *Pieds-Noirs* settlements in rural areas. It also collected taxes from villagers and carried out a campaign of terror against those Muslims who actively opposed them. The French military, fresh from defeat in Indochina, was doggedly determined not to lose again and spared no quarter in the forcefulness of its responses. It sent the bulk of its professional forces to Algeria, especially elite paratroop units (the paras), and ramped up the military draft to ensure sufficient numbers. Eventually, 400,000 French troops were in Algeria.

Heggoy describes French forces as divided into two task groups. The first group, made up of a mix of professional soldiers and draftees, had the mission to protect important facilities, keep the roads open, and guard Algerian villages from FLN visits. The second group consisted of professional soldiers, including the paras, who were tasked with capturing or killing FLN insurgents.

A key event changed the complexion of things. In response to the execution of FLN prisoners and violence by *Pieds-Noirs* in the capital city of Algiers, FLN refocused its guerrilla efforts there. Algiers's Muslim quarter, the famous Casbah, had been fertile FLN territory: its unemployment rate among Muslims was 50% and most of its males were under the age of 20, all in a port city where Europeans lived the good life. On 31 September 1956, FLN operatives set off three car bombs in Algiers, including one at the offices of Air France airlines. French General Jacques Massu of the 10th Parachute Division was ordered to bring the violence to an end "by any means necessary."

General Massu cordoned off the Muslim section and systematically purged every street of the Casbah. To facilitate the search for suspects, he authorized "hard interrogation" – in plain language, *torture*. The 10th paras, hardened by their Indochina experience, assumed these chores with grim efficiency. The operation was effective but costly. Heggoy summarized the impact of the torture[12]:

> A complete change in public attitudes among Arabs and Berbers occurred … as they reacted to the tactics by Massau's men. … Most of the FLN leaders and their specialists in urban terrorism were captured or killed, and the rebel cells in [Algiers] were nearly eradicated, but Massau's methods led nevertheless to a tightening of the rebels' hold on the Algerian population.

Buoyed by their military gains in the Casbah, the French increased Massau's methods in rural areas. The high command also urged territorial-control units to ratchet up "pacification" efforts. However, most territorial commanders either were unsure of exactly what this meant or thought this a hare-brained idea. Into this breach stepped then-Captain Galula, one for which he was uniquely prepared.

David Galula was born in Sfax, Tunisia, in 1919 to a Jewish family of well-to-do olive-oil merchants.[13] His father, educated in France, obtained French citizenship in 1924 for the whole family. Though smart, young David did poorly in school, preferring play to study until he became enthralled by one of his uncles, a military officer. Galula buckled down, gained admission to the French military academy at St. Cyr, and was commissioned in 1939. Just months later the Germans invaded France, driving segments of the French Army, especially its Jewish officers, into exile for the remainder of WWII. Galula operated as an undercover agent in Tunisia but in 1942 joined the *résistance* against the Germans on the mainland.

After the war, Galula was assigned as a military attaché to the French embassy in Beijing. "Fiercely anti-communist," he closely followed Mao's insurgency.[14] In 1948 he was a United Nations observer in Thessaloniki during the Greek Communists' unsuccessful insurrection. And, while an intelligence officer in France's Hong Kong diplomatic office, he studied the successful application of Mao's principles by the Viêt Minh against the French Army in Indochina. When he volunteered in 1956 for duty in Algeria, he was, in his words, "ready to test certain theories I had formed."

Captain Galula assumed command of the 3rd Company, 45th Colonial Infantry Battalion, and his company was responsible for two hamlets. FLN had a secret three-man political-administrative cell in such hamlets with one person coordinating military attacks, one directing political propaganda, and a third collecting taxes. Galula estimated there were two or three dozen FLN regulars carrying out such tasks in his area, avoiding direct confrontation. A small number of operatives in hamlet after hamlet was able to sustain the insurgency through "terror and persuasion" with tacit support of the locals, especially, Galula noted, if French troops terrorized and alienated the population.

Most territorial-control units like Galula's set up fortified positions outside villages from which to conduct patrols and searches. Instead, Galula rented domiciles from the locals and placed two-thirds of his troops inside his two hamlets. At night, the remaining one-third of the company put out ambush patrols to discourage night-time movement by the locals. He then conducted a house-by-house census.

He instructed his soldiers to treat locals with respect and use a minimum of force. They greeted and talked with villagers every day so they knew by sight who lived there and who might be a drop-in FLN worker. Galula put out word that FLN operatives who voluntarily turned themselves in would be pardoned to make amends, but any that 3rd Company captured would be turned over to higher command. And, he instructed the locals to select a mayor and council through which he could direct services such as a medical aid station and village-improvement projects.

All this did not take place smoothly. Villagers often refused to cooperate and FLN operatives tried to undo progress through night-time visits of intimidation. One of his own platoon leaders asked to be transferred, and adjacent French units ridiculed Galula's troops as "pacifiers." His battalion commander, doubtful at first, took pride in 3rd Company's curious but highly effective operational style and afforded visiting dignitaries a first-hand look at how it all worked. Still, some visitors reacted adversely – Galula's *counts of insurgents killed or arrested* were too low to suit their tastes.

By 1958, though using such kill-arrest metrics, it appeared the French had things in Algeria militarily well under control. However, FLN's bare-bones insurgency still held strong sway with most Algerians. Further, the military's policy of torture had become a divisive, hot-button issue within France itself,[15] and any talk of an *Algérie algérienne* sparked fiery debate between political factions. Amidst this political maelstrom, General Charles de Gaulle, hero of WWII for directing the *résistance* against the Germans, was elected President. To the shock and dismay of some in the French military, de Gaulle opened discussions for a negotiated settlement. Several generals staged a coup, quickly aborted, to seize control from de Gaulle. Things took another disastrous turn when a *Pieds-Noirs* reactionary group, *Organization armée secrète* (OAS), initiated terrorist attacks against both Muslim Algerians and the French army. Negotiations ultimately led to full independence for Algeria in 1962.

By that time, now-Lt. Colonel Galula had come to the U.S.A. to study at Harvard University. He wrote, in English, *Pacification in Algeria* and became recognized in this country as a counterinsurgency guru. (Only years later was his book translated into French.) It would be a mistake though to think of Galula as a dreamy idealist. He strongly supported an *Algérie française*. He recognized "a strong wind of independence was blowing across Algeria," but FLN's goal of complete victory, and its refrain of *"la valise ou le cercueil"* (a suitcase or a coffin) for the *Pieds-Noirs*, that is, *leave or die*, convinced him an independent Algeria was not an option.

Further, he shared his military's dislike for the French media, to whom he attributed much of the blame for the uproar over torture. Galula conceded that the posture toward torture of detainees in some units was "talk or else," but opined that the press's coverage was "90 percent nonsense."[16] And, finally, he did not seem terribly upset by the military's coup against President de Gaulle. Of this he said, "I decided to let my superiors fight it out among themselves. Paris finally won."[17]

Illustrative Case: French Indochina and America in Vietnam

While *Algérie française* in some ways is an extreme case, it is in others typical. Commonplace were its basic social structure – *colonial rulers at the top, a small, complicit local elite in the middle, and a mass of impoverished, disenfranchised locals at the bottom* – plus its heavy reliance on the military to ferociously hold the system in place. Recalling Robert Merton's typology of goals-and-means, there was no game in town worth playing for those at the bottom. In such situations, outright *rebellion* is just a matter of time.

In Algeria, the French military killed the insurgent but not the insurgency. Here too there is typicality. Military analysts Ben Connable and Martin Libicki have shown in a study of the 89 Small Wars since WWII, this has been the fate experienced most often by modern militaries and embattled governments they supported.[18] Insurgencies occur where people have genuine grievances and when a government and military defending the status quo attempt a purely military solution, this typically exacerbates the situation. To deepen our understanding as to why this has been so, we now turn our Spotlight to America in Vietnam.

Spotlight on Social Science Thinking and Research

Trúóng Nhú Táng, *A Viet Cong Memoir*[19]; and Lt. Colonel John A. Nagl, *Learning to Eat Soup with a Knife*[20]

The French had seized power in Vietnam in the 1880s. They took advantage of the area's resources to export for a handsome profit tobacco, coffee, rubber, rice, and lumber. Trúóng Nhú Táng was born in 1923 to a wealthy Sàigònnese family in a status between the French colonial elite and the local laborers and peasants – the usual *three-tiered colonial structure*.[21] He acquired traditional Confucian ethics, above all the "unshakeable" principles of loyalty to family and nation. Following his father's direction that he become a pharmacist, he went to Sàigòn's Collège Chassel-Laubat for children of French administrators and the most privileged of Vietnamese. He was taken aback to learn there he was a *mite* (short for *Annamite*, the pejorative word used by the French for all Vietnamese).

Still, at the end of WWII, he traveled to Paris to study pharmacy. Many of the Vietnamese elite were educated in France, among them Nguyên Sinh Cung, son of a mid-level, government official. As a young man he went by several aliases before settling on Ho Chi Minh, "Bringer of Light." Educated at Paris's Sorbonne Université, Ho's eyes had been opened there by the gap between France's ideals of democracy at home and its colonialism abroad. This led him to Mao's revolutionary thinking.

Ho was in France when Trúóng arrived.[22] He met with the newly arriving Vietnamese students and urged them to take advantage of their studies to prepare for regaining their country's freedom. "Nothing," he told them, "is more precious than independence and liberty."[23] Trúóng to this point had never heard of Ho Chi Minh and knew nothing about how the French colonial arrangement in Indochina had come about. He abandoned pharmacy, obtained a master's degree in Political Science and licentiate in Law at the Sorbonne, and returned to Vietnam in 1951 ready to take up the cause. *A Viet Cong Memoir* is both the story of his personal journey and his analysis of what transpired.

In 1951, as in the years before WWII, there were numerous nationalist insurgent groups in Vietnam, but the struggle had become caught up in the Cold War. No insurgent group was stronger than *Ho's Viet Minh* (Vietnam Independence League), supported by Red China and the Soviet Union. Communist or not, the insurgent groups readily shared with each other Mao's thinking about how to do revolutionary warfare. Cold War politics dictated that the U.S.A. back and finance the French.

In 1954, a Viêt Minh-led force shocked everyone when they overran a heavily fortified French outpost at Dién Bién Phú. However, victory did not bring independence. The principals and their Cold War sponsors met in Geneva and divided Vietnam into two nations, North and South. The resultant Accords granted Ho Chi Minh's faction control of the North and, in the South, Vietnam's emperor and vassal of the French, Báo Dai, appointed Ngô Dình Diêm. In backing Diêm, Trúóng noted, the U.S.A. was "blithely assuming ... the mantle of the newly departed French colonialists."[24]

Diêm proved to be an enormously unpopular president. He was from a mandarin but staunchly Catholic family, very unusual in an overwhelmingly Buddhist Vietnam. He did not help matters by retaining tax breaks set up by the French for the Catholic Church (a large landowner) and giving Catholics preference in hiring for government positions. He also authorized the flying of the Vatican flag alongside the flag of South Vietnam. Insurgents of all stripes in the South who had fought against the French were still about and Diêm considered them mortal enemies. He thus alienated insurgents loyal to "Uncle" Ho, nationalists comfortable with the two-state solution, and even former pro-French Vietnamese. The table was set for disaster.

Trúóng himself was a non-Communist nationalist willing to accept either a strong, democratic South or a multi-party, united Vietnam. While holding a top position with the Bank of Vietnam, he gravitated toward a network of top leaders disenchanted with Diêm. In 1960 this group formed the National Liberation Front (NLF), a big tent of mostly non-Communist Diêm opponents.[25] North Vietnam extended military assistance through the NLF to any and all guerrillas who opposed Diêm, loosely termed, Viêt Công. On the other side, the U.S.A. provided Diêm's regime with vast military and economic support, including 15,000 military advisors. Wracked by discord, the Army of South Vietnam (ARVN) struggled against the Viêt Công. All this came to a head in November 1964 when several ARVN generals with tacit U.S. approval initiated a coup in which Diêm was assassinated.

With the generals and, by extension the U.S.A., now so visibly in control, the NLF zeroed in on one goal: reunification of North and South. The North's People's Army of Vietnam (PAVN) stepped up its infiltration of troops into the South to fortify Viêt Công efforts to overthrow the South's government. The U.S.A. began its first aerial bombing campaign of the North and, in March of 1965, the first American ground forces splashed ashore in Dà Nãng. By 1969, almost 600,000 U.S. troops would be in South Vietnam. Trúóng and other NLF leaders were arrested in 1967 and he himself was tortured and kept in solitary confinement for a year. Upon release in 1968, he and other NLF leaders moved from Saigon to the Cambodian border for the duration of the war, and from there made plans for a coalition government at the end of the war.[26]

But Trúóng noted a troubling change: "the blue half of the NLF flag had become too red." As the North had committed more and more troops to front lines in the South, NLF had become less and less autonomous. A frustrated Trúóng explained:[27]

> We began to hear more and more about our class deficiencies. … Many of us were from well-to-do families … Many of us had struggled against the French, and in moving to the jungle … were decisively committed against the Americans. … After the prisons, B-52s, diseases, and malnutrition, it was outrageous to suggest that we were somehow second-class revolutionaries. … [But after Ho Chi Minh's death in 1969], the ideologues were firmly in control, and they were not inclined to waste effort on their junior partners.

A main point of Trúóng's analysis is now apparent: the NLF began as a nationalist organization but became characterized by the South Vietnamese government and the Americans, and then the North Vietnamese, solely as a Communist insurgency. However, it is not unusual for the *moderates to become marginalized or eliminated* by a government under siege or the hard-core insurgents or both. That takes the violence to another level and limits the options for resolving the conflict.

We turn now to John Nagl's analysis. Born in 1966, Nagl grew up in Omaha, Nebraska, where he graduated from the Jesuit-run Creighton Preparatory School. He went on to the U.S. Military Academy at West Point, was commissioned in 1988 as an armor officer, and then went to Oxford University as a Rhodes Scholar. Nagl served as a tank platoon leader in Persian Gulf War I and thereafter returned to Oxford to work on his doctorate. For his dissertation, the title, of course, borrowed from T. E. Lawrence, he compared the British in Malaya and the U.S.A. in Vietnam, the former a success story, the latter, one of the many gone awry. It is a study of a military's willingness and ability to adapt: what are the features, Nagl asks, of militaries that are *learning organizations*?[28]

The British Army has had its own love affair with modern military doctrine that does not like messy, slow campaigns. However, Great Britain is an island nation whose Army historically has had to play second fiddle to its Navy. This was especially true in its long history as an expeditionary force policing the many far-flung colonies of the British empire, of which Malaya was a part. In this tradition, a relatively small force usually was left to fend for itself, albeit with superior firepower, but also by forming long-term relationships with local elites and power brokers. With notably violent exceptions, this contributed to a military culture that kept political objectives at the fore, encouraged the economical use of manpower, and reduced the Army's tendency toward the exercise of brute military prowess in places outside continental Europe.

The war in Malaya from 1948 to 1960, termed the *Malayan Emergency*, featured a two-pronged, Communist-inspired insurgency. One part contained factions of ardently anti-British ethnic Malays seeking independence, the other of ethnic Chinese fearful of independence but angered by conditions in the tin mines and rubber plantations where they worked. Between 1948 and 1951, the British lost sight of their expeditionary heritage and their strategy devolved into *jungle bashing*, that is, heavy-handed, battalion-sized, search-and-destroy missions that alienated local populations.

In 1952, the British conducted a top-to-bottom assessment of their aims and strategy. The new field commander, General Sir Gerald Templar, felt the locals should and could both run the government and resist the insurgents. The British recalibrated, identifying a "united, self-governing Malayan nation within the Commonwealth" as its goal. Several initiatives were put in place: a plan for securing "New Villages" to gain the trust of the Chinese subpopulation, suggestions for involving locals to better provide security and gain intelligence, and the use of *small-unit Ferret Forces* to seek out hard-core insurgents more surgically. These actions, plus talks about steps that might lead to Malayan independence, eventually cost diehard rebels the support of the population and created enough stability that Malaya – now, Malaysia – was granted independence in 1957. A newly elected non-Communist government declared an end to the Emergency in 1960.

Direct American involvement in Vietnam began in 1950 to assist the once-formidable French colonial army there. American military culture historically has been firmly grounded in the precept, "annihilating an enemy's military is the surest path to victory." An exception was the Marine Corps in South America and the Caribbean early in the 20th century, but WWII made that chapter a long-forgotten memory. After the division of Vietnam into North and South, American military advisors busied themselves providing architecture and instruction for an ARVN, creating a conventional military in their own image.[29] The Viêt Công and PAVN meanwhile retained the insurgency doctrine and organization that had brought them victory over the French.

General William Westmoreland assumed command of U.S. forces in South Vietnam in 1965. Admitting Vietnam "was a new ball game for us," he settled on using helicopters to create highly mobile battalion, company, or even platoon-sized units in search of Viêt Công and PAVN fighters, and to then "pile it on" with close-in air support and artillery. As a metric for gauging how well things were going, the U.S.A. settled on the *body-count ratio*, the number of Viêt Công and PAVN killed per encounter by the number of Americans killed.[30] Once the PAVN and Viêt Công caught on to this tactic, they adopted "hugging the belt" as theirs, that is, engaging American units in combat only in physical spaces so tight (ravines, for example) or in such close physical proximity that American use of close-in air support and artillery was less feasible. The result was a fierce, bloody little war.

Nagl's final tally of organizational learning in the two campaigns shows first, *for both the British and the Americans*, there was a great deal of bottom-up critical input, questioning of tactics, and pointed suggestions for superior officers. The key difference is *what happened to this information flow*. The British top command for the latter part of the Malayan Emergency set the tone for honest assessments. Nagl states[31]:

> [General] Templar was afraid of neither receiving bad news nor sending it back to London. ... Bottom-up input was welcomed; district advisors, British army privates, surrendered enemy personnel were asked for ideas on how to better

accomplish [the mission]. … [The result was] a rare example of a military adapting its organization, doctrine, and training during the course of a conflict.

Among the Americans too, there was no shortage of discussion about the necessity and value of doing things differently. Marine Lt. General Victor Krulak, among others, complained to Secretary of Defense Robert McNamara, "the raw figure of VC killed … can be a dubious index of success since, if their killing is accompanied by devastation of friendly areas, we may end up having done more harm than good."[32] And, the Marines devoted some of their forces to a Galula-style village pacification program, about which General Westmoreland remarked, "I believe the Marines should [be] trying to find the enemy's main forces and bring them to battle."[33]

Nagl's analysis showed that a "can-do" attitude and a norm of "loyalty to the chief is the first commandment" resulted in plunging ahead, even as top officers received bad news from below or themselves thought things were not going well. For example, Nagl quotes one U.S. general:

> I will say that once it became apparent that we were going to send in massive forces, I was of the opinion it was a major mistake. … There were others who shared that view … I had come to learn in my military career, once decisions are made … it is not the business of subordinates to criticize and complain, but to put all shoulders to the wheel.

Nagl concluded: "[As a result] suggestions were quashed by the high command. … Even General [Creighton] Abrams had great difficulty in implementing change when he assumed command in Vietnam in 1968."[34]

The U.S.A. made an enormous military effort in Vietnam. About 2.9 million troops deployed there, and the U.S. Air Force dropped more than 7 million tons of bombs over North and South Vietnam, Laos, and Cambodia, about twice the tonnage expended in the European and Pacific theaters during World War II. All told, more than 58,000 American troops died and an estimated 500,000 Viêt Công and PAVN perished, a ten-to-one kill ratio, along with 170,000 ARVN. Still, Army Colonel Harry Summers's well-known critique of the American effort in Vietnam credits the loss to the U.S.A.'s failure to apply military force more ruthlessly.[35] Maybe so, but as we have seen, this flinty-eyed position does not have a good track record in such settings.

Vietnam was a much tougher nut to crack than that faced by the British in Malaya. Malaya was a British colony, and part of their success came from the opportunity, then their decision, to seek a route for granting independence. The French could have done so in Vietnam, but chose otherwise. Toward the end of *A Viet Cong Diary*, Trúóng summarized the predicament as follows:

> [T]he only ally Vietnamese nationalism had even known was the Communist International. Hô Chi Minh had grasped this support with the fervor of a drowning man, and he had woven the fabric of independence out of the twin fibers of nationalism and communism. … To all appearances, it had never occurred to Western

leaders to try to accommodate Vietnamese aspirations for independence, decent government, and economic progress … [T]he French had given up control inch by bloody inch. And the Americans regarded the country as a pawn of strategy, turning a blind and ignorant eye to the motivations of its people.[36]

North Vietnam's PAVN captured Saigon on 30 April 1975. The exclusion of non-Communist NLF officials[37] and factions by the conquering regime left Trúóng disillusioned and heartbroken. He escaped the country as one of the boat people in August of 1978.

Illustrative Case: Northern Ireland

Northern Ireland provides our third case study. Thirty years of bloodshed known euphemistically as *The Troubles* ended there only recently with the 1998 Peace Agreement. Though they are "white" like their Protestant neighbors, Catholics in Northern Ireland historically have lived under a system of strict *ethnic/religious segregation* – not unlike Jim Crow codes in the U.S.A. – perpetuated by the larger Anglican and Presbyterian Protestant groups.

The Troubles have their roots in Irish colonial history. English barons seized land in the Gaelic Isle of Eire in the 12th century and by 1300 all of Eire was under English control. The English and Irish were both Catholic. However, in the 1500s, King Henry VIII adopted Anglicanism as the religion of England, and Scotland later named Presbyterianism its official religion. The English invited now-Protestant Scots to Ireland as a counterweight to the rebellious Irish, and they settled mostly around the northern province of Ulster. The result was the familiar *three-tiered system*: *Anglican-English nobles* and *Presbyterian-Scot settlers* in the top two layers, and *Catholic, Irish-Gaelic peasants* on the bottom. The defeat of Catholic King James's forces near the city of Dublin in 1690 by Holland's Protestant Prince William of Orange reinforced this hierarchy.

At the turn of the 20th century, Britain (the entity of England, Scotland, and Wales) considered extending "Home Rule," that is, allowing an Irish parliament to govern, answerable to London. Most Catholic Irish favored the idea while Ulster-Scots vehemently opposed it. Ulster-Scots in 1913 organized a militia, the *Ulster Volunteer Force* (UVF) and the Irish countered with the *Irish Republican Army* (IRA).[38] Mired in a bloody stalemate, the British in 1921 proposed a two-state solution. The 26 mostly Catholic provinces would comprise the Irish Free State with Dublin as its capital, and the six northernmost ones where Ulster-Scots were predominant would become Northern Ireland with Belfast as its capital city. Ulster-Scots moved quickly to cement their political and economic advantages in the North. Irish-Gaelic Catholics there, about a third of the population, acquiesced in their traditionally subservient role.

We turn our analysis now to two studies of the "modern" IRA and UVF during the Troubles.

Spotlight on Social Science Thinking and Research

Richard English, *Armed Struggle: The History of the IRA*[39]; and Aaron Edwards, *UVF: Behind the Mask*[40]

Richard English, Professor of Politics at Queen's University in Belfast, identifies himself as a "double outsider" in studying the IRA. He was born in Belfast in 1963 but grew up in England. His mother's family were Irish Protestants and his friends in Northern Ireland (hereafter, N. Ireland) cut across the sectarian divide. Professor English and fellow political scientist Aaron Edwards (whom we will meet shortly) draw upon personal interviews, archives, memoirs, newspaper articles, and previous accounts. Both professors contend that the slide into decades of murder and mayhem was not a foregone conclusion. Rather, they argue, actions on both sides, *purposely intended to politicize and solidify ethnic/religious power differentials*, channeled the course of events in that direction.

First, the Irish Free State slowly moved toward independence and formally became the Republic of Ireland in 1949 (hereafter, Ireland). Fearing an Irish takeover of the entire island, a new version of the UVF soon was revived. The UVF was a small, by-invitation-only group. Some were seasoned British Army veterans, and they were encouraged by several well-placed, hardline Protestant politicians. In 1966, UVF cells on the prowl fired "warning bullets" into the homes of "soft" politicians and conducted drive-by shootings in Catholic neighborhoods.

Meanwhile, Ulster-Scots remained in complete control in N. Ireland. Government agencies afforded Protestants special advantages in education, housing, and services. The economy, with Belfast's shipyards and proximity to Scotland, was more than viable, but bigoted practices held northern Catholics at the margins. A Protestant police force, the Royal Ulster Constabulary (RUC), was distrusted for good reason in Catholic neighborhoods. The Orange Order, a Protestant fraternal organization, conducted elaborate public parades flaunting William of Orange's victory in 1690 over King James. And, Reverend Ian Paisley, founder of the evangelical Free Presbyterian sect in 1951, vociferously and divisively ramped up the rhetoric of intractable doctrinal chasms between Presbyterianism and "Romanism" under the dictum, *No Surrender!*

In this crucible, the N. Ireland Civil Rights Association (NICRA) was formed, modeled after the U.S.A.'s Civil Rights Movement of the 1950s and early 1960s.[41] NICRA sought reform and hoped to duplicate in N. Ireland the U.S. movement's success through *non-violent action*. When they took their message to the streets in October of 1968, RUC policemen and Protestant bystanders in Londonderry attacked NICRA marchers with water cannons and clubs. Over subsequent months, intense rioting periodically broke out there and in Belfast. Thousands of Catholics were displaced from their homes, and several were killed by roving bands of Protestants. On 14 August 1969, N. Ireland's Prime Minister asked London to dispatch British troops to restore order – the beginning of *Operation Banner*, a military presence lasting until 2007.

The riots, purges, and killings prompted the Belfast wing of the IRA to call an emergency meeting in December of 1969. They voted to secede from the Dublin-based "Official IRA" and form their own organization, the *Provisional Irish Republican Army* (P-IRA, informally, the "Provos"). Vowing "Never Again," the Provos promised armed defense from Protestant attacks, "be it the British Army, RUC, or Protestant bigots." Their goal now was violent revolution to force the British out of N. Ireland and the reunification of N. Ireland with Ireland.

The P-IRA agenda was not an easy sell to Ulster's Irish Catholics. Most at first assumed British soldiers would stand as a buffer between them and Protestant violence, and many held out hope that reforms sought by NICRA might yet prove tenable. Soon, however, rough searches, curfews, and arrests of suspects by British soldiers – at times using control-and-intimidation techniques from colonial Cyprus and Yemen – radically changed the perception of them. Confrontations with angry Catholics became routine and, in 1970, six Catholics died in such clashes. In January 1971, the P-IRA approved the targeting of British soldiers and RUC policemen, and Gunner Robert Curtis, the first British soldier to die in Operation Banner, was shot and killed a month later.

The British reacted strongly. Under the Special Powers Act, they instituted *internment without trial*, that is, authorization to imprison suspects without trial or due process. That the policy was applied mainly to Catholics made it especially onerous to them. Martin McGuinness, who became a decades-long P-IRA leader in Londonderry, and Gerry Adams, long-time leader of the Belfast-based political arm of the P-IRA, *Sinn Fein*, later identified friction with British troops and the internment-without-trial policy with stimulating sizable increases in recruitment.[42]

Still, things were about to get worse. On 30 January 1972, NICRA called for a march in Londonderry to protest the internment policy.[43] Some marchers deviated from the approved route and threw rocks at soldiers from the 1st Battalion, Parachute Regiment. The paras broke ranks and, firing shots, chased down marchers, killing 14 of them. With *Bloody Sunday*, the last wisps of support among Catholics for NICRA-style *reform* vanished, and P-IRA-style revolutionary *secession* became the preferred, or at least tolerated, response.

Aaron Edwards, Senior Lecturer in Defence and International Affairs at the Royal Military Academy Sandhurst, was born in 1980 to a North Belfast, Protestant family. The neighborhood, he writes, was "deeply intertwined with the history of the modern Ulster Volunteer Force (UVF)." This personal connection provided him access in recent years to surviving UVF leaders.[44] In addition to the UVF, such neighborhoods by 1972 had produced a proliferation of vigilante and area-defense groups.[45]

For its part, the UVF pursued its own style of violence with its base philosophy, "If you can't get an IRA man, get a Taig."[46] A typical operation was no more complicated than an evening of drinking at the pub, interrupted by a trip into a Catholic neighborhood to shoot a random Catholic on the street. Sometimes a Catholic would be brought back to the pub, tortured and mutilated, with the body dumped back on the street. This focus on intimidation also led them to perfect homemade bombs for targeting Catholic pubs and shops, a technique soon copied and used horrifically by the Provos.

In these years, some UVF members saw the spiral of violence in religious terms. For example, Billy Mitchell, once a UVF luminary, told Edwards years later:

> The much loved phrase – "by all means under God" – was simply a synonym for "by force of arms" but with added thought that God himself would approve … [T]his … idea that the Ulster Protestant was a modern-day Israelite and the Irish Roman Catholic was a modern-day Philistine … gave me, and many more like me, the firm conviction that force of arms was legitimate in the struggle for Ulster's continued existence.[47]

Escalating P-IRA violence remained a potent recruitment tool for the UVF. On 21 July 1972, *Bloody Friday*, the Provos set off 20 explosions, most of them car bombs, within a half hour in central Belfast. Long-time UVF member, David Ervine, was 19 at the time and that day having a pint with friends. He told Edwards he had seen puffs of smoke rising over the Belfast skyline: "It was so brutal, so raw. [I] took up a long-standing invitation to join the UVF." That day pushed him and a flood of others "off the fence ... [I thought] the best means of defence was attack."[48]

The stage was set for decades of violence. One P-IRA member reflected later[49]:

> "[D]id the IRA ... succeed in protecting Catholics ... from [Protestant] attacks ...? I would say no. ... At least [we] were promoting [ourselves] as someone who would do something ... The IRA ... could never actually defend. There was no way to defend against these things. So the only way to appear to be defending ... was to take out [kill] other people.

By the early 1990s the conflict's intensity had declined due to the constant presence of the British Army in both communities and war weariness on all sides. The P-IRA signaled a willingness to move towards the "ballot-box," and the British government in London wished to reduce the huge costs of their involvement. The time was ripe for compromise.[50] The core *Good Friday Agreement* of 10 April 1998 stated: (1) N. Ireland shall remain part of Britain; (2) a N. Ireland Assembly shall have *equal representation* from Protestant and Catholic areas; and (3) N. Ireland's Executive shall be chosen by the N. Ireland Assembly.

The carnage described above is of particular interest because all the participants are of the same "race," White, and all are Christians. However, the *intersection of ethnicity/religion* in N. Ireland created *identifiers* denoting to all *which group was firmly in charge* and *which one was the underling*. Hence, every encounter between those in the two broad groupings came to reflect this divide, evidencing enjoined differences in *power*, *ethnicity*, and *religion*. The virulence of the Ulster Protestants' attempt to preserve complete dominance stemmed from their fear of being a similarly subjugated ethnic/religious minority in a united Catholic Ireland. All told, there were 3,500-plus assassinations of Catholic and Protestant civilians and of security-forces personnel. Standardized by population size, this would amount to about 500,000 political killings for the U.S.A. for a 30-year period.

Questions for Discussion

1   Successful insurgencies involve a population-centered approach. Using Robert Merton's structural-strain theory as a point of departure, what would form the core of an effective counterinsurgency strategy?
2   French troops in Algeria were ordered to end FLN-driven violence "by any means necessary." What challenges did this approach create when dealing with the Algerian insurgency? How did Capt. David Galula's approach respond to these challenges?
3   Apply Lt. Col. John Nagl's question, "What are the features of militaries who are learning organizations?" to the experiences of the French and U.S. militaries in Vietnam. How was their operational success (or not) shaped by their (in)ability to adapt?

4   Let us reframe the *Troubles* through the lens of Merton's theory. What were the goals
     of both the UVF and the IRA? Did either group have access to legitimate means to
     achieve these goals? Consequently, what strategies did each group use to achieve their
     goals?

Notes

1 *U.S. Marine Corps Small Wars Manual*, 1940, Washington, D.C.: Dept. of the Navy, p. 1.1. The
   manual reflects the operational requirements and guidelines from interventions by the U.S.
   military in Central America and the Caribbean between 1898 and 1934.
2 The most acclaimed biography of T.E. Lawrence is Jeremy Wilson's *Lawrence of Arabia: The
   Authorized Biography of T.E. Lawrence*, 1990, New York: Atheneum Books.
3 From T.E. Lawrence, *The Seven Pillars of Wisdom*, New York: Random House, p. 192, quoted
   in Robert L. Bateman, 2008, "Lawrence and His Message," *Small Wars Journal*, Blog, downloaded from http://smallwarsjournal.com/blog/lawrence-and-his-message.
4 David Kilcullen, 2006, *Survival: Global Politics and Strategy*, 48 (4): 111–130, p. 111.
5 For instance, *Mao Tse-Tung on Guerrilla Warfare*, Washington, D.C.: United States Marine
   Corps, PCN 140 121800 00, original text written by Mao in 1937; and *On Guerrilla Warfare*, at
   Selected Works of Mao Tse-Tung: https://www.marxists.org/reference/archive/mao/works/1937/
   guerrilla-warfare/.
6 See Thomas X. Hammes, 2006, *The Sling and the Stone: On War in the 21st Century*, St. Paul,
   Minn.: Zenith Press, Chapter 5, "Mao and the Birth of Fourth-Generation War".
7 Robert K. Merton, 1938, "Social Structure and Anomie," *American Sociological Review*, 3
   (October): 1172–1182.
8 Merton's use of this benign-sounding term, *innovation*, implies that people falling in this category might very well be conformists if legitimate means for achieving material comfort and
   success were available to them.
9 Literally, "black feet," a reference to the black boots worn by French colonial soldiers, later a
   term referring to all European Algerians, whether in the military or not.
10 Alf Andrew Heggoy, 1972, *Insurgency and Counterinsurgency in Algeria*, Bloomington, Ind.:
   Indiana University Press.
11 David Galula, 2006, *Pacification in Algeria, 1956–1958*, Santa Monica, Calif.: RAND
   Corporation, originally published in 1963.
12 Heggoy, op. cit., p. 234.
13 Ann Marlow, 2010, *David Galula: His Life and Intellectual Context*, Carlisle, Penn.: U.S. Army
   Strategic Studies Institute, monograph downloaded from http://www.StrategicStudiesInstitute.
   army.mil/, pp. 21–24.
14 Ibid., pp. 30–36. At one point, Galula accompanied a Chinese Army unit in the field and was
   captured and briefly held prisoner by Mao's forces. He later admitted, somewhat reluctantly,
   that his captors had treated him very well.
15 In part, this debate was spurred by soldiers themselves, who upon return from their tour of
   duty in Algeria, expressed shame and guilt for their participation in or tolerance of torture.
16 Galula, op. cit., pp. 183–184. The full quote is: "In most [areas], the policy was to interrogate suspects summarily and to dismiss rapidly those who had not readily confessed; in [other areas] … it
   was 'talk or else.' Under pressure of a press campaign on 'tortures' (in my view 90 percent nonsense and 10 percent truth), a special unit was created in the fall of 1957, *Détachement Operationnel
   de Protection (DOP)*, … to [take] over from amateurs all handling of prisoners, guerrilla and
   civilian." Of this change, he speaks highly: "If I were to state now the single most important
   improvement in our counterinsurgency operations in Algeria, I would flatly put the DOP first."
17 Ibid., p. 239.
18 Ben Connable and Martin C. Libicki, 2010, *How Insurgencies End*, Santa Monica, Calif.:
   RAND, National Defense Research Institute. Their analysis reveals that the government under
   duress has prevailed outright in less than 30% of cases. The percentage drops to about 20 in
   instances where a foreign military power becomes visibly involved to bolster the established
   government. In the bulk of cases, the insurgents win outright or agree to some sort of power-sharing to end the fight.

19 Trúóng Nhú Táng, David Chanoff, and Doàn Vãn Toai, 1986, *A Viet Cong Memoir: An Inside Account of the Vietnam War and its Aftermath*, Vintage Books: New York.

20 John A. Nagl, [2002] 2005, *Learning to Eat Soup with a Knife: Counter-insurgency Lessons from Malaya and Vietnam*, Chicago, Ill.: University of Chicago Press.

21 Trúóng et al., op. cit., pp. 1–6.

22 Ho Chi Minh, the leader of a resistance organization, the Viet Minh, was an operative for the U.S. Office of Strategic Services (OSS, forerunner of the Central Intelligence Agency) during WWII in Vietnam after the takeover by the Japanese. Ho assumed therefore the U.S.A. would grant Vietnam its independence following the war. The U.S.A. instead elected to reinstall its ally, the French, as colonial masters.

23 Trúóng et al., op. cit., pp. 12–13, 15.

24 Ibid., pp. 33–38, 39.

25 Ibid., pp. 65–71.

26 Life in the jungle along the Ho Chi Minh Trail – the major infiltration route from the North to South – was filled with hardships and tribulations, but all this paled in comparison to being on the receiving end of a B-52 strike:

> The terror was complete. One lost control of bodily functions as the mind screamed incomprehensible orders to get out. … Sooner or later, though, the shock of the bombardments wore off … [P]eople just resigned themselves – fully prepared to "go sit in the ancestors' corner.
>
> Ibid., pp. 167–170

27 Ibid., pp. 187–189.

28 Nagl deployed to Iraq in 2003 as the operations officer for the 1/34th Armor Regiment, 1st Infantry Division, where he had the opportunity to apply some insights from his study of the Vietnam war. His assessment: Difficult.

29 The actions by these U.S. advisors in Vietnam was consistent with a 1957 Army study that had concluded: "The required forces … for a small war appear to be much the same as those for the atomic war against the Soviet Union" (quoted in Nagl, op. cit., p. 126).

30 Ibid., pp. 154–156.

31 Ibid., pp. 105–107.

32 Ibid., p. 157. For an in-depth description of a Marine Combined Action Platoon conducting pacification operations, see Bing West, 2003, *The Village*, New York: Pocket Books.

33 Ibid., p. 179.

34 Ibid., pp. 115–116.

35 The U.S.A. did take some military options off the table. It did not invade the North nor target civilians in the bombing there as happened in Japan toward the end of World War II. Likewise, except for one brief incursion in 1970, the U.S.A. did not invade PAVN sanctuaries in Laos or Cambodia. In part, these decisions reflected geo-political realities. China (and the Soviet Union) supplied war materiel in support of PAVN and Viêt Công military actions in Vietnam, but neither deployed troops there as China did in the Korean war.

36 Trúóng et al., op. cit., pp. 190–191.

37 Trúóng was the sole exception: he was appointed Minister of Justice, a position he held only briefly. At first he supported the reeducation program as a way to heal "the bitter, often savage enmities that a generation of civil war had left in its wake." He resigned when he saw the program rather was a way of indiscriminately incarcerating and punishing South Vietnamese (ibid., pp. 274–277).

38 Richard English, 2003, *Armed Struggle: The History of the IRA*, Oxford: Oxford University Press, pp. 20–25.

39 Ibid.

40 Aaron Edwards, 2017, *UVF: Behind the Mask*, Newbridge, Ireland: Merrion Press.

41 English, op. cit., pp. 88–96.

42 This made for a very bloody year: 9 P-IRA members and 33 Catholics were killed by security forces, and 56 security-force members were killed by the IRA.

43 English, op. cit., pp. 148–155.

44 Ibid., pp. 27–38.

45 The estimated peak strengths of the UVF and P-IRA are "several hundred" and 1,500, respectively. The Ulster Defence Association (UDA), an umbrella organization connecting many neighborhood groups, claimed to have 20,000 "members."

46 "Taig" is a pejorative term in N. Ireland for "Catholic," equivalent in rancor to the "n-word" racial epithet. The pejorative equivalent for "Protestant" is "Hun."

47 Edwards, op. cit., pp. 35–36.

48 Ibid., p. 55.

49 English, op. cit., pp. 173–174.

50 Each of the individuals mentioned in the Spotlight ultimately played some role in reaching the Good Friday Agreement. The *British government* in the 1980s admitted it could not defeat the P-IRA militarily, and the *N. Ireland government* conceded it must meaningfully accommodate its Catholic minority. *Gerry Adams*, Sinn Fein's leader from 1983 to 2018, and *Martin McGuinness*, Londonderry's former P-IRA commander, realized they had to entrust a power-sharing arrangement with Loyalist Protestants if there was to be peace. Both *Billy Mitchell* and *David Ervine* worked behind the scenes to persuade their UVF comrades to accept the Good Friday Agreement. And *Rev. Ian Paisley* consented to the peace-process in formal declarations with Gerry Adams and Martin McGuinness. Newspaper headlines declared, "Hell Freezes Over!"

Recommendations for Additional Reading/Viewing

The Battle of Algiers (film), 1966, directed by Gillo Pontecorvo.
(Highly acclaimed, fictionalized but realistic account of Algerian resistance in the city of Algiers)

Edward Burke, 2018, *An Army of Tribes: British Army Cohesion, Deviancy and Murder in Northern Ireland*, Liverpool, U.K.: Liverpool University Press.
(Analysis of Operation Banner, British Army campaign in Northern Ireland; draws on scholarly literature in sociology, anthropology, and criminology to explain a not-so-pretty picture)

Colonel Thomas X. Hammes, 2006, *The Sling and the Stone: On War in the 21st Century*, St. Paul, MN: Zenith Press.
(A review of 20th-century asymmetric wars with an eye toward applications for ones in the 21st century)

William D. Henderson, 1980, *Why the Vietcong Fought: A Study of Motivation and Control in a Modern Army in Combat*, Westport, Conn.: Praeger Publishing.
(Revealing study of the Việt Công organizational structure and motivation to fight based on interviews during the war with Việt Công prisoners-of-war)

Harold G. Moore, and Joseph L. Galloway, 1992, *We Were Soldiers Once… and Young*, New York: Ballantine Books; and *ABC News – Vietnam: They Were Young and Brave* (documentary), ABC Day One, Moore and Galloway return to LZ-X-Ray, narrated by Forrest Sawyer.
(Description of the battle of LZ X-Ray, U.S. 2/7th Cavalry, 1st Air Cavalry Division vs. the North Vietnamese 66th Glorious Regiment; to research book, Moore and Galloway received permission to visit site of LZ-X-Ray, where their hosts were five People's Army officers who fought against them; together they re-walk the battlefield)

United States Marine Corps, [1940] 2009, *Small Wars Manual*, New York, NY: Skyhorse Publishing; General David H. Petraeus, Lt. General James F. Amos, and Lt. Colonel John A. Nagl, 2007. The U.S. *Army/Marine Corps Counterinsurgency Field Manual: U.S. Army Field Manual No. 3–24, Marine Corps Warfighting Publication No. 3-33.5*. Chicago: University of Chicago Press.
(Official U.S. Marine Corps and U.S. Army doctrine of the whys of small wars and how to fight them)

10 Spectrum of Conflict
"New" Wars

Reader's Guide: Two recent developments justify the designation of 21st-century wars as "New" Wars: the interconnectedness of societies' institutions in a dynamic global network, and the accentuation of this by new social media and computing technologies. By tapping into these, disgruntled actors and social groups gain access and power far beyond their means. By exploiting these, nation-states expand the frontiers on which to wield influence and coercion. Our Spotlights provide three examples. The first reviews two studies of the uprisings (intifadas) by the Palestinians against Israel, the first on the terrorist organization, Hamas, and its reliance on public-opinion data in timing its attacks, and the second on adaptations by the Israeli Defense Forces. The authors of the latter suggest new terminology to capture what took place. The second Spotlight reviews the capacities and legalities of using drones – remotely piloted aircraft – by the U.S.A. to kill terrorist kingpins and suspects in the Middle East. Finally, we turn to cyber- and information-warfare with examples and a review of Russian military doctrine for their use.

In the previous chapter, we indicated that some contemporary *Small Wars* deserve separate treatment as *"New" Wars*. The person most responsible for applying the label "new" to 21st- century wars is Mary Kaldor, Professor of Global Governance at the London School of Economics. Her many writings, culminating in her 1999 book, *New and Old Wars: Organized Violence in a Global Era*, touched off rounds of informative debate.[1] Are *New Wars* actually "new?" Are they actually "wars?" Why is it not enough just to assert, as Colonel Harry Summers did in another context, "a war is a war is a war." And so on.

In Kaldor's schema, 19th- and 20th-century *Old Wars* correspond to our use of the term *Big Wars*, characterized by trysts among conventional militaries (see Chapter 8). *Globalization*, she argues, gives 21st-century *New Wars* distinctive features. Game-changing patterns of cross-national trade and investment create a wholly interdependent system of global capitalism, with a consequent reordering of the winners and losers. Recent winners are nations embracing the openness and restructuring required to play this game; losers are states whose control over their populations rely on isolation. It is in these states that Kaldor's 21st-century *New Wars* are found.

Kaldor had in mind Bosnia-Herzegovina where she worked in the 1990s both as a researcher and an activist. However, her analysis easily could be extended to other regions. Without embracing the meaning of "new" wars in the exact way Kaldor does, we are

DOI: 10.4324/9781003282549-10

persuaded that something fundamentally new has been created and given form in 21st-century wars. We also find direction in how her thinking points to the necessity for political rather than purely military solutions.

The U.S.A. and 9/11

To illustrate the point that something new is afoot, consider the 21st-century's defining moment for people in the U.S.A.: 11 September 2001, in the popular vernacular, 9/11. The date marks the attacks on the 110-storey twin towers of New York City's World Trade Center and on Washington, D.C.'s military complex, the Pentagon. Note how this attack differs from the bombing of Pearl Harbor in December of 1941. In that incident, military pilots from the Imperial Japanese Navy conducted an air assault on U.S. naval ships and nearby military aircraft – a *Big War* encounter between two near-peer military forces leaving little doubt against whom to address a response.[2]

In contrast, 19 civilians, at the behest of Osama bin Laden's jihadist think-tank, al-Qaeda (see our discussion in Chapter 5), carried out the attack of 9/11 by hijacking four civilian passenger jetliners. They managed to crash three of them into two of their intended targets, killing 2,977 Americans, mostly civilians. Fifteen of these operatives were from Saudi Arabia, a key ally of the U.S.A. in the Middle East. The remaining four were from Egypt, the United Arab Emirates, and Lebanon. The U.S. response began a month later in Afghanistan to deprive al-Qaeda of its training facilities there. Following a regime change in Afghanistan, President George W. Bush ordered the overthrow of Saddam Hussein as the next step in what he termed the *War on Terror*.

Is there anything "new" about this? The use of *terror* – directing violence toward non-military targets by nonstate actors for political purposes – is not new, as we saw in the cases of Algeria, Vietnam, and Northern Ireland in Chapter 9. "Terror" though is a *tactic*. Hence, the designated enemy, as it were, in a *war on terror* is not another military entity or nation whose military is a target, but those, anywhere, who espouse this tactic. What does stand out in this instance is how new technologies and the extreme interconnectedness of 21st-century states make it possible for insurgencies to be truly crossnational and to play directly to international audiences.

The U.S.A. – rightly or wrongly – quickly proceeded with an invasion of Iraq. Secretary of Defense Donald Rumsfeld envisioned a powerful military thrust and quick departure. It would feature technological wizardry applied so overwhelmingly that Iraqi soldiers would give up in "shock and awe."[3] On 19 March 2003, the *shock* began, and a lightning-fast invasion followed as *awe* set in. As Iraqi celebratory gunfire turned into wanton violence, U.S. ground forces stood by as looters ransacked museums and government buildings, and high-level staff destroyed all they could to obscure past activities. Asked to comment, Secretary Rumsfeld said, famously, "Stuff happens."[4]

Iraq's ethnic/religious demography figured prominently in this. About two-thirds of Iraq's population are Arab Shi'as, 20% Arab Sunnis, and about 12% are ethnic Kurds, mostly Sunni. Saddam Hussein was Sunni and, though not religious, he followed the Arab custom of nepotism by intensely favoring fellow Sunnis. Hence, though Sunnis were a small minority, they were firmly in charge before the U.S. invasion, and Saddam's demise meant Shi'as would displace Sunnis in positions of power. Still, descent into a vicious civil war was not a given.[5]

The complete collapse of the government, capitulation by the military, and inaction by the U.S.A. created a vacuum in services from electricity to police protection. L. Paul Bremer, appointed by President Bush to preside over Iraq temporarily, made two momentous but ill-advised decisions: he barred the 25,000 government workers from Saddam's Ba'athist Party from future public sector employment, and he disbanded the 400,000-member Iraqi Army, most of whose officers and elite units were Sunni. This ousting of the Sunnis proved to be disastrous. Displaced, bitter Sunnis quickly became an armed and dangerous threat to the American occupation.[6]

Meanwhile, an experienced, hard-core Islamist from Jordan, Abu Musab al-Zarqawi, had arrived on the scene. A former *mujahideen* ("holy warrior") from the war in Afghanistan against the Soviets, he was a long-time bit player in al-Qaeda with a fondness for grisly violence in the name of Islam.[7] Soon, sophisticated and deadly suicide bombings announced that al-Zarqawi and his little group, eventually called *al-Qaeda in Iraq* (AQI), were well into his plan of fomenting civil war between Sunnis and Shi'as.[8] Most American units responded obliviously with hard-hitting, often indiscriminate, tactics that further angered warring factions and bystanders alike.[9]

We will relate two scenarios that reveal something elemental about 21st-century *New Wars*. In 2005, one of the authors of this book (Wilbur Scott) and two of his colleagues conducted interviews with troops from the 3rd Armored Cavalry Regiment (ACR) and the 4th Infantry Division's Iron Brigade at Fort Carson in Colorado Springs.[10] These units had deployed to Iraq in 2003, expecting to take on the Iraqi army's armored units. By the time they arrived, the Iraqi army already had collapsed. What they encountered instead looked to them like sullen Iraqis, unappreciative of Americans having liberated them from Saddam Hussein, surreptitiously triggering roadside explosives to harm them as they drove about in their armored vehicles. The soldiers carried on, tankers against elusive individuals. They returned home baffled and frustrated.

When the 3rd ACR redeployed to Iraq ten months later, now led by then-Colonel H. R. McMaster, they returned with a new game plan. Colonel McMaster had put into place a *full-spectrum* regimen, that is, one in which his troops in their redeployment training practiced scripts for approaching and enlisting the support of Iraqi locals, while confining their lethal edge to the hard-core insurgents among them. During its second deployment, the 3rd ACR successfully reclaimed the town of Tal Afar from the grip of AQI and became a model for how American units might more effectively operate in this type of battlefield.[11] A new *counterinsurgency field manual*, put together in 2006 under the direction of Army General David Petraeus and Marine Corps Lt. General James Amos, spelled out this *population-centered approach*.[12] General Petraeus installed this as the strategic doctrine for the war in Iraq (and, later, Afghanistan) when he became commander of Middle East operations in 2007.

Meanwhile, then-Major General Stanley McChrystal had been assigned as commander of Task Force 714, a supreme special operations team made up of Army Delta Force, Navy SEAL, and Army Ranger elements. It was given the task to apprehend or kill al-Zarqawi, an *insurgent-centered approach* under the rubric of *counterterrorism*. He analyzed this operation in a book written after his retirement[13] while at Yale University's Jackson Institute for Global Affairs.[14]

TF-714 found it could not pin down where al-Zarqawi was or prevent his group's attacks. AQI was *centralized in its decision-making*, that is, al-Zarqawi and two or three confidants decided what to do, but *decentralized in its operations*, that is, semi-autonomous cells carried out assignments. And, it attracted a continuing supply of new members by posting video footage of its beheadings and suicide bombings on the internet. McChrystal begrudgingly concluded AQI was, if nothing else, *highly adaptable*. Then it hit him: "the *new war's environment* favored *flexibility over efficiency*" (emphases added).[15] The many changes McChrystal instituted in TF-714 thus were designed to increase its *operational flexibility* by breaking down barriers among its elements, enhancing information flow among them, and giving them the autonomy to improvise more swiftly.

On 7 June 2006, two 500-pound, laser-guided bombs from a U.S. Air Force F-16 killed al-Zarqawi in a remote farmhouse as he met there with his spiritual advisor. The organization that pulled this off, McChrystal wrote, was "worlds apart from the one [I took over] in September 2004."[16] One might marvel, in a nonjudgmental way, at what is so 21st century about this. An elite special operations unit, supported by highly specialized intelligence and communications staff with an intercontinental network of telecommunications infrastructure, had devoted tens of thousands of hours to locate and surveil their target. To track him to that farmhouse, they had used drones flown over Iraq by pilots on the ground outside Las Vegas, Nevada. Finally, they had called upon a pair of F-16 fighter jets, on call over Iraq, to deliver a quick half-ton of ordnance, *all to kill one man.*

There are some hard lessons here for *counterinsurgency/counterterrorism* efforts. By 2007 AQI indeed was on the ropes: al-Zarqawi was dead, but it had badly overplayed its hand in Sunni provinces and some Sunnis now considered the American presence a lesser evil.[17] Soon, some 50 Sunni sheiks defected from AQI's orbit in what they called *Sahawa al-Anbar*, the Anbar Awakening. Things fell apart in 2008 as the date, arranged earlier by President Bush, approached for the U.S.A. to begin its drawdown. Prime Minister Nouri al-Maliki, leader of the Shi'ite-based Islamic Dawah Party, staunchly resisted pressure from the U.S.A. to share power with the Sunnis (and Kurds). An emboldened AQI targeted Sunni sheiks for assassination, and frustrated Sunnis abandoned The Awakening. A revitalized AQI morphed through several stages in western Iraq and eastern Syria to emerge in 2011 under a new title: the *Islamic State of Iraq and Syria* (ISIS).

The Israeli-Palestinian Conflict as "New Wars"

Any description of a violent conflict is fraught with political nuances. Such is the case with the Israeli-Palestinian Conflict. The title for this section, for instance, might suggest we favor the Israeli side, while an alternative term, Palestinian-Israeli Conflict, might imply implicit preference for the Palestinians. Our purpose though is to provide a balanced account of certain aspects of the conflict and an indication of what is "new" about them.

The area bounded by the Jordan River on the east and the Mediterranean Sea on the west is known as the "Holy Land" for good reason. It is the birthplace of two world religions, Judaism and Christianity. Arabs introduced Islam, a world religion which originated in Saudi Arabia, to the region, making it the religious homeland as well for Palestinian Arabs. The city of Jerusalem thus contains some of the holiest sites for all three. In ancient times, the area was part of the Israelite kingdom of Judah. The Romans invaded around the time of Christ, and Jesus was crucified by them along with the many others who took part in revolts against Roman encroachment. The Romans named the area *Palaestina* and expelled Jews from the area, the beginning of a large *Jewish diaspora*. In the 8th century, Arabs introduced Islam to the region and became the dominant group. There remained a small Jewish minority, and some Arab Palestinians were Christian. Muslims, Jews, and Christians mostly lived there with relative calm from the 16th to the 20th century as part of the Ottoman Empire.

An ostracized minority in many places, the 19th-century ideology of *Zionism* appealed to many in the Jewish diaspora because it held out the goal of a Jewish homeland in Palestine. An opportunity arose when the Ottoman Empire, which had sided with Germany in World War I, collapsed toward the war's end. The British had been meddling in the area (see our discussion of Lawrence of Arabia in Chapter 9) and took control of Palestine after the war. Mixed messages from British authorities promised both Arab independence and support for a "Jewish home" in Palestine.

World War II intervened. As the war unfolded, some Jews were able to escape Nazi Germany and its domains in Eastern Europe by immigrating to Palestine. At the end of the war, as the horrifying magnitude of the Holocaust began to become clear (see our discussion in Chapter 8), ever larger numbers of Jews migrated to Palestine with the approval of Britain and the U.S.A. These migrations engulfed the Arab Palestinians, who responded with violence, and Britain appealed to the United Nations for a solution. In November of 1947, the U.N. General Assembly passed Resolution 181 that partitioned Palestine into three parts. It put aside 45% of the territory for an Arab state, 55% for a Jewish state, and kept the city of Jerusalem a separate entity under U.N. supervision. Jews accepted the arrangement, but the Arab Palestinians rebelled, claiming with some justification they would be displaced from much of what they saw as their homeland.

In the war that followed, the Jewish residents and settlers, with military support from the U.S.A. and Britain, prevailed over the Arab Palestinians, who had been joined in the fray by neighboring Arab countries. The State of Israel now claimed 85% of what had been Palestine, leaving two totally separated geographical areas for the Palestinians, both under the occupational control of Israeli forces: the *West Bank*, stretching from the Jordan River to Jerusalem, and the *Gaza Strip*, a slice of land along the southeastern coast of the Mediterranean Sea. And, the Holy City now had two sections, *East* and *West Jerusalem*, under the control of Arab Palestinians and Israelis, respectively. A minority of Arab Palestinians remained in the State of Israel, but some 700,000 fled from their former homes to refugee camps, primarily in Jordan, Syria, and Lebanon.

We turn our Spotlight now to two studies of the conflicts which followed.

Spotlight on Sociological Thinking and Research

Richard Davis, *Hamas, Popular Support and War in the Middle East: Insurgency in the Holy Land*,[18] and Eyal Ben-Ari et al., *Rethinking Contemporary Warfare: A Sociological View of the Al-Aqsa Intifada*[19]

Richard Davis holds a Ph.D. in Political Science from the London School of Economics, is a Founding Fellow of the Centre for the Study of Intractable Conflict at Oxford University, and served as Policy Director of the U.S.A.'s Homeland Security Council from 2004 to 2006. He also is the Chief Executive Officer of Artis International. Artis's website describes the company as one that "conducts field-based scientific research across the globe, in all cultures under any circumstances." Indeed, that is what Davis and his Founding Fellow at Artis, Scott Atran, are known for. *Hamas, Popular Support and War in* stands as a good example. The book is based on extensive field research in which Davis conducted interviews with leaders of the Palestinian resistance groups, Hamas and Fatah, and key authorities in the Israeli security services, including Shin Bet, Israel's counterterrorism agency.

In addition, he makes use of an unusual trove of data: public opinion surveys tapping the attitudes of ordinary Palestinians collected by Dr. Khalil Shikaki, founder of the Center for Palestinian Research and Studies (PRS), located in the West Bank city of Ramallah, and the Jerusalem Media and Communications

Center, founded by two Palestinian journalists.[20] The book centers on Hamas, the premier and most violent Palestinian resistance group. The time periods addressed in the book are the First Intifada (Palestinian uprising) in 1987 to 1993, the Oslo Peace Accords of 1993, the Second Intifada in 2000 to 2005, and the 2006 Palestinian elections. Davis's intent is to study all this scientifically in hopes an understanding of it may be useful in reducing the violence.

Hamas was founded in Gaza in December of 1987. The triggering event was the killing of four Palestinians on their way home from work, ploughed into by an Israeli motorist in retaliation for the stabbing of an Israeli businessman days earlier. This was the beginning of the First Intifada, a revolt by ordinary Palestinians. The word *"hamas"* means "zeal" in Arabic. Its Palestinian founders, all university-educated in either Egypt or Kuwait (in engineering or the like), were well-steeped in the lore of the Muslim Brotherhood, a movement founded in Egypt in 1928 to resist Western influence. Hamas's central stated goal is the reclamation by force of the entire territory of what had been Palestine. Davis counts 260 high-profile terrorist attacks on Israelis by Hamas between 1989 and 2014, including 79 suicide bombings. These attacks, plus those of assorted other Palestinian groups, killed 1,596 Israelis and wounded another 8,120.[21]

In the classic style of 20th-century terrorist organizations, Hamas has separate political, social, and military wings which work independently as much as possible. Independence is necessary since Israel's Shin Bet maintains a vigorous Targeted Assassination and Apprehension Program (TAAP) to kill or arrest Hamas leaders. Israeli legal strictures constrain lethal retaliation, so TAAP conducts robust intelligence-gathering and surveillance activities in the Palestinian territories to *detect pending attacks*. Hamas's military wing, the Qassam Brigades, thus must operate in a very decentralized and covert fashion to preclude putting the leaders of the political and social wings at risk.[22]

The political wing, the Politburo, has representation from Gaza, the West Bank, those in prison, and leaders in exile. It is responsible for internal governance and relations with local Palestinians, Israelis, the press, and international audiences. The social wing carries out an important practical and public relations mission, in Arabic, *dawa*: providing food, medical care, and religious schooling for underserved Palestinians. Hamas can conduct these activities because of donated money from sympathizing Arab nations and other private donors. Its sphere of influence is strongest in Gaza, while its rival group, *Fatah* (meaning *the opening* in Arabic), founded in 1959 by well-known Palestinian leader, the late Yassir Arafat, is the predominant group in the West Bank.

The First Intifada lasted from 1987 to 1993. As indicated above, it was initially a grass-roots revolt by young Palestinians rather than organized groups. Their weapons in clashes with Israeli forces were stones, knives, and Molotov cocktails. The Israeli military, conscious of the *asymmetry of forces* and wishing to project *a public image of restraint*, responded with nonlethal weapons such as truncheons and rubber bullets, and used police-style tactics more in keeping with arresting criminals. This stance however was applied with enough brute force to merit the informal label, "breaking Palestinian bones."[23] Hamas encouraged the uprising and conducted its first vehicle-borne suicide bombing. The architect of the actual bomb used in this attack and many subsequent ones was Yahya Ayyash, an engineering graduate of Birzeit University in the West Bank. Dubbed "The Engineer" by Shin Bet, he soon rose to the top of TAAP's most wanted list.

The 1993 Oslo Accords brought the violence to an end as Fatah agreed to recognize the right of the State of Israel to exist and committed itself to a two-state solution. This put Hamas in a bind. While its commitment to violence is intractable, acceptance among Palestinians themselves for such violence is not. To frame this theoretically, Davis draws on a lecture about political action given by one of sociology's founding fathers, Max Weber, at the University of Munich in 1918 (see our discussion of Weber in Chapters 2 and 4). Weber placed actions by political groups on a continuum, with those advancing their *preferred endgame* at one end, and those addressing the *wants of their broader constituency* at the other. A political group must balance these sometimes-contradictory demands.[24]

By 1993, Dr. Shakili had set up PRS and was conducting public opinion surveys of the Palestinian populations of the West Bank and Gaza, and Hamas was busy trying to undercut the Oslo Accords. Davis learned in a 2012 interview with Ami Ayalon, the former head of Shin Bet, that he (Ayalon) had noticed in 1996 a correlation between support by Palestinians for violence, reflected in Dr. Shakili's polls, and Hamas's military operations. Davis therefore put together a dataset to test the link between the *ebbs and flows in Palestinian support for violence* and the *timing of Hamas's use of violence*. Davis's subsequent multivariate, time-series analyses for the years 1993 to 2015 confirmed Ayalon's suspicions. Davis states: "[O]ver the last 22 years, when Palestinian *Support for Violence* increases, as reported by PRS, Hamas violence increases in the following period."[25]

Why does support for violence vary in the Palestinian population? It appears that *violence directed against them*, whether by lone-wolf Israelis, Israeli security forces, or Palestinian resistance groups, increases willingness to see violence used on their behalf. To cite one example, in September of 1995, 18.3% of Palestinians surveyed by PRS indicated they supported the use of violence. By December, TAAP had assassinated The Engineer and, as the percentage climbed to 39.1, Hamas initiated a wave of retributions. Retributions and retaliations make the lives of ordinary Palestinians not only difficult but perilous, and there is a limit to what they are willing to tolerate. As their support for violence decreases, Davis's data show, Hamas would shift some funds from military operations to the *dawa* program. Interestingly, since the Israelis were tracking the PRS polls as well, a similar balancing act came into play for them in figuring out the judicious use of TAAP.

Keeping fingers on the pulse of the population, by either the insurgents or the state's security forces, is not "new," although the availability of systematic public opinion data is somewhat novel. What is "new" is more evident in the Second or Al-Aqsa Intifada. Eyal Ben-Ari and his associates[26] observed Israeli military operations during this uprising from September of 2000 to March of 2005. Its name derives from Palestinian protests after a visit by Ariel Sharon, leader of Israel's right-of-center Likud Party, to the Temple Mount complex, also the site of Dome of the Rock and the Al-Aqsa Mosque, holy sites for Jews and Muslims, respectively. Accompanied by Israeli soldiers, Sharon made the provocative visit to assert the complex would always be under Israeli control. More than 4,000 Palestinians and 1,000 Israelis perished in the ensuing four and a half years of violence.

At the time of the research, Ben-Ari was Professor of Sociology and Anthropology at the Hebrew University of Jerusalem. Zeev Lerer, Uzi Ben-Shalom, and Ariel Vainer were officers in the Israeli Defense Forces (IDF), had doctoral degrees in Military Psychology, and were serving as researchers and advisors in the IDF's Behavioral Science sections. The initial focus of the research was not the intifada, but one they shifted to as the violence unfolded. They collected data through formal interviews and more informally by accompanying and talking with troops on missions in the West Bank or near the Gaza Strip, a method they called *focused journeys*.[27]

The Al-Aqsa Intifada began with street demonstrations but morphed rapidly into something "not quite war." Ben-Ari and associates noted the tactics of the Palestinians, from throwing rocks and Molotov cocktails to firing handguns and light machine guns, laying roadside bombs, and conducting drive-by shootings and suicide attacks in Israeli urban centers. Under the onslaught, the Israeli population favored a strong response. The IDF reacted with tear gas, rubber-coated steel bullets, plane and helicopter-fired missiles, tank patrols, and the destruction of Palestinian outposts and strongholds.[28] Ben-Ari's team suggest some additions to the lexicon of military sociology to capture what was taking place.

The research team noticed a prime characteristic, present too in the First Intifada but accentuated greatly in this second uprising: the ubiquitous presence of *social media* and the *internet*. Rioters and others routinely posted pictures of encounters on the internet, thereby placing the conflict under a kind of constant *global surveillance*, and created an atmosphere "in which 'every bullet' may have wide-ranging implications."[29] Adaptations to this had begun before the uprising. IDF leaders already had articulated *dignity of man* principles and *in-between measures* for decision-making to reduce the likelihood of events going awry. Now, *humanitarian officers* were added to some units as arbitrators in sensitive interactions, and the IDF allowed the advocacy group, *Machsom* (Checkpoint) Watch, a volunteer organization of Israeli women, to provide observers at troublesome checkpoints.

This emphasis on human rights, Ben-Ari and his associates note, was consciously reflected in the vocabulary of *precision warfare* and considerations designed to reduce civilian casualties. Modern artillery, aircraft, and drones were portrayed as capable of *surgical strikes* and *pin-point accuracy*. Drawing up *most-wanted lists* for carrying out *targeted assassinations* were justified to introduce exactness in identifying and killing violent leaders of the insurrection. Unintended bystanders sometimes were killed, but their deaths often were characterized as unavoidable.[30]

At first, the Israeli military used existing regular units, what Ben Ari and his associates call *textbook units*. Soon the IDF began to disassemble and reconfigure some of these into tailor-made *instant units* to carry out specific tasks. Such instant units combined clumps of soldiers with different specialties and operated in highly decentralized command structures. This fragmentation and recombination of textbook units into instant units raised concerns about their *unit cohesion*, the key to binding individual soldiers into solid fighting groups (see our discussions in Chapters 1 and 5).

The research team found cohesion is not produced here in the same way as in textbook units. For example, they observed the camaraderie in an instant unit having breakfast after an overnight mission. Present were its two tank crews, a handful of infantry types, and some female soldiers in charge of surveillance instruments. The team observed: "It seems everyone is aware of who is in the dining hall but the social communication takes place primarily *within* the groups" (emphasis added).[31] They characterized what they saw as substantial *internal cohesion* within the *instant unit*'s subgroups, but *loose couplings* among its constituent parts. Is this sufficient? The team theorized these soldiers "proceed under the assumption that the proper way to act is to cooperate, take others into consideration, and be identified and committed to the new social framework." This would promote *swift trust* among subgroups and heighten *cohesion* despite limited interaction.

Ben-Ari and his associates expected some *structural decoupling*, that is, gaps between the IDF's formal rules and the informal practices developed by its soldiers. They thus were on the lookout for how the limits of permissible action were conveyed to and accepted by rank-and-file troops. The team on occasion did encounter transgressions, especially following attacks in which Israelis were killed. The incidence of *structural decoupling* appeared leadership-driven: where commanders worked to make conforming behavior part of their unit's image of a "good soldier," and demonstrated the utility of such during operations, troubling gaps between formal rules and informal practices were less likely.

The Legality and Ethics of Individuated Warfare

At the start of the War on Terror, the idea of deploying aerial, armed robots would have been, in the words of security analyst, P. W. Singer, "the stuff of Hollywood fantasy." But by 2012 more than 7,000 "drones" were in operation, most prominently two remotely piloted aircraft (RPA)[32] in the U.S. Air Force inventory, the MQ-1 *Predator* and MQ-9 *Reaper*. The use of Pred/Reapers in the Middle East by U.S. forces began in earnest in 2006 and won quick acceptance. In 2005 95% of U.S. military air missions were manned. By 2012, that figure had dropped to 59%.[33]

RPAs first were developed as an intelligence/surveillance/reconnaissance (ISR) platform,[34] but the U.S.A. soon employed them for *kinetic* missions as well, that is, ones involving lethal force. Although the ratio of missions varies in response to several contingencies, most of the time is spent in passive ISR activities, while a smaller fraction is devoted to killing people and breaking things. The U.S. Air Force in 2015 fired more ordnance in Afghanistan from RPAs than from conventional warplanes.

Pred/Reapers are extremely cost-effective. Their price tags and dollar costs of deployment are miniscule compared to manned warplanes, and they are considered low risk for use in difficult, challenging areas, allowing planners to radically reduce the number of "boots on the ground." Because of fuel considerations, fighter jets do not spend much time over the target area, perhaps 20 minutes or so. Fuel-efficient Pred/Reapers can loiter at an altitude of 10,000 to 20,000 feet for hours and their high-resolution cameras give a startlingly clear look at the target area.

Pred/Reapers are disassembled and moved by air transport to locations across the globe near the actual areas over which they will fly. There, once airborne, control is switched through a satellite datalink to a Mission Control Element (MCE) located elsewhere, most famously at Creech Air Force Base about 40 miles north of Las Vegas, Nevada. An MCE crew consists of a pilot (an Air Force officer), a sensor operator (an enlisted Airman), and an intelligence specialist who may be co-located or at another station altogether. The three-person crew is only one part of a larger network of command, intelligence, and communication entities, all of whom talk by radio or internet texting to sort through a blizzard of tactical considerations and data points.

The Air Force has very strictly defined rules and procedures for bringing its lethal weapons into play. A multi-layered oversight network must be negotiated to reach the point where the pilot has the go-ahead to release ordnance, in this case, AGM-114 *Hellfire* missiles. MCE crew members, and several others on integrated radio nets and video feeds, move together down a checklist to verify the target and legality of engaging it. In the absence of time-sensitive occurrences, this process could be drawn out over hours, days, or even weeks.

This use of lethal force in the New Wars of Afghanistan and Iraq is not directed against adversarial militaries. The targets are *individuals* confirmed or suspected to be part of an al-Qaeda network and so have drawn the term, *individuated warfare*. We turn now to an analysis of the concept.

Spotlight on Social Science Thinking and Research

Jack McDonald, ***Enemies Known and Unknown: Targeted Killings in America's Transnational War***[35]

Jack McDonald received his PhD in Security Studies and Theory at King's College, London. In his doctoral work and research since then, he has focused on relations among ethics, law, technology, and war. Which brings us to *individuated warfare*. In establishing what makes this brand of warfare *new*, McDonald begins by analyzing several incidents. In the first, John Yoo, a legal advisor in President George W. Bush's Department of Justice from 2001 to 2003, defended the Administration's emerging policy of targeting civilian "combatants" by citing the case of Japanese Admiral Isoroku Yamamoto in WWII.[36]

Here, the U.S. military sought to take advantage of an intelligence leak revealing Admiral Yamamoto's exact upcoming travel plans. While it had little compunction bombing Japanese cities, the U.S. military did not have an operational plan for assassinating high-ranking enemy commanders. Admiral Chester Nimitz, commander of the U.S. Pacific fleet, thus sought authorization from Secretary of the Navy, Frank Knox, who consulted with several clergy about the morality of such an act. President Roosevelt gave the final approval. On 18 April 1943, American pilots in P38G *Lightnings* tracked and shot down the small plane in which Admiral Yamamoto was traveling.

Was this a legal killing? International law under the Law of Armed Conflict (LOAC) spells out the conditions for a state's legal use of military violence. That law requires *distinction* (separation of military from civilian targets), *proportionality* (minimal damage to nonmilitary life and property incidental to military operations), *military necessity* (violence is required to achieve the military objective), and *unnecessary suffering* (weapons causing unnecessary suffering are prohibited). LOAC was established in 1947 in response to the excesses of WWII, but McDonald asks: Would the killing of Admiral Yamamoto have been legal under LOAC? His answer is "yes": Japan's military had attacked the U.S. naval station at Pearl Harbor, LOAC allows nations an inherent right of self-defense (here, for the U.S.A. *to declare war on Japan*), and defines the antagonists' military forces (here, members of the Japanese military) as legitimate targets under the above conditions.

What about *targeted killings* in what President Bush called the War on Terror? There are two problems with the latter label. The term "war" in LOAC legally refers to lethal conflict between the militaries of two or more states and, two, the term "terror" refers to a tactic. It makes no legal sense to talk about a *war on a tactic*. However, President Bush explained[37]:

> Our enemy is a radical network of terrorists and every government that supports them. Our war on terror begins with *al-Qaeda*, but it does not end there. It will not end until every terrorist group of global reach has been found, stopped and defeated.

In 2001, the U.S. Congress extended an *authorization* to the President *to use any military force deemed necessary and appropriate* (AUMF) against those responsible for 9/11. This is a slippery slope from a legal point of view, because al-Qaeda is transnational and because President Bush's declaration implies defeating it and every other such group everywhere.

McDonald examines other examples. What of the killing of al-Zarqawi orchestrated by McChrystal's Task Force? Yes, legal, McDonald concludes. Iraq was an officially recognized *zone of war* between the Iraqi and the U.S. militaries, and then, after the collapse of Saddam Hussein's military, *a zone of armed conflict* between the new Iraqi state and the U.S.A. on one side and several armed, paramilitary groups, including AQI, on the other. Al-Zarqawi was a Jordanian citizen, but one who publicly identified himself as the leader of AQI – a group he himself created – and claimed responsibility for numerous car bombings, assassinations, and other incidents of lethal violence in Iraq. And he was killed in Iraq. These qualify him as a legitimate target.

Now a third case, that of Anwar al-Awlaki, an American citizen killed in Yemen on 30 September 2011 by a Central Intelligence Agency (CIA) Pred/Reaper as his car sped down a country road. Al-Awlaki was born in Las Cruces, New Mexico, of Yemeni parents in 1971 and held dual American and Yemeni citizenship. He condemned the 9/11 attack on the World Trade Center but became known in the U.S.A. as a Muslim cleric publicly critical of the American invasion of Iraq. Al-Awlaki returned to Yemen in 2004 to preach at a university and was imprisoned in 2006 by Yemeni authorities without trial, perhaps at the instigation of the U.S.A., for encouraging Muslims everywhere to kill Americans and Yemenis who tolerated their presence. Following his release after 18 months, he openly affiliated himself with al-Qaeda in the Arabian Peninsula (AQAP). In 2010, the Obama Administration, citing the 2001 AUMF, put him on a *find-and-kill* list.

A legal killing? McDonald reviews the debate. Yemen was a *zone of armed conflict* – there was an internal struggle between the Yemeni government and AQAP – and the U.S.A. had a small number of military advisors there on the ground. However, al-Awlaki's status as an American citizen suspected of criminal activities (a legal evidentiary issue) normally carries with it the *right of due process* spelled out in the U.S. Constitution and Bill of Rights. Would not the correct legal procedure then have been to arrest al-Awlaki and subject him to a formal trial? Republican Senator Rand Paul of Kentucky thought so. While agreeing that al-Awlaki was a "despicable human being who wanted to harm the United States," Senator Paul found the legal justification an overreach of authority frightening in its implications: "the Obama administration has established a legal justification [for killing] that applies to every American citizen, whether in Yemen, Germany, or Canada."[38]

On the other hand, the authority cited by the President was not "self-derived" but extended legally by Congress. The 2001 AUMF allowed the President to define "lawful targets," and arguably his intent was to stay within the law. Further, one could plausibly argue that al-Awlaki forfeited his constitutional protections by advocating and supporting attacks on Americans while affiliated with AQAP, thereby shifting his status to that of a *civilian combatant* under LOAC. Still, McDonald concedes, "on a gut level, the idea that a military force might be assessing us on an individual basis is both new and difficult to comprehend. This is especially true since governments make mistakes," doubly worrisome when "acting in relative secrecy as judge, jury, and executioner."[39] That the killing was carried out by the CIA added to the controversy. The rules for initiating and justifying such a kill are not always transparent, and those of the CIA especially tend to be enshrouded.

Targeting individuals for execution has not been restricted solely to top al-Qaeda leaders. The search for intelligence required to expose and track terrorist networks uncovered many mid- and lower-level personnel whose activities also were seen to merit adding them to a hit-list too. Though Pred-Reapers have a loitering capacity that makes it possible to eliminate *collateral damage* (the incidental killing of civilian bystanders), targeted killings also have sometimes challenged LOAC's principle of *distinction* when "target-rich" opportunities have presented themselves. For example, a Pred-Reaper strike in 2013 was carried out on a wedding convoy near Radda, Yemen. Intended to kill a mid-level AQAP operative, Shawqi Ali Ahmad al-Badani, and several AQAP associates, it also took the lives of eight to sixteen *civilian noncombatants* as well.[40]

We can see from this it has been a short step from *targeted killings* to *signature strikes*, that is, lethal hits on "unknown" individuals whose *pattern of life* activities sufficiently matched those of "known" terrorists in past encounters (hence the title of McDonald's book). This practice amounts to a form of *pre-crime justice*, since persons not previously known to be terrorists might be targeted in anticipation of their becoming known as terrorists. The problem is these considerations apply most aptly to *obvious warfare* with militaries wearing different-looking uniforms. The U.S.A.'s pattern of life calculations, in current non-obvious wars, are derived from statistical analyses of "big data."

These enormous datasets, available and used for both commercial and military purposes, are made possible by the routine collection of millions of bits of personal data from internet, social media, and smart phone technologies. If these truly are valid, McDonald notes, *signature strikes* are in principle like any other use of lethal force (putting aside the issue of using them pre-emptively). The caveat is, we are talking about civilians. Such strikes, McDonald concludes, "draw attention to the uncertain nature … [of] secret judgements that render [nonmilitary] people vulnerable to death."[41]

Cyber and Information Warfare

A hundred years ago, sociologist W. F. Ogburn noted (and lamented) that *material culture* and its technologies change more rapidly than do the *social norms* for adjusting to the consequences.[42] New Wars provide apt illustrations of Ogburn's *cultural lag theory*, perhaps nowhere more startling than in the areas of Cyber and Information Warfare. The idea of sabotaging an enemies' communication network or using propaganda to dispirit a foe is not new. What is new is the domain and reach made possible by a highly computerized, interconnected world.

For starters, consider the meaning of the word *cyberspace* – a word for which people 30 years ago would have had no reference. Israeli security analyst Lior Tabansky provides a basic definition: "all the computerized networks of the world, as well as all the endpoints that are connected to [them] and are controlled through commands that pass through these networks."[43] Note that cyberspace is not only the largest global network (the *internet*), but also includes other networks *outside the internet* that are not supposed to be accessible to others. Some of these are *private* (certain companies and countries have their own restricted-use networks), while others are *transactional* (like ones used to do online banking or stock-market trading), or *automated/control systems* (like ones that regulate the operation of machinery such as electric-power grids).

What makes *cyberwar* possible is the penetration of these networks by *hackers* – those not having authorized access – to control, subvert, or crash their functions.[44] American cybersecurity specialist Robert Clarke gives two recent examples. On the night of 6 September 2007, the calm over Syria along the Turkish border was shattered. An Israeli Air Force jet strike-force destroyed in a matter of seconds a "facility" being constructed in remote southeastern Syria by North Korea. Here is the cyberwar part: in the minutes before and during the bombing, Syria's highly sensitive, Russian-built radar system had shown blank screens, nothing in the area, even as Israel's attack formation approached and did its thing. There are several ways the radar's computer system could have been compromised and tricked into seeing "nothing." No one, Israeli, Syrian, or Russian, is admitting to this – but informal consensus is that Israeli hackers paved the way for a successful strike.[45]

A second example shows a variation, this time in the city of Tallinn, Estonia. Estonia celebrated its present-day independence with the collapse of the Soviet Union in 1989. During its reign, the Soviets had erected in Tallinn a giant bronze statue of a Red Army soldier over a mass grave of Soviet soldiers who had died "liberating" Estonia. In February of 2007, nationalist Estonians sought to remove the offending statute from their midst. A smaller Russian ethnic minority of Estonians objected. In the standoff on the evening of 27 April, known as Bronze Night, the statue was moved to another less visible part of the cemetery.

Now, the cyberwar part. Estonia is one of the most "wired" of nations in its use of the internet, more so than even the U.S.A. On the day after Bronze Night, the servers of Estonia's most used internet sites were jammed by hundreds of thousands of "pings," bringing all online transactions to a halt in an attack that lasted several weeks. A *botnet*, a robotic network, had activated thousands of computers that had been contaminated in the weeks or months before with a hidden attack message, waiting only to be activated and directed at some crucial moment. The country had been incapacitated and brought to its knees without any military action. Estonia later traced the origin of the controlling botnet to Russia and hackers using Cyrillic-alphabet keyboards.

Another variation of cyberwar that has been observed in several countries is *information warfare*, the use of the internet to sow *disinformation*. The most common techniques rely on *trolls* (human instigators) and *bots* (computers), to saturate cyberspace, in the words of Brookings Institution analyst Ben Nimmo, with the four D's: *distort* the facts, *distract* the main issue, *dismiss* the critics, and *dismay* the audience. Prime examples include the Russian campaigns to cover its invasions of Crimea and eastern Ukraine in 2014, and to sway the 2016 Brexit election in the U.K., the 2016 Presidential election in the U.S.A., and the 2017 French Presidential election.[46]

Information warfare on the internet can be especially effective because it capitalizes on some of the vulnerabilities in how humans easily "know" the world. Social media platforms are designed to steer viewers toward sites or content that conform to what they already know or like to believe. This is done through algorithms that track responses to *clickbaits*, that is, photos or headlines that serve as teasers enticing the viewer to click on the link to a particular webpage. The effect of this is *homophily* and *confirmation bias*, the funneling of like-minded viewers to sites presenting information they already agree with or are open to accept. If not careful to diversify one's sources of information, a viewer can become susceptible to *fact-deprived messages*. Skilled trolls and bots can spin shady marketing schemes and false political narratives. Foreign trolls often use local social media accounts, and bots can be used to give a sense of widespread support for a message.

P. W. Singer (mentioned in the previous section) recounts a case from the U.S. 2016 Presidential election in his analysis of cyber "like-war."[47] On 4 December 2016 (a month after the election) 28-year-old Edgar Welch entered a Washington, D.C. establishment, Comet Ping-Pong Pizza (so-called because the pizzeria has ping-pong tables for its diners' amusement). Armed with a Colt AR-15 assault rifle, he had driven from his home in North Carolina to save children he believed were held there in the basement for abuse by Hilary Clinton and others of the Democratic elite. After firing shots and thoroughly casing the restaurant (amid terrified customers), he surrendered to police, admitting later, "intel on this wasn't 100%." He was sentenced to four years in prison for the incident.

Here is the information war part. The hoax had originated on a Twitter account known for white supremacy material in the final days of the election with pictures of children at the pizzeria, actually a fund-raiser for St. Jude's Hospital (which specializes in the treatment of childhood cancer). The story gained life on conspiracy-theorist Alex Jones's InfoWars YouTube channel with its 2 million subscribers. Russian trolls picked up the story, and their bots gave it further exposure. The issue generated 1.4 million mentions on Twitter, and Josh Posobiec, a naval reserve officer, pushed the story to his 100,000 online followers. The case indicates a sizable audience for this kind of tale, and a marvelous infrastructure manipulated by a small number of social media accounts.

We turn now to the case of Russian information warfare.

Spotlight on Social Science Thinking and Research

Keir Giles, *Handbook of Russian Information Warfare*[48]

Keir Giles is a respected expert on security issues in Russia and Eastern Europe and on armed forces in the Russian Federation. He has been a Senior Consulting Fellow at the U.K.'s Royal Institute of International Affairs and Fellow at the Conflict Studies Research Centre, formerly affiliated with the Ministry of Defence. He was asked to write the *Handbook* to provide a basic orientation for NATO officers working in information-warfare and influence operations. He derived material for the book from non-classified sources, which he says, lend sufficient credibility but do limit its scope. Most of those sources were in Russian, which he translated for the book himself.

Giles states that much of the strategic thinking and tactics in the source material are said by their Russian authors to be ones used against Russia itself by the West. If so, these still represent what has become Russia's policy. However, this is not the only basis. Rather, Giles argues, the current policy reflects an updated version of enduring principles for managing interstate relations dating back to the Soviet Union. This is not surprising since the renewal of this thinking, set aside in the first years of *glasnost*, has occurred under the direction of Vladimir Putin, a former Komitet Gosudarstvennoy Bezopasnosti officer (KGB, in English, Committee for State Security) from 1975 to 1991.

The first central difference to notice between the Russian approach and that of the West is the *timing of information warfare* (IW). In Russian strategic thinking, IW is something that is always carried on in relations with other countries, not just in time of conflict or its initial phases. Further, the range of IW activities and methods of implementation is wide and runs the full gambit of options. These include operations directed not only at foreign militaries and their populations, but at the Russian people as well, as indicated by incorporating within IW the Russian state's presentation of the news and current affairs on the state-run media, *Russia Today* (RT) and *Sputnik*. In contrast, government-directed IW and influence operations are not permitted legally in peacetime by Western nations.[49]

As such, terms of "cyber-something" do not appear in the Russian lexicon, except to refer to Western methods, and there is no formal equivalent of the 24th U.S. Air Force Cyber or the U.K.'s 77th Security Assistance Brigade. Thus, the key word, Giles states, is *information*. In peacetime, it is manipulated covertly, in time of war, more overtly. Since 2014, Russia's IW essentially has been one of *distributing disinformation*. Giles's analysis of documents reveals the Russian belief that today's "means of information influence" provide an *asymmetric, nonkinetic avenue* to achieve strategic political and social goals, including the defeat of an enemy's armed forces. Russian Chief of Staff General Valery Gerasimov summarizes (in a document translated by Giles)[50]:

> Of great importance here is the use of the global internet network to exert a massive, dedicated impact on the consciousness of the citizens of states that are [our] targets ... The extensive deployment [of information resources] *enables the situation in a country to be destabilized from within.*
>
> (emphasis added)

As such, Russian operatives send polluted information especially attractive to the dissatisfied at either end of an adversarial nation's political spectrum, including that nation's own disruptive social media pundits (referred to in another context as *useful idiots*) who, for their own reasons, are in the business of spreading similar themes. In any case, the purpose is to *amplify dissension* with the goal of *undermining institutions* and *limiting viable policy options* in Russia's favor.[51]

Ironically, Russia is a latecomer to accepting the internet. Its leaders, like those of China and North Korea, were extremely suspicious of giving their own citizens unimpeded access to a permissively open and free communications network, a problem each has solved by creating their own internal networks and approved software. However, inefficiencies observed by Russian leaders in the 2008 armed conflict in Georgia led to the formation of an Information Troops unit within the Russian military, consisting of hackers, linguists, and specialists in psychological warfare. On the civilian side, extensive funding provided specialized educational and technical training to grow a legion of *state-supported troll farms* and *botnets. Cyber criminals* operating outside Kremlin control are given leeway if they prove occasionally useful.[52] As a result, Russia's reputation for now having world-class cyber capabilities for the *full-spectrum of IW*, Giles asserts, is well deserved.

Like most cyber-security analysts, Giles warns that cyber- and information-warfare are already here and about to take a quantum leap with another technology, *artificial intelligence*. Ignore this, he states, at your own risk.

Questions for Discussion

1   During the First and Second Intifadas, both the Israelis and Palestinians apparently believed that keeping informed of public opinion and then tailoring their actions to it was a strategic and tactical necessity. Using the lens of New Wars, why was this so important?

2   Eyal Ben-Ari and associates recount features of what they see as New Wars and even provide new concepts for accounting for them. Review two or three of the most salient of these. With reference to these, are you persuaded there is something fundamentally new about 21st-century wars?

3   "Big wars" are known for their extreme death tolls, aka multicides. Now with New Wars, we have individuated warfare with targeted killing. What are the ongoing ethical concerns for civilians with this capability? Would you argue this creates new, pressing ethical questions? Explain.

4   How would you summarize Russia's strategic thinking on information warfare? Do you think this is of special concern, especially in light of Russia's reputation for world-class cyber capabilities? Explain.

Notes

1   Mary Kaldor, 1999, *New and Old Wars: Organized Violence in a Global Era*, Palo Alto, Calif.: Stanford University Press. See also Mary Kaldor and Basker Vashee (eds.), 1997, *New Wars*, London: Pinter Publishers; and Kaldor, Ulrich Albrect, and Geneviève Schméder (eds.), 1998, *The End of Military Fordism*, London: Pinter Publishers. Present material is from Kaldor, 2013, "In Defence of New Wars," *International Journal of Security & Development* (March), downloaded from http://www.stabilityjournal.org/articles/10.5334/sta.at

2   A similar account of these early days of the American response to 9/11 appears in Wilbur J. Scott, 2020, "Organizational Adaptations in the Hunt for Abu Musab al-Zarqawi," pp. 45–63 in Thomas Vladimir Brond, Uzi Ben-Shalom, and Eyal Ben-Ari et al. (eds.), *Military Mission Formations and Hybrid Wars: New Sociological Perspectives*, London and New York: Routledge.

3   Harlan K. Ullman and James P. Wade, 1996, *Shock and Awe: Achieving Rapid Dominance*, Washington, D.C.: National Defense University.

4   Sean Loughlin, 2003, "Rumsfeld on Looting in Iraq: 'Stuff Happens,'" CNN.com/U.S., War in Iraq, 12 April, downloaded from http://www.cnn.com/2003/US/04/11/sprj.irq.pentagon/.

5 Ali A. Allawi, 2007: *The Occupation of Iraq: Winning the War, Losing the Peace*, New Haven, Conn. and London: Yale University Press, pp. 1135–1138; General Stanley McChrystal, 2014, *My Share of the Task: A Memoir*, New York: Penguin Group, pp. 121–122.

6 Allawi, op. cit., p. 240.

7 Joby Warrick, 2015, *Black Flags: The Rise of ISIS*, New York: Doubleday.

8 McChrystal, op. cit., pp. 112–113.

9 Thomas E. Ricks, 2006, *Fiasco: The American Military Adventure in Iraq*, New York: Penguin Press.

10 Wilbur J. Scott, David R. McCone, and George R. Mastroianni, 2009, "The Deployment Experiences of Ft. Carson's Soldiers in Iraq: Thinking About and Training for Full-Spectrum Warfare," *Armed Forces & Society*, 35 (April): 460–476.

11 Cf., George Packer, 2006, "The Lesson of Tal Afar: Is It Too Late for the Administration to Correct Its Course in Iraq?", *The New Yorker*, 10 April; Major General David H. Petraeus, 2006, "Learning Counterinsurgency: Observations from Soldiering in Iraq," *Military Review*, January–February: 2–12.

12 General David H. Petraeus (U.S. Army) and Lt. General James F. Amos (U.S. Marine Corps), 2007, *The U.S. Army, Marine Corps Counterinsurgency Field Manual*, Chicago and London: University of Chicago Press.

13 General McChyrstal was dismissed for insubordination in 2010 by President Barack Obama when McChrystal was commander of Multi-National Forces in Afghanistan. In April of that year, he and his staff had entertained a reporter from *Rolling Stone* magazine for several days, which included bouts of drinking together. The resulting story cited derogatory remarks by them about the commander-in-chief, vice president, the president's national security advisor, and the ambassador to Afghanistan. McChrystal admitted he had "messed up." At his retirement ceremony, Secretary of Defense Robert Gates assured those gathered that McChrystal left "with his place secure as one of America's greatest warriors." Cf. Elizabeth Bumiller, 2010, "McChrystal Ends His Service with Regret and a Laugh," *The New York Times*, 23 July, downloaded fromhttps://www.nytimes.com/2010/07/24/us/24mcchrystal.html.

14 General Stanley McChrystal, with Tatum Collins, David Silverman, and Chris Fussell, 2015, *Team of Teams: New Rules of Engagement for a Complex World*, New York: Penguin Random House.

15 Ibid., pp. 80–82.

16 Ibid., p. 242.

17 Michael Totten, 2007, "Anbar Awakens, Part I: The Battle for Ramadi," *Michael J. Totten's Middle East Journal*, 10 September, available at https://www.michaeltotten.com/archives/001514.html.

18 Richard Davis, 2016, *Hamas, Popular Support and War in the Middle East: Insurgency in the Holy Land*, London and New York: Routledge.

19 Eyal Ben-Ari, Zeev Lerer, Uzi Ben-Shalom, and Ariel Vainer, 2010, *Rethinking Contemporary Warfare: A Sociological View of the Al-Aqsa Intifada*, Albany, New York: State University of New York Press.

20 Davis, op. cit., pp. 6–10.

21 Ibid., pp. 35–36.

22 Ibid., pp. 44–46.

23 James Ron, 2000, "Savage Restraint: Israel, Palestine and the Dialectics of Legal Repression," *Social Problems*, 47 (4): 445–472.

24 Davis, op. cit., pp. 10–13.

25 Ibid., pp. 95–96.

26 Ben-Ari et al., op. cit.

27 Ibid., pp 17–18.

28 Ibid., p 33.

29 Ibid., pp 53–61, 152. They derive *global surveillance* as a characteristic of 21st-century wars from military sociologist Martin Shaw, 2005, *The New Western Way of War*, London: Polity Press.

30 Ibid., pp. 165–168.

31 Ibid., pp. 72, 79–84.

32 The U.S. military eschews the term "drones." The Army prefers the label Unmanned Aerial Vehicles (UAVs) and the Air Force calls them Remotely Piloted Aircraft (RPAs).

33 Mark R. Rose, Richard D. Arnold, and William R. Howse, 2013, "Unmanned Aircraft Systems Selection Practices: Current Research and Future Directions," *Military Psychology* 25: 413–427, p. 413.

34 In 2001, the military added two laser-guided AGM-114 Hellfire Missiles to the Predator. The term "Hellfire" originated as an acronym for "Helicopter-Launched, Fire-and-Forget Missile." The Reaper carries four Hellfire Missiles.

35 Jack McDonald, 2017, *Enemies Known and Unknown: Targeted Killings in America's Transnational War*, Oxford and New York: Oxford University Press.

36 Ibid., pp. 16–20.

37 Ibid., p. 38.

38 Ibid., p. 65.

39 Ibid., p. 187.

40 Ibid., pp. 112–113.

41 Ibid., pp. 190–191.

42 W.F. Ogburn, [1922] 1964, *On Culture and Social Change*, Chicago: University of Chicago Press.

43 Lior Tabansky, 2011, "Basic Concepts in Cyberwarfare," *Military and Strategic Affairs*, 3 (May): 75–92, p. 77.

44 Richard A. Clarke and Robert K. Knake, 2010, *Cyber War*, New York: HarperCollins, pp, 69–70.

45 Ibid., pp. 1–6.

46 For an analysis of Russian tactics in the Ukraine, see Holger Mölder and Vladimir Sazanov, 2018, "Information Warfare as the Hobbesian Concept of Modern Times – Russian Information Operations in the Donbass," *Journal of Slavic Military Studies*, 31 (3): 308–328; for analysis of Russian tactics in the 2016 Presidential election, see Darren L. Linville et al., 2019, "THE RUSSIANS ARE HACKING MY BRAIN!, Investigating the *Russian Internet Research Agency* Twitter Tactics during the 2016 United States Presidential Election," *Computers in Human Behavior*, 99: 292–300, and Adam Badawy et al., 2018, "Characterizing the *Russian Internet Research Agency* Influence Campaign," *Social Network Analysis and Mining*, available for download at 1812.01997.pdf (arxiv.org).

47 P.W. Singer and Emerson T. Brooking, 2018, *LikeWar: The Weaponization of Social Media*, Boston and New York: Houghton Mifflin Harcourt, pp. 127–130.

48 Keir Giles, 2016, *Handbook of Russian Information Warfare*, Rome, Italy: NATO Defense College Fellowship Monograph.

49 Ibid., Part 2.

50 Ibid., p. 18.

51 Ibid., pp. 36–41.

52 Ibid., pp. 49–57.

Recommendations for Additional Reading

Martin C. Libicki, 2016, *Cyberspace in Peace and War*, 2nd ed., Annapolis, Md.: Naval Institute Press.

(Comprehensive introduction to cyber-warfare and cyber-security by a noted expert and Rand Corporation scientist)

Lt Colonel Wayne Phelps, US Army, with Lt Colonel Dave Grossman, US Army, 2021, *On Killing Remotely: The Psychology of Killing with Drones*, New York: Little, Brown, and Company.

(Description and analysis of drone – unmanned aircraft systems – warfare and the psychological adaptations required of those operating these systems)

Paul Scharre, 2018, *Army of None: Autonomous Weapons and the Future of War*, New York: W.W. Norton.

(Pentagon security analyst describes the future warfare, already here in nascent form, of autonomous "killer robots")

Wilbur J. Scott, David R. McCone, and George R. Mastroianni. 2009, "The Deployment Experiences of Fort Carson's Soldiers in Iraq: Thinking and Training for Full-Spectrum

Warfare." *Armed Forces & Society*, 35 (April): 460–476; Anthony King, 2022, "Why Did the Taliban Win?" *Armed Forces & Society*, June, https://doi.org/10.1177/0095327X221096702.
(Two articles to help understand the U.S. military's difficulties in Iraq and Afghanistan)

P.W. Singer, 2009, *Wired for War: The Robotics Revolution and Conflict in the 21st Century*, New York: The Penguin Press; and P.W. Singer and Emerson T. Brooking, 2018, *Like War: The Weaponization of Social Media*, Boston: An Eamon Dolan Book, Houghton Mifflin Harcourt.
(Two books at the center of New Wars by an analyst known for his insightful writings on 21st-century warfare)

11 Military Families

Reader's Guide: Civilian family patterns changed mightily in the second half of the 20th century, one of which is the entry of many married women into paid careers. These patterns have affected military families too, along with other factors unique to them, especially expected absences by the military member risking injury or death. Our first Spotlight details two studies focusing on the intersection of the military and family as "greedy" institutions, a greediness compounded by working military spouses when the military still prefers the "two-person, one-career" model. We review adaptations to all this by military couples, and by the military itself. The second Spotlight contains a study of couples in the Israeli Defense Forces, an interesting case because Israel is the only country in which both men and women are subject to conscription and subsequent service in the reserves. We conclude with two studies, the first assessing the effects of deployments on Dutch children of military families, the second on painful adaptations by adult children of U.S. military Missing-in-Action from the war in Vietnam.

The *family* is a fundamental *social institution* in all societies. At a minimum, for a society to exist over time, two basic events must somehow occur conjointly: adults defined as related to each other, usually by marriage or kinship, must derive the essentials of daily living – food, clothing, and shelter – *and*, from a societal point of view, sexual activity must be regulated to produce enough babies, for whose care parents and related kin are responsible. Families optimally might also be a key source of emotional and social support for their members.

Despite this generic mandate, there is a great deal of *diversity in how societies have done and now do family*. This is due to two sets of contingencies. The first is demographic: unless willing to replace its numbers largely by acquiring members from other societies, a society's *birth rate* must be high enough to comfortably offset its *death rate*. (See our related discussions in Chapters 3 and 5.) Throughout most of human history, that is, until the advent of industrialization in the late 18th century in England and Holland, death rates in most all human societies were quite high, in standardized terms, around 50 deaths a year for every 1,000 people in a society's population. This rate varied in response to several factors, but important for our purposes is the pressure that high death rates create for producing birth rates high enough to compensate for them. A comfortable margin is around 5 more births per 1,000 each year than the number of yearly deaths per 1,000.

DOI: 10.4324/9781003282549-11

A birthrate of 55 per 1,000 population requires sexual activity and family life oriented toward producing an average of eight to nine live births per woman. This calls for a substantial amount of *gender polarization*, that is, the organization of gender roles, within both the family and the society in which it is located, to *emphasize childbearing*. With industrialization and ensuing declines in the death rate (due to advances in disease prevention and treatment), there has been a consequent decrease in the number of births necessary for societal survival. For instance, most industrial societies now have death rates of around 10 deaths per 1,000, and a birth rate required to offset this number, 15 per 1,000, can be attained with one to two live births per woman. Thus, industrial societies have had the option of reorganizing gender roles and family structures accordingly.

The second contingency is the interplay among social institutions themselves, as the substance of any particular social institution constrains and shapes the possibilities within others. Thus, the economic structure or dominant religious fabric of a society will impact the structure and purpose of its families, and families will also have an influence on these institutions. For example, as noted in our discussion of social institutions in Chapter 2, *nonindustrial societies* and *extended family structures* go hand in hand, that is, the demands of hunting/gathering and horticulture/herding favor family arrangements that include elaborate kinship networks and lots of babies. Agrarian families typically include other kin living together in the same compound who, in addition to taking part in economically productive work, fulfill essential tasks like childrearing and elder care. And, since youngsters work within the family enterprise at an early age and are considered adults in their early teenage years, having babies contributes child workers who are valuable sources of labor.

By contrast, a *nuclear family arrangement*, consisting of parents and offspring living for the most part independently of extended kin, is more compatible when economic production moves from agriculture to industry. Now, an extended family is no longer the basic unit producing the essentials – factories, then businesses and corporations are. One of the parents, typically the father in a heterosexual marriage-, leaves home daily to go to work elsewhere, while the mother maintains the household and cares for children, the so-called *breadwinner model*. Restricting the family structure to the conjugal unit favors serial relocations for economic advantage. Further, having babies now means introducing *consumers* into the family, not workers, often until they are well beyond their teenage years. This expense provides a stout incentive for having fewer children.

Recent Changes in the Civilian Family

Not only are birth rates declining, but marriage is decreasing as well in industrialized countries. To produce high birth rates, young people in nonindustrial societies must marry at a very early age, the age of 16 would not be too early, to provide married women enough time before menopause to have a sufficient number of babies. As the prerequisites and payoffs of an industrializing society invade the calculus for having children, the *age at first marriage* begins to rise, first to the late teens, then into the early twenties, and, on to the late twenties.[1]

This seemingly innocuous change has important repercussions. For starters, there is a widening number of years between *sexual maturity*, something that normally occurs in the early to mid-teenage years, and age at first marriage. Requiring celibacy before marriage when the gap between the two is a decade or so becomes problematic. Hence, this trend triggers ever more permissive attitudes toward *pre-marital sex*, and surveys of married couples since 1950 have shown a steady increase in the percentage who partake in such.[2] Although celibacy before marriage was the ideal in 1950, still about a third of married couples said they very discreetly enjoyed some form of sexual intimacy with each other before marriage, usually something short of intercourse. The most recent surveys show that percentage now exceeds 90%, and typically includes intercourse.

Another consequence of declines in the birth rate and delayed age at first marriage is an expansion of options available for women outside of childbearing and childrearing, evidenced most vividly by married women in the paid labor market. Before 1950, young single women who entered the labor market typically quit their jobs when they got married. Those who did not were those who had little or no choice, usually poorer married women, or those choosing to have a career, often as a nurse or schoolteacher, and who therefore often remained unmarried over the life course. But, starting in about 1950, married women increasingly remained in the labor force after marriage until the birth of the first child. However, by the 1970s, many employed married women continued to work after the birth of children, a family pattern known as the *dual-earner model*.

Table 11.1 shows the *employment status of women and men* in the U.S.A. for the year 2020. It reveals that over half of women in the U.S.A. over the age of 16 are in the paid labor force, regardless of marital status (never married: 56.8%, married: 53.8%, and separated or divorced: 55.1%). The percentages are similar for men over the age of 16, except that more than two-thirds of married men currently work for pay (67.9%). This includes

Table 11.1 Employment Status[a] by Gender, Marital Status, and Education for the Non-Institutionalized U.S. Adult Population,[b] March 2020 (%)[c]

	Working for Pay (%)	Seeking Work (%)	Not Working for Pay (%)		Working for Pay (%)	Seeking Work (%)	Not Working for Pay (%)
Women age 16+				**Women** age 25–64			
Never married	56.8	11.4	31.8	< High school	39.4	13.8	45.8
Married	53.8	6.3	39.9	High school	56.0	9.8	34.2
Sep./divorced	55.1	8.6	36.3	Some college/voc	66.0	8.2	25.8
				BA/BS or +	76.2	5.0	18.8
Men age 16+				**Men** age 25–64			
Never married	58.2	12.3	29.5	< High school	64.0	10.6	25.4
Married	67.9	4.9	27.2	High school	72.2	8.6	19.2
Sep./divorced	60.3	8.4	31.3	Some college/voc	77.3	7.3	15.4
				BA/BS or +	86.6	5.4	8.0

a Employment categories are: Employed in Labor Force (Working for Pay), Unemployed (Seeking Work), Not in Labor Force (Not Working for Pay). Percentages sum to 100 across rows.

b Non-institutionalized adult population is made up of persons 16 years of age or older living in any of the 50 U.S. states or the District of Columbia who are not in institutions, such as nursing homes or prisons, and are not on active duty in the U.S. military.

c Table 11.1 constructed from U.S. Bureau of Labor Statistics (BLS), 2000, March, *Women in the Labor Force: A Databook*, Tables 4 and 8, downloaded from https://www.bls.gov/opub/reports/womens-databook/2021/home.htm. The U.S. Bureau of the Census conducts regularly national probability samples of 60,000 U.S. households for BLS to estimate labor force participation and unemployment.

those in both full- and part-time work, and the percentages of women employed part-time is slightly lower than for men.

Another key trend is the linear relationship between education and labor force participation, that is, the more years of schooling women or men complete, the more likely they are to be in today's labor force. Among women aged 25 to 64, 39.4% with less than a high school education are in the labor force. That percentage rises to 56.0 for high school graduates, 66.0 for those who complete some college coursework or attend vocational school, and 76.2 among those who completed baccalaureate or graduate degrees. A similar but more elevated pattern exists for men. This represents both a changing labor market where positions require more education to qualify for job entry, and an incentive on the part of individual women and men to capitalize on educational attainment when more is invested in it. Still, in part because of how they are distributed in occupational categories, and in part because of lingering gender discrimination, women, regardless of marital status or educational attainment, earn less than men.

Educational attainment also has become a factor in the likelihood to marry. Figure 11.1 depicts the percentage of women since 1950 who are married for each of three educational groups, those who did not complete high school, those who completed high school or some college, and those who are college graduates. In 1950, the percentage of women who were married in these three groups was nearly identical, at about 60. Since then, these three groups have gone separate ways in terms of marriage. The percentage of women who have not completed high school that marry has dropped steadily over the last 65 years to 27, and for the interim education group, the percentage who marry has declined steadily since 1970 to 45. Meanwhile, the percentage of college-educated women who marry has held steady, registering at 59 in 2016.

Related to education as well is the likelihood that a marriage will end in divorce. The incidence of divorce in the U.S.A. rose most dramatically between 1970 and 2000: it was three times higher in 1980 than it was in 1950. Since 2000, both marriage and divorce rates have declined slightly.[3] In 2020, the marriage rate in the U.S.A. – the number of marriages per 1,000 people in the population – was 5.1 per 1,000. The rate of divorce, which is reflective of marriages from previous years coming undone, in 2020 stood at 2.3 per 1,000. This would mean, if the two rates held steady over time, almost one-half of the

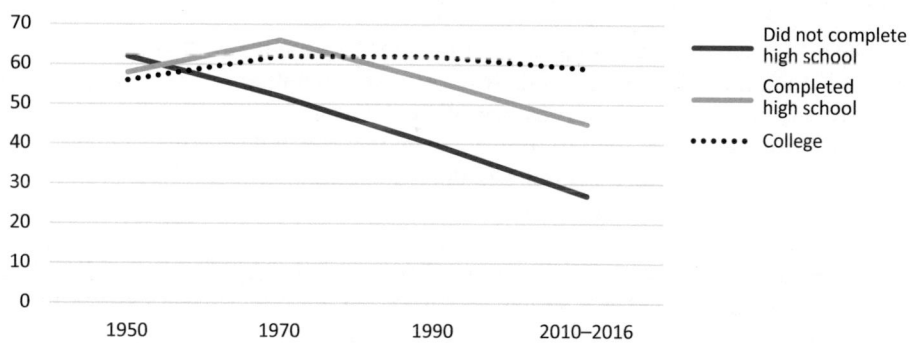

Figure 11.1 Percentage of U.S. Married Women by Educational Attainment, U.S.A., 1950–2016

Source: Authors' Creation.
Note: Adapted from Colette Allred, 2018, "Variation in Percentage Married According to Educational Attainment, 1940–2016," in *Marriage: More than a Century of Change, 1900–2016*, Family Profiles, FP-18-17, Bowling Green, Ohio: Center for Family & Research Marriage, downloaded from https://doi.org/10.25035/ncfmr/fp-18-17.

2020 marriages would end in divorce (the average length of time between marriage and divorce is eight years). Most likely to end in divorce are marriages among persons under the age of 25 and for those with less years of education.

Studies show a similarly patterned increase in *cohabitation*, defined as unmarried couples living together in an emotionally and sexually intimate relationship. Cohabitation used to be the realm of the poor and the rebellious. The term "living in sin" expressed the moral judgment directed at couples who chose to live together outside of marriage. However, the rates of cohabitation have increased greatly over the last 50 years. As Figure 11.1 might suggest, there are class differences at play here. Middle- to upper-class couples are more likely to cohabit as a *prelude to marriage*, while working- and lower-class couples increasingly are using it as a *replacement for marriage*.[4]

Unsurprisingly, cohabiting households often include children, either those born to the cohabiting couple or those coming into such an arrangement, usually when the biological mother and a male partner move in together. Family researcher Wendy Manning estimates that 40% of children in the U.S.A. will have lived in a cohabiting household by the age of 12, and it is now the norm for women without college degrees to have children outside of marriage rather than within it.[5]

Studies of unmarried low-income women with children show these women do view marriage as a noble institution that, in principle, they wish they could join.[6] However, many grew up in poverty, and their views of marriage are marred by the economic and hence marital instabilities of their elders. Many of these women thus find it important to establish their own economic independence, a tough requirement when living in poorer neighborhoods with limited opportunities. Still, they consider children a necessity whose presence adds status and a powerful identity: *mother*.

Finally, up until 2015, *same-sex marriage* was illegal in parts of the United States due to the influence of religious institutions. Change began in 2011 with the repeal of "Don't Ask, Don't Tell" in the U.S. military (discussed in Chapter 7), followed by the 2015 Supreme Court decision, *Obergefell vs. Hodges*, which legalized same-sex marriage nationally. In 2019, the Census Bureau estimated that approximately 1%, or around 1,012,000, of U.S. households were same-sex families.[7]

Military Families and the All-Volunteer Force

We have spent some time reviewing civilian families because many of the trends among them apply to military families as well. Still, being in the military incurs unique circumstances. Just as the 1970s were a critical period of change for civilian families, that decade also marked big changes in military policy for its families. Before the 1970s, the U.S.A. made do with a small professional military augmented in time of war by a draft (see our discussion in Chapter 5). Accommodations by the military for spouses and children largely were restricted to officers and senior non-commissioned officers (NCOs), and the breadwinner model was the taken-for-granted order of the day. Drafted enlisted men were single or simply treated as if single. For them, the informal maxim was, "if the military wanted you to have a wife, it would have issued you one."[8]

This posture changed with the advent of the All-Volunteer Force (AVF) in 1973. The military now had to compete with civilian employers by extending more competitive pay and benefits, including those for the service member's family. And, studies soon revealed a prime factor affecting the likelihood of re-enlistment: spousal and family satisfaction with military life.[9] For insight into the dynamics of military families in the early AVF, we turn our Spotlight now to one of the all-time classics in military sociology, and a follow-on analysis of it 30 years later.

Spotlight on Sociological Thinking and Research

Mady Wechsler Segal, "The Military and the Family as Greedy Institutions,"[10] and Karin De Angelis and Mady Wechsler Segal, "Transitions in the Military and the Family as Greedy Institutions"[11]

Mady Wechsler Segal, a renowned sociologist in her own right, also is the spouse of David Segal (see our introduction of them in Chapter 1 and the Spotlight on David Segal's work in Chapter 5). They both were born in New York City, by coincidence in the same hospital, he in 1941 and she four years later. She obtained undergraduate degrees at Queens College of the City University of New York and continued her studies at the University of Chicago where they met and married.[12] Both completed doctorates in Sociology under the tutelage of Morris Janowitz. The Segals first joint employment was at the Army Research Institute in 1973, where they studied the shift to an AVF and its viability as a combat force. Eventually they settled into the Sociology Department at the University of Maryland, where they established a premier program in Military Sociology.

It was at the University of Maryland that Segal established herself as a foremost social science authority on military families. Prior to the "Greedy Institutions" article in 1986, her research had focused on the social psychology of small group processes. However, "Greedy Institutions" shifted her research trajectory to the study of the military family, and the Department of Defense routinely sought her expertise to guide relevant policy decisions. The article's importance lies in its clever examination of the intersection of two major social institutions, and its exploration of the consequences for service members and their families.

Segal derived the concept of *greedy institutions* from one of well-known sociologist Lewis Coser's lesser-known works published in 1974 with the same title.[13] Coser had observed that social institutions – the family, the economy, the polity, etc. – require high levels of commitment by society's members, but, since individuals participate simultaneously in multiple institutions, members must find ways to juggle or limit their competing demands. Laws and informal norms can facilitate this. The 40-hour work week in industrial societies for instance is a cultural and legal expectation for many civilian jobs, a provision that makes it difficult for employers to impose on their hourly-workers' time without their consent and overtime pay.

The military does not follow the same restrictions. Long days and weekend work are typical, to say nothing of demands on service members in the event of a national defense emergency, all of which are expected to override non-military obligations, making it an especially greedy institution. Segal, however, argues that what really makes the military unique is the *pattern of demands* it makes on its members and, by extension, on their families.[14] These include: the *risk of injury or death*, frequent *geographical relocations*, frequent *separations*, occasional *residence in foreign countries*, and *normative constraints* on spouses and children. Although other occupations experience some of these demands – police officers, for example, routinely incur risk of injury or death, and the spouses of religious leaders often experience normative constraints on their behavior – it is the combination of these demands that is unique to the military.

The family too is greedy. While the all-encompassing nature of the family in nonindustrial societies is especially greedy, nuclear families can be greedy as well. Here, the demands of childrearing and household fall on two parents without extended kin on the premises. This customarily has been resolved by dividing up the greediness unequally between the parents. In the *breadwinner* model, being wife and mother becomes the woman's full-time, unpaid job, and, in the *dual-earner* model, working mothers typically retain a disproportionate share of parental and household work.

So, what happens when greedy institutions intersect? In 1978, Segal notes, about 60% of all military personnel were married and the percentage of military married males was higher than among civilians for every age group. Most military marriages were between a *military man and a civilian wife* in the breadwinner model – in military parlance, the *two-person, one-career* model. This intersection was quite greedy for military wives, who were expected to take up any slack created by distracted or absent military husbands, and to perform volunteer work for the military in support of their husband's career. For its part, the military, greedy anyway, was more so during overseas operations – in those days, short-lived regional conflicts or positionings – and command tours or unit duties for officers and senior NCOs. These times created impositions for both spouses at critical life stages, such as the birth of a child, or key events, such as Christmas or a child's birthday.

However, notable changes in the military family were beginning to take place. With a greater acceptance of women working in the paid labor market, military wives became more likely to seek jobs or careers of their own. One variation of this was an increase in *dual-service couples*, in which both spouses were military members. Also, just as familial arrangements in the larger society have proliferated, *sole parents* have become more visible in the ranks. Military wives' organizations thus became more active in advocating for the needs of evolving military families, and the military began expanding its family support services to include family advocacy programs, job counseling, and child development centers.

In 2015, Karin De Angelis and Segal provided an updated assessment of the military and family as greedy institutions. De Angelis, one of the authors of this book, completed her undergraduate degree in Sociology at the University of Chicago before serving six years in the U.S. Air Force, including a deployment to Al Udeid Air Base in Qatar in 2004. Following military service, she obtained her doctorate with an emphasis on Military Sociology at the University of Maryland. De Angelis now is Professor of Sociology in the Department of Behavioral Sciences & Leadership at the U.S. Air Force Academy, where she has been on the faculty since 2011. The bottom line of their update is straightforward: the *greedy-intersection* thesis is more relevant than ever.[15]

For starters, the tempo of military operations since 1986 has shot up dramatically. In 1991–1992, the U.S. military deployed about 697,000 troops to the Middle East for the Persian Gulf War, which thankfully was of very short duration. Smaller operational deployments in the 1990s were carried out in the Sinai Peninsula, Somalia, Bosnia, and Kosovo, as well as dozens of other contingency missions. Much more salient, the terrorist attacks on the World Trade Center in New York City and the Pentagon in Washington, D.C., on 11 September 2001 led to armed conflict in Afghanistan from 2001 to 2021 and in Iraq from 2002 to 2011. More than 1.7 million U.S. service members have deployed since 2001 in support of operations there. Deployments were longest for the Army and Marine Corps (typically

12 months) and shorter for the Navy and Air Force (6 months). Many Active-Duty units deployed multiple times, and, under the *Total Force* concept, these deployments included a substantial number of Reserve and National Guard units as well.

Likewise, there have been changes in the military family. We have constructed Table 11.2 (not included in De Angelis and Segal's update) to help sort through the changes. There are many moving parts: the numbers of short-term enlistments vary between the two end-years, 2000 and 2020, in response to the wars in Afghanistan and Iraq; the branches of the military differ in these arrangements (the Army and Navy are very similar, while the Marine Corps and Air Force often diverge in opposite directions); and, the numbers of military women, rising before 2000, continued in a slow but steady increase. Fourteen percent of Active-Duty parents were women in 2020. In 1960, the number of military dependents (spouses, children, and adult dependents) exceeded the number of service members, a figure that dipped during the war in Vietnam but returned as a fixture with the advent of the AVF. In 2000, that ratio was 1.41 to 1, 1.33 to 1 in 2015, and 1.18 to 1 in 2020.

Next, the military marriage rate has declined around 10 percentage points since the "Greedy Institutions" article in 1986, but only 2 percentage points of that are since 2000. This rate is still higher than for comparable age groups among civilians. While individuals volunteering for military service are less likely to be married than age-comparable civilians, once in the military they are more likely to marry than civilians of the same age group and have a lower average age at first marriage. Thus, the increase in those service members who are single with no children in 2020 may reflect short-term enlistment trends. In any case, the percentages of married couples remained steady from 2000 through 2015, meaning few left the military, remarkable given the U.S. military's wartime operational tempo and correspondingly greedy demands.

Another significant change is in the number of *military spouses (mostly wives) who are working.*[16] In 2015, a reported 40% of "civilian" spouses were in the paid labor market and another 12% were looking for work. Also, a total of 6.4% were *dual-military couples*, both of whom obviously work outside the home. These figures are more pronounced in 2020, where fully two-thirds of military spouses are employed (49%), dual-military (6.8%), or seeking employment (12%). Such employment incurs a double bind. Employed military spouses are hampered by being both *tied-stayers* and *tied-leavers*, that is, their movement and longevity for advancement is bound to the geographical location and relocations of a military spouse, and the military member, in an institution assuming *two-persons but a single career*, lacks a full-time person to manage the home front.

The increase in dual-military couples in Table 11.2 seems quite small, rising from 5.7% in 2000 to 6.4% in 2015 and 6.8% in 2020. However, these numbers mask an important dimension of this arrangement. In 2020, women in the military were five times more likely than military men to be in a dual-military marriage, 19.7 vs. 4.1%.[17] Dual-military marriages face unique challenges in maintaining successful military careers and satisfying family lives. If they wish to be on *fast-tracks* for advancement, their jobs must routinely be given first priority. It is possible but difficult to arrange *collocation* (assignment to the same geographical area), so assignment to distant military installations is likely, and they can expect to have training

Table 11.2 Family Status for Active-Duty U.S. Military, 2000–2020[a]

Family Status	2000	2005	2010	2015	2020[b]
Single[c]	**47.0%**	**45.4%**	**43.7%**	**45.6%**	**50.0%**
Single, no children	40.8	40.0	38.3	41.1	46.1
Single, with children[d]	6.2	5.4	5.4	4.5	3.8
Married	**52.9%**	**54.7%**	**56.4%**	**54.3%**	**49.9%**
Married, civ. spouse, no children	10.7	12.8	13.8	14.0	13.6
Married, civ. spouse, with children	36.5	34.9	35.9	33.9	29.5
Dual-military, no children	3.2	4.1	3.8	3.8	4.4
Dual-military, with children	2.5	2.9	2.9	2.6	2.4
Ratio, mil. dependents to service members	1.41 to 1	1.36 to 1	1.39 to 1	1.33 to 1	1.18 to 1
Married, civ. spouse work status[e]					2019
Employed	—	—	35%	40%	49%
Seeking work	—	—	12%	12%	14%

a Table constructed from Defense Manpower Data Center, *2020 Demographics, Profile of the Military Community*, Washington, D.C.: U.S. Department of Defense, Table 5.12, available for download at https://download.militaryonesource.mil/12038/MOS/Reports/2020-Demographics-Report.pdf

b In 2020, there were 1,569,841 military dependents and 1,333,822 active-duty service members.

c Single includes never married, annulled, divorced, and widowed.

d Children includes those under age of 21, and dependents under age of 23 enrolled as full-time students.

e Based on surveys of military families in 2010, 2015, and 2019. Third work status category is "Not seeking/wanting employment." Survey data not available for 2000 and 2005.

assignments and deployments twice as often between them. And, if they have children, each spouse must maintain a Family Care Plan (FCP) with their unit commanders. The FCP details exactly who and how children will be cared for in event of their absence, a chore that usually falls to one of their mothers or female siblings.

The full impact of the wartime deployments since 2001 has weighed heavily on military families. The deployment of military members to hostile war zones creates a stress of its own, and the conflicts in Afghanistan and Iraq have led to thousands of fatalities and serious injuries, including loss of limbs, traumatic brain injuries, and post-combat stress disorders (see our discussion in Chapter 12). The family is the focal point of recovery, and hence typically the primary caregiver in the case of such injuries. During recent conflicts in the Middle East, the U.S. military has encouraged military families to remain on or near military installations where they could be incorporated into formal support networks, but that is not always possible. Increasingly, parents of service members have become sources of support or even the primary caregivers for their military sons and daughters.

Finally, the U.S. military allows LGBQ individuals to serve openly with equal pay and opportunity for promotion and provides benefits to family members. The U.K. and European Union countries also have legalized same-sex marriage and same-sex military families have been afforded inclusion.

We have underscored many problems faced by military families, but there are important benefits. The military provides good, stable employment, especially for those with only a high school degree, along with opportunities for education and training to enhance advancement. Further, the military has developed a reputation for extending access to these and familial resources regardless of race, so differences in marriage rates found in the civilian population between Whites and African Americans disappear among those in the military.[18] And, military members can retire after 20 years of service in their early 40s with a significant *defined benefit* (rather than defined contribution) *pension* plan.

Cross-Cultural Patterns

Zeynep Aycan, Professor of Psychology and Management at Koç University in Istanbul, Turkey, has provided an especially helpful guide for cross-cultural analyses of work–family conflicts.[19] She begins with the observation that perceptions of work–family conflicts (WFCs) are the sum of two normative evaluations: the extent to which *family obligations are seen as interfering with work* (FIW) and/or that *work expectations are infringing upon family life* (WIF). Both FIW and WIF, she continues, are conditioned by culturally based ideas about the types of demands and support mechanisms in each society's family and work domains. While she identifies several cultural dimensions on which societies vary, we will restrict our discussion here for the sake of brevity to *collectivism* vs. *individualism*. In the former, the family is more extended and takes precedence over the individual, while in the latter more value is placed upon an individual's autonomy.

These are ideal types and most cultural differences are relative, and what is dominant in one culture simply receives less emphasis in another. However, comparative studies of societies have shown that *family and work domains* tend to be viewed as more *segmented* in individualist cultures, but more *integrated* in collectivist ones. The former view is more likely to result in higher perceptions of both FIW and WIF, the latter in lower levels. For instance, in the case of women in the labor market, time devoted to work and family may be seen as zero-sum when individuals devote themselves to time-intensive careers, but are left to resolve competing parental demands on their own. In collectivist settings, going to work may be seen as providing support for the family and, in turn, other family members assist in childrearing and household chores.

The cross-cultural lens also can alert us to the exact nature of role conflicts. Aycan notes WFCs are viewed in some countries as stemming from *employee–parent* tensions, but as *employee–spouse*, *employee–household*, or *employee–social woman* ones in others. These differences reflect the salience of these roles across cultures and understanding the source of these conflicts may point both the individual and the organization to solutions. If the wife is expected to routinely organize and stage events for the extended family, for instance, much of the WFC may be *employee–social woman* related. If the woman more values the *social woman* role, she likely sees the conflict as WIF and might temper work involvement; if she is devoted to a career, she probably thinks of it as FIW and might hire a catering company to fulfill that obligation. And, sensing a chance to assist valued workers, an organization might help stage such social events for employees' families.

We turn now to an analysis of work–family arrangements in Israel, the only contemporary society in which military service is compulsory for both men and women.

Spotlight on Sociological Thinking and Research

Meytal Eran-Jona, "Married to the Military: Military–Family Relations in the Israel Defense Forces"[20]

Meytal Eran-Jona is known for her research on the intersection of gender, family, and the military in the Israel Defense Forces (IDF). She completed her doctorate in Sociology at Tel-Aviv University under the guidance of Hanna Herzog, a trailblazer of gender studies in Israel. Eran-Jona served as Senior Sociologist for the Civil-Military Project at the IDF's Behavioral Sciences Center, and as Head of the Israeli Defense and Intelligence Research and Analysis Division's research center. She currently is Chair of the Diversity and Inclusion Office of the Weizman Institute of Science, and consultant for its consortium of 30 European organizations and universities to promote the participation of women in physics and other STEM fields.

The IDF provides an interesting case study since, as noted above, Israel requires compulsory military service of both men and women and doing so is a badge of honor for most Israelis. Israeli Jewish men are expected to complete three years of active-duty military service, and Jewish women about two years. Exempted are married women with children, religious males studying in accredited Jewish Law institutes, and those with physical disabilities. Religious women may do alternative service. Israeli Arabs, just over 15% of the population at the time of Eran-Jona's article, may volunteer to serve but are not required to do so. After the initial active-duty service, most men and women soldiers transfer to reserve units, where their obligation extends to age 45 for men but a shorter time for women. Reserve units train steadily and must be prepared to be at full operational capacity on 48-hours notice. A small number of soldiers volunteer to serve long-term, forming the core of the IDF.

As in other industrial and industrializing societies, family arrangements in Israel have changed and proliferated in the last decades of the 20th century. Prior to this, Israeli Jews and Arabs lived most often in extended families who were the center of economic activity for its members. Increasingly, this has given way to *dual-earner, nuclear families* with attendant attitudinal shifts and debate about the proper roles for men and women in the family and in the paid labor force. "Married to the Military" examines the extent to which the attitudinal and structural changes taking hold in Israeli society are shared by service members in the IDF.

Israel geographically is a small country, about one-fifteenth the size of Italy or one-sixth the size of the state of Florida in the U.S.A. Except for one Air Force installation, there are no military bases with housing for families. Most service members live in their own domiciles and commute daily to military duties. Married soldiers tend to live in a nuclear arrangement, with parents and children under one roof, although there frequently are kinfolk living in proximity. The percentage of dual-military couples is increasing but is in the single digits of all military families in the IDF.

Eran-Jona's study focused on three groups: *wives of military men in the entire IDF, wives of combat officers,* and *husbands of female service members.*[21] The study employs a mixed-methods design, with survey data from 965 respondents (728 wives and 237 husbands), and in-depth interviews with 20 combat officers and 20 of their wives. All respondents had at least one child. The military men of the first group (husbands of wives in the sample) are representative of married male soldiers in the IDF, while combat officer-husbands represent a highly specialized, elite subgroup. Female soldiers, wives of husbands in the sample, mostly are in support positions and low in rank. About 75% of the soldiers' wives and 90% of the husbands of female soldiers in the study are employed at least part-time.

Turning first to how these military couples share parental and household chores, Eran-Jona observes that her respondents' *attitudes* reflect those of the broader Israeli society that spouses should share household duties, but their *behaviors* do not match these expectations. More than 90% of the wives of male service members said they were the ones responsible for childcare and household chores. About one-half of the soldiers' wives and 70% of the combat officers' wives said they were less than "highly satisfied" with this arrangement. In contrast, 22% of the husbands of female service members reported they were the primary caregivers, and 48% said they share childcare and household duties equally with their soldier-wives. To some extent, the military shapes this pattern, as married women soldiers with children may take at least one hour a day off from work for childcare. Among husbands of female soldiers, 70% were highly satisfied with this arrangement.[22]

Eran-Jona then addresses the extent to which work impinges upon family time (WIF, in Aycan's terminology). Virtually all the husbands of female soldiers reported their wives were in the home for at least some time each day, and two-thirds said their wives were usually home by 6 p.m. Eighty percent of the wives of military men reported the same for their husbands, with about half saying their husbands routinely arrived home between 6 and 8 p.m. However, over half of the wives of combat officers said their husbands were in the house only one day a week or less, and on days when they did arrive home, it usually was after 8 p.m. Most military wives, combat officers' wives in particular, pointed to their husbands' *extensive working hours* as their main discontent with military life. This feature, they said, more than any other created marital strain and curtailed the scope of their own employment and chances for advancement.[23]

Finally, across the board, most spouses in the survey indicated that they relied upon nearby family members, or even friends and neighbors, to assist them with childcare and household chores. Without such help they would hardly be able to balance family and work. When they paid for these services, most commonly it was for babysitting or house cleaning. These percentages, unsurprisingly, were lowest among those spouses who were less likely to be employed outside the home.

Eran-Jona's research has been impactful. She showed how the wives of IDF soldiers, mostly employed outside the home themselves, have compensated for the greedy demands of military life. However, the limited provisions available for married female soldiers with children, but not male soldiers with families, also pointed to policy changes that might alleviate the strains. Eran-Jona thus presented these and other similar findings to the IDF's Chief of Human Resources. In response, the IDF has instituted changes in the *combat-officer career path* to allow for more family-friendly options and has created additional *family services* for all its military families.

The Children of Military Families

War is hard on all living things, especially children who reside in the geographical areas of conflict. Since we have just examined family issues in Israel and the IDF, a relevant study of note is one reported in 2009 by a team from the Hebrew University of Jerusalem and Al-Quds University in Jerusalem's Palestinian West Bank.[24] The team collaborated to collect standardized data on rates of exposure to violent incidents and subsequent health issues to allow direct comparisons between Israeli and Palestinian adolescents. Both groups of youngsters have witnessed or been subjected to a substantial amount of violence in the years during and after the Second Intifada (see our discussion in Chapter 10). Predictably, the team found significant numbers of youth in both groups with troubling *somatic anomalies*, *mental health symptoms*, and *functional impairments*; the incidence was higher for Palestinian youth and for girls more so than boys. The most frequent coping strategy for both groups was *accepting reality*, an indication of their sense of powerlessness in the face of violence.

While this topic is deserving of much further commentary, the remaining focus in this chapter is the *children of military families*. By nature of their parents' deployment to zones of conflict, they typically reside outside the zone of violence, but are subjected to different sets of stressors, some unique to military life, others not. In Table 11.2, we noted that there were about 1.57 million military dependents in the U.S. active-duty military, a ratio of 1.18 dependents for every service member. Two-thirds of those dependents are children. Almost 40% of these military children are 5 years of age or less (38.9%), another one-third are between the ages of 6 and 11 (33.8%).[25]

In the U.S.A., military service members and their families are concentrated in areas where most military installations are located: the deep South, Eastern seaboard, Texas, and California. Military families can expect to relocate from one area to another about two-and-a-half times more often than civilian U.S. families, and nearly all will move outside the continental U.S.A. and experience the deployment of the service member-parent at least once.[26] These moves and occasional absences of a parent are not unique to the military, as indicated by sociologist Morten Ender's compilation of families in organizations with operations abroad: governmental foreign service, international business corporations, and missionary groups.[27] However, military deployments of parents may incur a particular source of stress for their children.

We turn our Spotlight now to two studies that explore this issue in some detail.

Spotlight on Sociological Thinking and Research

Manon Andres and René Moelker, "There and Back: How Parental Experiences Affect Children's Adjustments in the Course of Military Deployments,"[28] and Edna Hunter-King, "Children of Military Personnel Missing in Action in Southeast Asia"[29]

Manon Andres and René Moelker are Assistant Professor and Associate Professor, respectively, at the Netherlands Defence Academy. Andres obtained her doctorate in Social and Behavioral Sciences from Tilburg University and specializes in the study of military families, especially work–family conflicts and familial well-being. Moelker holds a doctorate in Sociology from Erasmus University in Rotterdam. He has published dozens of studies and treatises on the military as a profession, military technologies, military families, military education, and the Dutch approach to irregular warfare. The purpose of the present study, begun in

2006, was to examine the attitudes and behaviors of Dutch families during the deployment of a parent to the conflicted Uruzgan province of Afghanistan or for post-conflict missions in Bosnia.

The study is unusual in two ways.[30] It employs a *longitudinal* design, that is, Andres and Moelker found themselves in the rare but highly advantageous position of being able to interview the parents *before, during, and after* these deployments. In 2006, Dutch service members who were about to deploy to Afghanistan or Bosnia were informed of the study and invited to participate. This allowed documentation of changes, rather than having to infer so from retrospective measures in the more typically occurring *ex post facto* designs.

Second, they accumulated a sample of 911 couples, for whom they devised pre-, mid-, and post-deployment questionnaires for both the service members and their partners. In addition, they conducted 120 semi-structured in-person interviews with service members' partners. More than half of the respondents had children, and for these participants the deploying parent was the father. The average age of their children was 9.6 years, and 82% of the mothers worked for pay outside the home, mostly in part-time positions. Three-fourths of the fathers deployed to Afghanistan and the average deployment lasted five months.

Andres and Moelker note that military deployment families fall in the category of families experiencing *temporary separations*, as might be found in businesses with overseas operations, jobs that call for being away from home weeks or months at a time, or the institutionalization or imprisonment of one of the parents. These require a *transition to a single-parent family* mode, with the substantial restructuring of roles and routines for many of the respondents. Unique for all in this study was the deploying fathers' risk of injury or death.

In preparation for the deployment, participants devised scripts for informing their children about their father's pending absence. The ages of the children most worried about the deployment were those 6 to 11. As a seven-year-old stated when his father's pending deployment to Afghanistan was explained to him, "You can tell me my daddy is coming back, but if they shoot him I have no daddy anymore."[31] The reasons for the absence were more vaguely described for younger children, while many teenagers stoically exuded a sense of assuming greater responsibilities. About 30% of the service members and their partners worried about the length of the deployment.

The cycle of deployment was similar for both service members and families left behind: an initial phase of anxiety and unsteady adjustment, then the settling into a routine for the long haul, and finally an anticipation near the end of the tour. While most mothers reported that their children adjusted quite well, about a quarter said their children had difficulties. The ages of those most affected again were 6 to 11, consistent with findings in similar studies. Mothers indicated children in this age group were more likely to encounter talk about the war in Afghanistan among classmates at school, prompting fears among them about their fathers' safety. One mother stated, "My daughter really feared her father would die." Parents of teenagers reported them more likely to watch the news and be aware of the risks, but to cope effectively. Very helpful for most children (and their mothers) were the occasional phone calls, email, and Skype sessions.[32]

Speaking of their own experiences, most mothers indicated only moderate amounts of stress and work–family conflicts, mostly at levels like those prior to deployment. A vast majority of fathers reported only stresses associated with the deployment itself, although these were exacerbated for the few who felt that things were not going so well back on the home front. Interestingly, the pending and actual homecoming was more of a stressor than respondents had anticipated. Though a joyful event, it triggered a transition to family life as structured prior to deployment. One mother summarized: "After the reunion, [the children] have to settle down once more ... It doesn't end when [the service members] have just returned home, it takes weeks again." And, some parents said their children worried that their fathers would leave again soon. One mother explained: "My son ... felt strange when my husband went to work and was afraid he would stay away [for months] again [before returning home]."

Andres and Moelker's multivariate analysis of the longitudinal data revealed two significant predictors of difficulties: child's age (in the curvilinear pattern discussed above) and mother's psychological stress. Perhaps because of its relatively short duration, length of the deployment was insignificant and there were no significant differences between deployments to Afghanistan and Bosnia. Their findings point to the importance of support networks and services for mothers (or fathers) left to manage work and family during such absences – a finding corroborated by other studies and experiences of other families.

In 1998, Edna Hunter served on the U.S. Department of Veterans Affairs Advisory Committee on Former Prisoners of War. She holds a master's degree in Clinical Psychology from San Diego State University and a Ph.D. in Human Behavior from the United States International University. Her affiliation with the Navy Medical Research Unit in San Diego began in 1967. She joined the Unit's Center for Prisoner of War Studies (CPWS), created in 1971, and later served as head of its Family Studies Branch. She has been the author or co-author of at least a dozen studies on American Prisoners of War (POWs) in Vietnam, their families, and the families of those listed as Missing in Action (MIA). In this article, she reflects on 25 years of research on these issues, especially the long-term effects on the *now "adult" children of American MIAs.*

The first American POWs of the Vietnam war were captured in 1964, the year before the introduction of Marine and Army ground troops in South Vietnam. The U.S.A. was not officially at war and North Vietnam categorized American POWs as "war criminals," so families were advised not to discuss their plight publicly. The same advice was given to MIA families. Fearing repercussions for those held in captivity, families mostly carried on in silence and isolation. Growing dissatisfaction with this situation led two POW wives in San Diego to form an action group for discovering and connecting POW/MIA families. In 1968, a news story about their efforts brought national attention and hundreds of new members. In 1969, the group inundated the North Vietnamese delegation at the Paris Peace Talks with cablegrams expressing concern for the POWs. The San Diego group went national in May of 1970 with a charter in Washington, D.C. under the name, National League of Families of American Prisoners and Missing in Southeast Asia.

By this time, the U.S. military estimated the number of American POWs to exceed 600 and the number of MIAs to be more than 2,500. Also in 1970, the U.S.A. began its drawdown of troops in South Vietnam, so the Department of Defense initiated planning for the return of the POWs someday. The Navy received funding to set up CPWS, whose task it was to study and provide policy recommendations for all Navy, Marine Corps, and Army POWs, MIAs, and their families.

The Air Force established its own center for assessing its POWs but elected not to include their families. CPWS's rationale for including families was that families were both *stress producers* and *stress alleviators*, but that well-functioning families would more likely be alleviators. In 1972, CPWS dispatched staff to all parts of the continental U.S.A., Hawaii, and Puerto Rico to interview POW/MIA wives, as well as those POW/MIA parents who were financially dependent upon their sons in captivity.[33]

CPWS had completed about half of their planned interviews when word arrived in January of 1973 that release of the POWs was imminent. The first POWs were brought home on 14 February, and by 29 March, 591 had returned to the U.S.A. Fifty-five percent (325) of these were Air Force pilots and 23 percent (138) Navy pilots. Twelve percent were Army (77) and Marine Corps (26) personnel. The handful of others were civilian contractors. They had spent an average of more than five years in captivity. More than 2,500 missing pilots, soldiers, and Marines did not come home. Of these, 658 were Navy, Army, and Marine Corps MIAs (the other 2,000 or so were Air Force pilots and air crew). The first CPWS study in 1974, and follow-up study five years later, included the children of both POWs and MIAs for the Navy, Army, and Marine Corps, a total of 874 children. Of these 567 were children of MIAs, who mostly were between five and ten years of age when their fathers disappeared.[34]

In 1988, the National League of Families mailed out a questionnaire to all children of MIAs who were members of the organization, including those whose fathers were in the Air Force, to assess how they were doing 15 years after the POWs' return home. By now all were adults. Fifty-one percent of their mothers had never remarried. The percentage of daughters and sons was virtually the same; about two-thirds were married and 40% had children of their own. The fathers of about three-fourths of them were Air Force MIAs.

All the respondents indicated that the singlemost problem marking their lives was the *prolonged but ambiguous loss* of their fathers. One daughter wrote on her questionnaire: "My father never came home! I'm always waiting for Daddy … I resent him for leaving – yet I'm so proud of him. I've gone on with my life, but every day I wonder about him." And a son expressed his "despair and frustration and hope. There will always be hope." Many revealed that *unrequited grief* had damaged their ability to form close relationships: "[I] have trouble becoming close to people"; [my brothers, sisters, and I] have had difficulties in romantic relationships"; "I have an irrational fear of my spouse dying at an early age"; and "I put off having children for 11 years because I wanted to make sure I had a career, should anything happen to my own husband." And, a small percentage expressed that they had lost faith in the government and other institutions: "Don't trust anyone!"

However, many mentioned some positive consequences. Comments included: "An experience like this really brings a family together … [Being a child of an MIA,] you develop an eternal hopefulness in all things." And a daughter wrote:

It is ironic that the most painful event of my life has also brought about the most changes of a positive nature. I have learned to value and appreciate relationships and life itself, for it can be so short. I have pondered the existential questions of life at a very early age, and in doing so it has enriched me with wisdom to help me through life's ups-and-downs … [Nevertheless], it is difficult at times to live with unanswered questions.[35]

Questions for Discussion

1   How have families, particularly in Western countries, changed the most in the 20th and 21st centuries? What are the implications of these changes for building and maintain a military force?

2   How might characteristics such as race, gender, or rank (officer vs. enlisted) change the *greediness* of the military and the family? How might these vary for two-person/ one-career, dual career, and dual service member couples?

3   The Israeli Defense Forces provides limited opportunities for married female soldiers with children to tend to family needs but has not done the same for other family types (including, male servicemembers with children). What additional family-friendly options would you recommend?

4   Children of military families experience unique stressors when a family member is deployed. What support networks and services do you see as most needed? How do your recommendations change for children of Prisoners of War and Missing in Action?

Notes

1   Philip Cohen, 2014, *Family Diversity is the New Normal for America's Children*, Briefing paper prepared for the Council on Contemporary Families.

2   See Lawrence B. Finer, 2007, "Trends in Premarital Sex in the United States, 1954–2003," *Public Health Reports*, 122(1): 73–78.

3   Valerie Schweitzer, 2020, *Divorce: More than a Century of Change*, FP 20-22, Bowling Green, Ohio: National Center for Family & Marriage Research, downloaded from http://doi. org/1025035/ncfrm/fp-20-22.

4   Elizabeth H. Pleck, 2012, *Not Just Roommates: Cohabitation after the Sexual Revolution*, Chicago: University of Chicago Press.

5   Wendy D. Manning, 2013, *Trends in Cohabitation: Over Twenty Years of Change, 1987–2010*, FP-13-12, National Center for Family & Marriage Research. Retrieved from http://ncfmr.bgsu. edu/pdf/family_profiles/file130944.pdf.

6   See, for example, Kathryn Edin and Maria Kefalas, 2005, *Promises I Can Keep: Why Poor Women Put Motherhood Before Marriage*, Berkeley: University of California Press.

7   Current Population Survey, *2019 Annual Social and Economic (ASEC) Supplement* conducted by the Bureau of the Census for the Bureau of Labor Statistics. Washington, D.C: U.S. Census Bureau [producer and distributor], 2019.

8   Sondra Albano, 1994, "Military Recognition of Family Concerns: Revolutionary War to 1993," *Armed Forces & Society* 20 (Fall): 283–302.

9   Chris Bourg and Mady Wechsler Segal, 1999, "The Impact of Family Supportive Policies and Practices on Organizational Commitment to the Army," *Armed Forces & Society*, 25 (Summer): 633–652.

10   Mady Wechsler Segal, 1986, "The Military and the Family as Greedy Institutions," *Armed Forces & Society*, 13 (Fall): 9–38.

11   Karin De Angelis and Mady Wechsler Segal, 2015, "Transitions in the Military and the Family as Greedy *Institutions*: Original Concept and Current Applicability," pp. 22–42 in Rene Moelker, Manon Andres, Gary L. Bowen, and Philippe Manigart (eds.), *Military Families and War in the 21st Century: Comparative Perspetives*, London and New York: Routledge.

12   Mady Segal reports that her mother's first response was, "You had to go all the way to Chicago to meet a nice Jewish boy from New York?" 2012, "Faculty Spotlight" in *Imagine: Sociology News, University of Maryland*, 6 (Spring), available for download at https://umdsocy. wordpress.com/05/15/833/

13   Lewis Coser, 1974, *Greedy Institutions: Patterns of Undivided Commitment*, New York: The Free Press. Coser is best-known for his book, *The Functions of Social Cnflict*, [1956] 1964, New York: Simon and Schuster.

14 Wechsler Segal, op. cit., pp. 15–22.

15 De Angelis and Wechsler Segal, op. cit., pp. 22–27.

16 De Angelis and Wechsler Segal, op. cit., pp. 36–39.

17 Table constructed from Defense Manpower Data Center, *2020 Demographics, Profile of the Military Community*, Washington, D.C.: U.S. Department of Defense, Table 5.12, available for download at https://download.militaryonesource.mil/12038/MOS/Reports/2020-Demographics-Report.pdf, Table 5.24.

18 See Jennifer Hickes Lundquist, 2004, "When Race Makes No Difference: Marriage and the Military," *Social Forces*, 83 (2): 731–757.

19 Zeynep Aycan, 2008, "Cross-Cultural Approaches to Work-Family Conflict," pp. 353–370 in Karen Korabik, Donna S. Lero, and Denise L. Whitehead, *Handbook of Work-Family Integration: Research Theory and Best Practices*, Amsterdam: Elsevier.

20 Meytal Eran-Jona, 2011, "Married to the Military: Military–Family Relations in the Israel Defense Forces," *Armed Forces & Society* 37 (January): 19–41.

21 Ibid., pp. 22–25.

22 Ibid., Table 1, p. 26.

23 Ibid., Table 2, p. 28.

24 Ruth Pat-Horenczyk, Muhammad M. Haj-Yahia, Mohammed A.M. Shaheen, et al., 2009, "Post-traumatic Symptoms, Functional Impairment, and Coping on Both Sides of the Israeli-Palestinian Conflict: A Cross-Cultural Approach," *Applied Psychology: An International Review*, 58 (September): 488–508.

25 Defense Manpower Data Center, op. cit., Table 5.64.

26 See Molly Clever and David R. Segal, 2013, "The Demographics of Military Children and Families," *The Future of Children*, 23 (Fall): 13–19, 26–30.

27 Morten Ender, 2002, *Military Brats and Other Global Nomads: Growing Up in Organization Families*, Santa Barbara, Calif.: Praeger Publishers.

28 Manon D. Andres and René Moelker, 2011, "There and Back: How Parental Experiences Affect Children's Adjustments in the Course of Military Deployments," *Armed Forces & Society* 37 (Summer): 418–447.

29 Edna J. Hunter-King, 1998, "Children of Military Personnel Missing in Action in Southeast Asia," pp. 243–256 in Yael Danieli (ed.), *International Handbook of Multigenerational Legacies of Trauma*, New York: Plenum Press.

30 Andres and Moelker, op. cit., 424–427.

31 Ibid., pp. 419–421.

32 Ibid, pp. 429–431.

33 Hunter-King, op. cit., p. 247. See also Iris R. Powers, 1974, "The National League of Families and the Development of Family Services," pp. 25–34 in Hamilton I. McCubbin et al. (eds.), *Family Separation and Reunion: Families of Prisoners of War and Servicemen Missing in Action*, San Diego, Calif.: Naval Health Research Center.

34 Edna J. Hunter and James D. Phelan, 1974, "Army, Navy, and Marine Corps Returned Prisoners of War," pp. 35–44 in McCubbin et al., op. cit.

35 Hunter-King, op. cit., pp. 249–251.

Recommendations for Additional Reading/Viewing

Margaret C. Harrell, 2000, *Invisible Women: Junior Enlisted Army Wives*. Santa Monica, Calif.: RAND.

(Respected study based on in-depth interviews with three young women married to junior enlisted soldiers)

Stacy Ann Hawkins et al., 2018, *What We Know about Military Family Readiness*, Monterey, Calif: Office of the Deputy Under Secretary of the Army.

(Assessment of research on preparedness of U.S. Army families to navigate challenges of military service)

Hamilton I. McCubbin et al., 1974, *Family Separation and Reunion: Families of Prisoners of War and Missing in Action*. San Diego, Calif.: U.S. Naval Health Research Center.

(Important study of U.S. Vietnam POWs' and MIAs' families and one-year follow-up of POWs' reunion with their families)

René Moelker, Manon Andres, Gary Bowen, and Philippe Manigart (eds.), 2015, *Military Families and War in the 21st Century: Comparative Perspectives*, Part III, London and New York: Routledge.

(Contributions on military families in the U.S.A., Slovenia, Belgium, Portugal, Turkey, Australia, and Japan)

David G. Smith and Mady W. Segal, 2013, "On the Fast Track: Dual Military Couples Navigating Institutional Structures," *Contemporary Perspectives in Family Research*, 7: 213–253.

(Highly informative research article based on interviews with 23 U.S. Navy dual-career couples)

While Time Stands Still (documentary film), directed by writer and filmmaker Elena Miliaresis, 2013.

(Follows the story of two military wives to document the impact of combat on military families)

12 Veterans and Veterans' Issues

Reader's Guide: Veterans of various wars do not enjoy equal gratitude for their service. We advance a good-war/bad-war distinction for explaining why some enjoy adulation and others denigration, and author Scott provides a personal account of rejected – and dejected – French veterans of the war in Algeria. The first Spotlight provides theoretical context with a review of the literature on the politics of war memory and commemoration. Turning to the demographics of U.S. veterans, veterans of the draft-based wars of WWII, Korea, and Vietnam are old and rapidly declining in number. Those who served in the Persian Gulf, Afghanistan, and Iraq in an All-Volunteer Force are more likely to be women and are more racially and ethnically diverse. Our second Spotlight contains two studies, one comparing the public's views of soldiers and veterans in Great Britain, the second a statistical assessment of "thank you for your service." The final Spotlight provides a diagnostic evaluation of "moral injury" and an innovative program by the Veterans Administration to address it.

An author of this textbook, Wilbur Scott, spent the 1991–1992 academic year at Université Blaise-Pascal, in Clermont-Ferrand, France. His official job was faculty advisor for a group of University of Oklahoma students during their semester abroad. His own research agenda was to meet and interview French veterans of the war in Algeria (see our discussion of the Algerian war in Chapter 9). Scott at the time was in the final stages of his own book on American veterans of Vietnam[1] and was looking for cross-cultural comparisons. The French fought the war in Indochina largely with Legionnaires, elite paratroopers, and West African colonial troops (see our discussion of the latter in Chapter 6). In Algeria, however, they bolstered these forces with more than a million conscripts from the homeland, so veterans of this latter war provide a more suitable comparison.

Scott spoke only rudimentary French, so was enrolled in a total immersion language program. Scott's faculty sponsor taught Spanish in Blaise-Pascal's foreign languages department but spoke very little English. They thus were a good match, learning each other's language with gusto. After the first month, the sponsor suddenly made an astonishing revelation: he himself was an Algerian war veteran. His reticence, Scott would learn, was not unusual. French Algerian war veterans lived under a shroud of what French military psychiatrist, Bernard Sigg, called "silence and shame" (*le silence et la*

DOI: 10.4324/9781003282549-12

honte).[2] Over the course of the year, the sponsor introduced Scott to a half dozen or so other faculty members whom he now discovered also were veterans of the Algerian war.

A similar story unfolded in the neighboring village of Romagnat where Scott rented a small apartment. When his landlord, a carpenter, learned of Scott's research interests, he approached men in the village who had been part of a local but defunct chapter of *Fédération nationale des anciens combattants d'Algérie* (FNACA).[3] FNACA was a grass-roots veterans' organization formed late in the Algerian war to advocate for a negotiated withdrawal and, after the war, for the rights of Algerian war veterans, including the *carte du combattant* (the war veterans' identification card) and its attendant benefits. The French Ministry had ruled that veterans of Algeria were not eligible for the *carte* since it had been a "civil war" without defined fronts. The Ministry reluctantly reversed its course in 1974, 12 years after war's end. Still, Scott was told, by 1991 only about 10% of eligible veterans of the war in Algeria had applied for and been issued the *carte*.

Scott's landlord urged its former members in Romagnat to revive the chapter so they and Scott could talk. The men decided to have a meeting in January of 1992 to which Scott would be invited. The January meeting convened in the town's wine cellar. It began with *kir* (cassis and white wine), then introductions all around, small talk, and, eventually, an agreement to meet again. The second and third meetings began much like the first but took a much different turn as several began cautiously describing their time in Algeria. What followed were intense outpourings of raw, unprocessed emotion – anger, hurt, confusion, regret, and resentment about what had taken place, both during and after the war. Scott, himself a Vietnam veteran, had seen and been part of similar episodes in sessions with Vietnam veterans in the early 1980s. In this vein, Benjamin Stora, a recognized authority on the history of Algeria (where he was born), recounted a conversation with journalist Jean-Pierre Farkas, who said to him[4]:

> So, at the market, I have a friend who sells fish; the other day, he suddenly said he that he served in the war [in Algeria]. That's all that was said. ... [O]ur fathers ... certainly did talk to us a lot about their experiences in World Wars I and II. I have never met an Algerian war veteran who has recounted his war experiences. Why have the French acted like this? I don't know why.

Romagnat's chapter was still having meetings when Scott departed in May.

Memory and Commemoration

In 1998, sociologists Jeffrey Olick and Joyce Robbins took stock of a resurgent field, the sociology of memory.[5] The field may be traced to the 1925 work, *Social Framework of Memory*, by one of Émile Durkheim's protégés, Maurice Halbwachs (see our discussion of Durkheim in Chapter 2). Central to Durkheim's thinking was his concept of the *collective conscience*: values, beliefs, and norms shared by members of a society that in turn unify diverse individuals into a collective. When these shared conceptions are accepted as a matter of course, the collective conscience becomes *reified*, that is, takes on an existence of its own beyond its individual beholders. *Social facts* such as the collective conscience thus are not reducible simply to the psychology of individuals.

It is this thinking that inspired Halbwachs's concept of *collective memory*. In the 1920s, Sigmund Freud was the dominant force in psychology. Freud argued that all a person's memories are stored routinely in the *unconscious mind*. A person's mind, he continued, selectively screens these memories and recalls only those approved to see the light of day, a protective psychological process he called *repression*. In rebuttal, Halbwachs placed the process *outside the mind*: "There is no point ... in seeking ... where [memories] are preserved in my brain or in some nook in my mind ... It is in society that people normally acquire memories. It is also in society that they recall, recognize, and localize [these] memories."[6]

Traditional historians sometimes make a distinction between memory and history. Memory, most of them agree, is very selective, and history seeks to right the tilt. The study of *war memory and commemoration* remained a tangential concern in both sociology and history until the last two decades of the 20th century. In 2000, historian Timothy Ashplant and his associates put together a volume to assess the state of this developing area. We now turn our Spotlight to their analysis.

Spotlight on Social Science Thinking and Research

T. G. Ashplant, Graham Dawson, and Michael Roper (eds.), *The Politics of War Memory and Commemoration*[7]

In 2000, Timothy Ashplant was a member of the Research Centre for Literature and Cultural History at Liverpool John Moores University, Graham Dawson a member of the Popular Memory Group at the University of Brighton, and Michael Roper a social and cultural historian in the Department of Sociology at the University of Essex. All three have authored books on the politics of memory with a special application to war and its aftermath. This edited book presents case studies in almost a dozen different countries and their wars. This Spotlight focuses on their introduction to the field.

The recent interest in war memory has been sparked, they say, first by the growing visibility in the 1980s and 1990s of Holocaust remembrances, and of demands for recognition by war victims and survivor groups (see our discussion of the Holocaust in Chapter 8). Also, several war anniversaries, for example, the day marking 75 years since the beginning of World War I or the 50th anniversary of World War II's end, underscored and reinforced these movements. Finally, the collapse of the Soviet Union in 1989 brought an end to the Cold War, but ushered in an era of irregular wars and "an explosion of ethnic strife" among peoples banding together around linguistic, tribal, religious, or cultural differences. All this caught the attention of academic researchers.

The bulk of this academic work, Ashplant and his associates observe, has followed one of two approaches.[8] One emphasizes the *political* nature of these movements, the other the *psychological* essence of them. The former focuses on the power of the state to exert, from the top down as it were, an official version of what a war or conflict was all about and to create public memorials and rituals for reliving its important symbolic themes. In contrast, the latter approach views memory and commemoration at the micro-level as a manifestation of the mourning process. Independent of all formal machinations, its existence is created as ordinary people seek to make sense of and to accept what has taken place. The two views potentially intersect. Individual mourning and meaning-making could take place within official eulogies and events, and official commemorations could make some effort to reflect, even if imprecisely, individuals' experiences associated with the war.

Despite this, Ashplant et al. contend the two approaches typically were not used in tandem. They present an edited volume by British historians Eric Hobsbawm and Terence Ranger, *The Invention of Tradition*, as representative of the political approach.[9] What comes through in this book, and works in the same paradigm, is the diligent, almost cynical, effort on the part of the state, as the book's title suggests, to "invent traditions" designed to characterize and legitimate a war effort. This is accomplished most directly by embedding the war in that culture's enduring themes and practices. The state has at its disposal an awesome array of tools to do so, for example, formally created and sponsored rituals, parades, speeches, memorials, platitudes, and "true" war stories. And should these efforts at persuasion fail, there is always coercion.

American historian Jay Winter's *Sites of Memory, Sites of Mourning* illustrates a critique of this political emphasis offered by the second approach.[10] Winter does not attempt to dispel the political perspective, but rather refocuses the spotlight on underlying sentiments welling from the bottom up, despite, or in spite of, whatever is being orchestrated from the top down. The volume contains a touching and moving study of *collective remembrance* of WWI through accounts of grieving survivors and family members. Here a common culture of memory and commemoration is stitched together through the cumulative efforts of ordinary people. This usually is carried out through the *informal networks of family, local community, and fictive kin* – others to whom the actors are not related by blood or marriage but through the shared experience of a war. All this occurs, Winters maintains, independently of manipulation by the state.

Ashplant and his associates offer several observations. One, a fully satisfying exploration of memory and commemoration should not fall neatly into either a political or psychological treatment of the material. Activity at both levels most likely is taking place, perhaps even in reaction to each other. They offer as an example David Lloyd's historical account of the construction of the World War I Cenotaph in London.[11] The original Cenotaph – generically, an empty tomb to honor a person or group of persons whose remains are elsewhere – was of wood and plaster, one of several temporary memorials built in 1919 for a parade by the British military to triumphantly celebrate the war's end. However, this make-shift Cenotaph became an enormously popular focal point for pilgrimages by former service members and others seeking solace by publicly grieving lost loved ones. In response, the government agreed a year later to construct a permanent one made of stone to replace it.

There are several solutions to this disciplinary inertia. Ashplant and his associates offer suggestions, in particular the use of oral histories, such as in *Anzac Memories* by Australian historian Alistair Thomson.[12] In this highly acclaimed work, Thomson analyzes the oral histories of three World War I "Anzac diggers" – slang for soldiers of the Australian and New Zealand Army Corps who fought in the fierce, deadly campaigns on the Isle of Gallipoli in western Turkey and in the battle of the Somme in France. The work documents the life-long struggles of these veterans, all in their 80s when Thomson recorded their oral histories, to find meaning in the carnage they had experienced. Thomson distinguishes between *personal memories* and *public versions*, the gulf between them that often led the diggers to suffer privately, and the interplay between them that led the veterans to adapt their private memories to wider popular, public versions. Ashplant and his associates write:[13]

> The complex entanglement of public and private memory is key to Thomson's approach. ... [He] identifies ... agents in this process, from official war historians and returned-servicemen's groups of left- and right-wing persuasion [who] literally fought in the streets in 1919, through to the film-makers, anti-Anzac Day feminist protestors, oral historians and elderly diggers of the 1980s. The private memories of the veterans are ... understood as ... shifting ... in relation to changing forms ... of the public legend, and to the identities of the men themselves at different stages in the life course.

Taken seriously, this perspective and its corresponding methodology call for focusing on the *competing versions of truth* made by the state and its attendant ministries, veterans and groupings of them, and other agents of public persuasion. The purpose is to document *who introduced a war's dominant narratives* into the public domain, and *how these were advanced, or not*, from contested claims to widely accepted facts.

"Good" Wars vs. "Bad" Wars

American psychologists Joel Brende and Edwin Parson have offered a distinction between "good" wars and "bad" wars. The terms reflect the orientation of the state and segments of the population about the necessity and justness of a war. In "good" wars, the state's narrative is widely viewed as legitimate and enjoys strong support. Although good wars do not necessarily produce contented veterans, they can ease their readjustment. A clear and moral purpose provides a compelling context for wartime service and for making peace with war experiences. "Bad" wars however are ones gone awry that governments choose not to sanction fully and for which the public splinters in its support. This makes coming to terms with the hostilities problematic. Brende and Parson write[14]:

> [I]n bad wars, what the soldier does in the war falls squarely on his shoulders. He alone bears the burden ... [and hence is left] adrift without a useful perspective that would aid him in formulating and weaving his own "tapestry of meaning."

There are many differences between the wars in Algeria and Vietnam.[15] Though both featured *uprisings against colonial powers*, Vietnam was never a colony of the U.S.A. – the original revolt was against French colonial rule. Still, there are many similarities. Each was an *undeclared war* where insurgents fought when they had the advantage and hid among the population when they did not. It was difficult for French and American troops to distinguish friendlies from enemies and establish a coherent plan for winning the support of the population. Opposition on the home front became increasingly strident. In France, leftist political parties became hotbeds of criticism of the war in Algeria, as did American college campuses during the Vietnam war. War critics also raised questions about atrocities, and, in France, about its brutal, deeply entrenched use of torture.

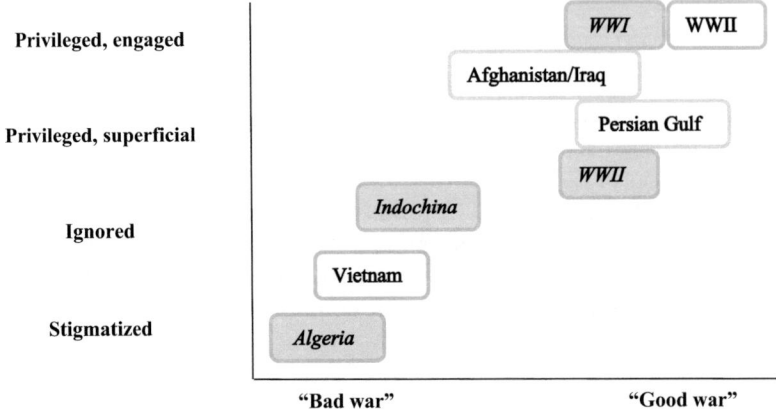

Key: French wars in italics; U.S. wars in Roman.

Figure 12.1 French/U.S. Veterans' Status by the Good War/Bad War Distinction

Source: Authors' Creation.
Note: Figure was constructed using Brende and Parson's description as a point of departure.

Figure 12.1 depicts the relationship between veterans' status, ranging from stigmatized to privileged and engaged, by the "good/bad" war distinction. For France and the U.S.A., World Wars I and II, respectively, provided high watermarks for their veterans. Both embodied defense of the homeland with almost universal public support and subsequent adulation. In each, the origin of significant largesse in veterans' benefits may be traced to these wars.

However, during the wars in Algeria and Vietnam, veterans formed organizations to challenge state policy, evidence these wars fell toward the "bad" end of the good war–bad war spectrum. In France, the two major organizations were FNACA and *Union nationale des combattants d'Afrique du Nord* (UNCAFN). FNACA during the war was most like a later organization in the U.S.A., Vietnam Veterans Against the War (VVAW). Each called for a truce and a return of the troops home, and focused on torture and atrocities, respectively, to reduce public support for war policy, a risky strategy. VVAW's dogged emphasis on atrocities, for example, aroused the sentiment within a portion of the American public that Vietnam veterans themselves were worthy of contempt. Back in France, UNCAFN took the opposite tack. Founded by veterans on the political right, it favored victory by any means, and if torture might in any way contribute to that, they were fine with it.

Following the war, FNACA shifted its focus to veterans' rights. Politically unaffiliated but to the left because of this focus, it became the single largest of that war's veterans' organizations. But, as we have seen, Algeria has remained a taboo topic, and FNACA's successes have been modest. UNCAFN, on the other hand, turned away from veterans' issues. Its principal figure, Jean-Marie Le Pen, in 1972 founded the French fascist party, *Front national* (FN), drawing strong support from the 340,000 former *Pieds-noirs* who resettled after loss of the war along France's Mediterranean coast. The FN has focused on measures directed against non-European immigrants from its former colonies under the banner of *La France aux Français* (France for the French).[16]

In the 1970s, the U.S. veterans' organizations from the World Wars, the American Legion and the Veterans of Foreign Wars (VFW), lobbied against bills before Congress that would have provided Vietnam veterans counseling, treatment, and compensation specific to that war.[17] Consequently, very few Vietnam veterans joined their ranks, and so formed their own initiatives and organizations. For example, Vietnam veterans led by Jack Smith, with no background in psychology other than personal efforts to deal with his own war experiences, linked with psychiatrists Chaim Shatan and Robert Lifton to justify a new diagnosis, post traumatic stress disorder (PTSD), formally adopted by the American Psychiatric Association in 1980.[18]

Meanwhile, VVAW's nosedive in popularity left a vacuum filled by Vietnam Veterans of America (VVA), founded in 1978 by Bobby Muller. A former Marine Corps lieutenant, Muller was paralyzed from the chest down after being wounded in a North Vietnamese assault on his platoon's position in 1969. Radicalized by frustrations about poor medical care at Veterans Administration (VA) hospitals, he was not shy about leading VVA's efforts to secure improved care, psychological counseling, and recognition for Vietnam's veterans.[19]

The divisiveness among the U.S.A.'s veteran groups thawed a bit in the 1980s. A Vietnam veteran-led campaign, with key donations from the Legion and VFW, culminated in a stunning Vietnam Veterans Memorial located in the heart of Washington, D.C.[20] The Legion and VFW now have more Vietnam veteran members than does VVA. The Legion and VFW (who at the time rated Senators and Congressional Representatives in terms of how conservatively they voted on *non-veteran* issues) lean decidedly to the right.

In 1989, a VVA lawsuit against the VA broke a ten-year standoff between the VA and veterans seeking treatment and compensation for cancers suspected of being caused by exposure to the herbicide, Agent Orange.[21] Because VVA challenged the two traditional veterans' organizations, and because it favored *diplomatic recognition* of what is now called the Socialist Republic of Vietnam, it was regarded during Muller's era as a left of center group. Since his departure, it has become more right of center.[22]

In France, literally every town has in its *centre ville* an impressive monument commemorating the more than 1.3 million Frenchmen who gave their lives in WWI. To most of these monuments has been added a plaque or maybe an additional statue in memory of WWII deaths and, sometimes, of those who died in Indochina. Rarely does one see any addition in honor of veterans of Algeria. In Clermont-Ferrand, for instance, one typically finds fresh wreaths and flowers in front of 40-foot, larger-than-life statuary commemorating the two World Wars, concentration camp victims, and Indochina. It is a city landmark known to all. A block away, in an easily overlooked, small park, stands a five-foot marker erected by FNACA in honor of service in Algeria.

U.S. Veteran Characteristics

In 1989, President George H. W. Bush elevated the Veterans Administration to cabinet-level status and redesignated it the Department of Veterans Affairs (DVA). DVA had an operating budget in 2021 of $241 *billion*, the second-largest in dollars of federal agencies behind only the Pentagon, and it runs the U.S.A.'s largest health-care system.[23] A "veteran" is defined as: one who served in the active-duty military, Coast Guard, uniformed Public Health Service, or uniformed National Oceanic and Atmospheric Administration; reservists and National Guardsmen/women who were called to active duty; or, those disabled while on active-duty training. Excluded are those dishonorably discharged. The definition incurs eligibility for medical treatment, benefits such as a G.I. Bill (stipends to attend college or vocational school), and compensation for disabilities. Only war-time veterans qualify for many of these benefits.

Table 12.1 Number and Demographics of U.S. Veterans, 2000 and 2018[a]

Period of Service	2000[b]	%[c]	2018[b]	%[c]	Median Age	Female (%)	Non-White (%)
Post-9/11	—		3,764	19.0	36.6	16.8	34.9
Gulf War 1990–91	3,025	10.9	3,804	19.2	49.6	14.6	31.0
Vietnam War era 1964–75	8,380	30.2	6,384	32.3	70.8	3.7	17.4
Korean War era 1950–55	4,046	14.6	1,306	6.6	86.1	2.9	11.4
WWII era 1941–46	5,720	20.6	485	2.5	92.6	4.6	9.1
Peacetime only	6,556	23.6	4,034	20.4	68.0	9.5	24.0
"Total"	27,727	100	19,777	100	65.0	6.7	76.7
Actual number of veterans	26,404		17,960				
			-32%				

a Jonathan E. Vespa, 2000, "Those Who Served: America's Veterans from World War II to the War on Terror: American Community Survey Report," United States Census Bureau, available at census.gov. Table constructed from material excerpted from Tables 2 and 3, pp. 8–9.
b Number in thousands
c % of "Total"

Table 12.1 contains U.S. veteran characteristics for 2000 and 2018. There is no register of veterans per se, so their numbers are estimated through surveys commissioned by DVA or done by the U.S. Census Bureau. A survey by the latter estimates there were 17,960,000 U.S. veterans in 2018, approximately one-third less than the 26.4 million of 2010. The distinction between "total" and "actual" number of veterans comes from the enumeration procedure in which a veteran of more than one war, for example, one who served in Korea and Vietnam, is included in the numbers for both wars. The number of veterans in 2018 indicates that about 7% of the adult population in the U.S.A. have served in the military.

The sharply decreasing numbers of veterans may come as a surprise since the U.S.A. has been at war in the Middle East for the past 20 years. However, U.S. wars since Vietnam have relied upon an All-Volunteer Force (AVF), which by design is much smaller than draft-based militaries of the past. Second, the large numbers of WWII, Korea, and Vietnam veterans in 2010 were old and, in 2018, the median age of veterans of WWII was in the mid-90s, and in the upper 80s and low 70s for those of Korea and Vietnam, respectively. Hence, of the 480,000 WWII veterans in 2018, the National World War II Museum reports only 240,000 remained alive in 2021 and they were dying at the rate of about 250 per day.[24] Thus, the projected total number of veterans by the year 2040 is around 10 million and will consist almost entirely of post-AVF peacetime veterans and those who served in the Persian Gulf war, Afghanistan, and Iraq.

Two other characteristics should be noted. Consistent with trends in the AVF, women and Non-Whites in the U.S. veteran population are noticeably increasing. Of those veterans who served in WWII, women and Non-Whites made up 4.6 and 9.1% of those remaining in 2018, but 16.8 and 34.9% in 2018 of those in the post-9/11 subgroups. This shift will intensify for women veterans in particular as the number of draft-based, 20th-century war veterans continues to decline.

Sociologists Alair MacLean and Glen Elder have systematically studied *how being a veteran who is male* impacts *employment and earnings* over the life course. Their work has shown there is no simple relationship: the effect of veteran status is mitigated by era (WWII vs. Vietnam vs. post-9/11) and other individual characteristics. However, more than any other variable, *whether the veteran was in combat* is an overriding consideration: combat veterans of any era do not fare as well as those who did not see combat.[25] Since female veterans of the AVF have served mostly in noncombat slots, their military experience should prove to be more consistently beneficial.

Studies of noncombat male veterans have reported another twist: *those who come from disadvantaged backgrounds* often gain desirable skills in the military while incurring *fewer opportunity costs*, such as interruptions in certain civilian career progressions. This also should favor many current women veterans, especially those who are Non-White and Hispanic. A recent analysis by sociologists Irene Padavic and Anastasia Prokos has confirmed these expectations. They demonstrate that women veterans earn on average about $8,000 more per year than corresponding nonveteran women, much of this attributable to educational differences between the two groups – an advantage accrued mainly during time in the military. This effect is stronger for women veterans of color than for White veteran women and for male veterans.[26]

We place our Spotlight now on two studies of the public's perceptions of veterans, one in Great Britain and the second in the U.S.A.

Spotlight on Sociological Thinking and Research

Rita Phillips, Vincent Connelly, and Mark Burgess, "Exploring the Victimization of British Veterans: Comparing British Beliefs about Veterans with Beliefs about Soldiers,"[27] and Meredith Kleykamp, Crosby Hipes, and Alair MacLean, "Who Supports U.S. Veterans and Who Exaggerates Their Support?"[28]

Rita Phillips completed her doctorate in Psychology at Oxford Brookes University in 2019 and currently is Lecturer in Psychology at Robert Gordon University in Aberdeen, Scotland. Her research interests revolve around social psychological phenomena in everyday life, especially how individuals experience and make sense of their worlds. Vincent Connelly is Professor of Psychology at Oxford Brookes. He has devoted the last decade to the study of military personnel and their families, especially how British Reservists balance the demands of family and both civilian *and* military work expectations. A Ph.D. in Psychology (from the University of Alberta, Canada) and Reader in Psychology at Oxford Brookes, Mark Burgess focuses his research on transformative experiences. In this vein, he has interviewed terrorists, peace protestors, victims of violence, and military personnel.

The present study analyzes how people of the U.K. characterize *soldiers* and *veterans*. Phillips and her associates note that public opinion polls reveal a peculiar finding: while soldiers and veterans are both highly thought of in the U.K., the perceptions of *veteran-ness* also carry a distinctly negative connotation.[29] Some polls indicate people associate physical and mental health problems with being a veteran, along with attendant issues of unemployment and alcoholism, this despite the incidence of these ailments among veterans being far lower than public perceptions. Being a soldier incurs no such negative connotations, though soldiering might incur some of these same problems and be a forerunner of them for when they become veterans themselves. So, why does being a veteran incur negative connotations?

To answer this question, Phillips and associates consider these perceptions as *social representations*. A term introduced into the literature by French social psychologist Serge Moscovici, social representations establish sets of ideas which individuals use to orient themselves to the social order and thus provide a code for communicating with each other. Unlike Leon Festinger's *cognitive dissonance* theory – which specifies that contradictory beliefs or inconsistencies between attitude and behavior induce stress and hence pressure to move toward consistency – Moscovici's theory shows how such contradictions may come about and be accommodated. Social representations tend to have two interacting systems, a *central* one of unambiguous beliefs, and a *peripheral* one whose entries are more contextual and situationally dependent. The perceptions of veterans thus might contain some opposing elements, with both beliefs existing side-by-side.

To test how this works, Phillips and associates used a *free word association task*.[30] A sample of subjects, working at a computer screen, were first asked to list, as quickly as possible, the three words that came to mind when thinking of a soldier and then of a veteran. In the second part of the task, subjects were allowed time to be more reflective, and then to indicate how important each was on a five-point Likert-type scale ranging from *extremely important* to *not so important*. In the last part of the task, subjects listed their demographic information. A convenience sample of 234 residents of the U.K. participated in the study. The sample was disproportionately female (64.5%), young (median age 24.2), and above average in education, so care must be taken in generalizing the findings.

Content analysis first was used to group together semantically similar answers, and then the hierarchical evocation method (HEM) – commonly used in social representational research – to identify central and peripheral elements. Finally, principal components factor analysis was employed to determine how these elements cluster for respondents' images of veteran and soldier. The raw data contain 1,404 word associations that the content analysis reduced to 14 themes. HEM revealed that seven of these were part of the central core of representations. These were *characteristics of war, victimizing associations, heroizing associations, associations with war experience and aging, positive personality associations, military human resources*, and *"other" contrasts*. The factor analysis pointed to a two-factor matrix. Elements in the first component are correlates of *victimizing associations* and of *heroizing associations* in the second component. These elements load differently, that is, have different response patterns, for respondents' thoughts about veteran and soldier.

The findings confirm the earlier trends reported in public opinion polling. The core elements for both veteran and soldier contain images of *warfare* and *being a hero*. However, for veteran, the core also contains *being a victim* along with *war experience and aging*, while, for soldier, the core includes *positive personal traits*, for example, trustworthy, reliable. Heroizing associations for both soldier and veteran seem anchored in the cultural notion in the U.K. that "military service represents a sacred duty ... and those who carry out this sacred duty are sacred people."[31] Victimization, war experience, and aging, on the other hand, are more closely related to "war as a reason for suffering," something exacerbated after time in the armed forces. Applied to veterans, subjects seemed to associate the image of "veteran" with stereotypes of WWI and WWII veterans – *heroes, but flawed ones*, exhibiting the infirmities of a lifetime of dealing with war's consequences.

Perhaps for this reason, Phillips and associates note, some younger, post-9/11 U.K. veterans dismiss the label "veteran" in referring to themselves, but use ones such as, I am "a former soldier (or Marine)," "former military," or an "ex-service member." The first two are especially preferred, since they seem to denote dashing combat service rather than time in support roles.

In our second spotlighted study, Meredith Kleykamp and her associates examine attitudes towards U.S. veterans and the willingness to provide support for them. Kleykamp is Associate Professor of Sociology at the University of Maryland and the Director of its Center for Research on Military Organization. She has devoted much of her research to the study of veterans' issues, especially public opinion towards veterans and the consequences of military service for veterans and their families. Crosby Hipes is a newly minted sociologist who has served on the U.S. Department of Defense's Recovering Warrior Task Force. Alair MacLean is Associate Professor of Sociology at Washington State University. Her research has focused on how trajectories originating in early adulthood affect employment over the life course, especially in the case of veterans.

The research team begins by noting the acclaim afforded U.S. military members since 1990. Figure 12.1 depicts the status of U.S. veterans of the Persian Gulf war, Iraq, and Afghanistan. The Persian Gulf war is an exemplary "good" one – honorable, quick, decisive, and resolved with overwhelming force. However, it lasted only 100 hours once it got rolling, robbing its veterans of a more storied place in modern memory.[32] The wars in Iraq and Afghanistan have hardly been "good" ones, but their veterans enjoy near universal adulation. They would say, however, the "thanks for your service," though well-intentioned, is rarely followed by further interest.[33]

Kleykamp and her associates wonder how much of this expressed support might be due to *social desirability bias*, that is, a response that is inflated (or deflated) by an awareness there is a socially approved answer.[34] A second problem may be that respondents *exaggerate support or disapproval in the abstract* but feel quite differently when presented with *specific scenarios*. A respondent might express strong support for veterans when asked, "How do you feel about our veterans who served in Iraq?" but answer negatively if asked, "How would you feel about having a mental health clinic for veterans in your neighborhood?" In this study, there are two specific outcome variables: *social distance*, that is, the amount of intimacy respondents would permit between themselves and veterans, and, *deservingness*, that is, the willingness to see more tax dollars used for veterans' programs.

To address social desirability bias, respondents are *randomly divided* into control and treatment groups; the *control group* is asked to indicate the number of items they agree with from a list of three non-sensitive issues; the *treatment group* is given the same instructions, but their list contains four issues, the additional issue being the one thought to elicit a socially biased response. The *difference in number of items agreed with* between treatment and control group provides an *unbiased estimate* of the proportion of respondents who "actually" agree with the sensitive item.[35]

Kleykamp and associates used a treatment group and two control groups, a recommended variation. Their traditional control group (Group C) was presented with the following for *social distance*:

Below is a list of different kinds of people who might move into a home near you. How many of these types of people would you be pleased to have move in next door to you? I don't want to know which ones, just how many in total?

1. A professional musician
2. Someone convicted of driving under the influence of alcohol
3. Someone who teaches kindergarten

The treatment group (Group A) was given the same question and three items, plus a fourth item:

4. A recent veteran of the wars in Iraq/Afghanistan

An expanded control group (Group B) was given the same question and four items as the treatment group but asked to express agreement or disagreement with each of the four items separately. This provides evidence of the *overt sentiment* attached to veteran-sensitive items. The difference between the unbiased estimate and the overt responses gives a robust estimate of the social desirability effect.

To measure *deservingness*, the same procedure for the three groups was used with the following question: "Your taxes support numerous programs, including the ones listed below. *How many of these programs should the government spend more to support?*" The three non-veteran items were: *Improvements in elementary school education, Regular pay raises for members of Congress*, and *Repairing infrastructure such as roads and bridges*. The added fourth item for Groups A and B was: *Health care for veterans of the wars in Iraq/Afghanistan.*

Data for the study were derived from a national probability sample of U.S. households through the Time-sharing Experiments in the Social Sciences (TESS) program. TESS conducts web-based surveys and, to reduce bias that certainly otherwise would result, provides temporary internet services for households without such access to ensure fuller participation in the survey. The survey was carried out during 15–21 July 2011 and netted a sample size of 4,142. Households were assigned randomly to Groups A, B, and C to determine which questionnaire the respondent would be filling out. Demographic and other relevant data were collected to complement analyses of the test questions.

In reporting the initial findings, Kleykamp and associates note the following. One, when asked directly if they would like having a veteran live next door, 87.7% of Group B respondents answered affirmatively. However, the unbiased estimate derived from the responses of Groups A and B indicates that only 85% actually would be pleased, meaning this support would be overstated by 2.7%. The corresponding figures for deservingness are 92.2 and 82%, meaning support for more health care dollars for veterans is overstated by at least 10%. So, on the one hand, support for veterans for these two measures is very high, that is, more than 80%, but, on the other, that support is overstated, especially when the issue is tax expenditures for veterans.

How do these patterns of support vary once social desirability bias is removed? Kleykamp et al. find those over the age of 45 are two to three times more likely than 18-to-29-year-olds to want veterans as neighbors, while Blacks and Hispanics are about three times less likely than non-Hispanic Whites to express such an inclination. When the issue is stated in terms of increases in tax expenditures for veteran health care, Democrats are two and a half times more likely than Republicans to support such expenditures (a difference of 11%), while Blacks and Hispanics are less likely to favor them.[36]

These data cannot tease out why these findings come about. The effect of race is interesting since, as we noted above, military service is notably beneficial for Non-White veterans, especially women. An earlier survey by MacLean and Kleykamp showed ambivalent views towards veterans based upon concerns about potential mental health and behavioral problems.[37] Perhaps this is a factor in the perceptions of veterans by many Non-Whites. Other studies have found a positive relationship between age and support for U.S. war policy in the Middle East, and a negative relationship between being Non-White and support for that policy. This would imply a bifurcated view of the military (and war policy) among Non-Whites, some viewing the military with suspicion, others embracing it as a career path. As for Republicans, the findings simply may reflect their proclivity for supporting purely military expenditures, but being cool to other types of government spending.

The Trauma of War

We begin this section with some grim statistics. The military, when all is said and done, is the central institution in society entrusted with the use of lethal force. One way to measure the impact of war is to note the number of military deaths, keeping in mind, as we discussed in Chapter 8, a war's full impact encompasses the entire range of military *and* civilian deaths, that is, *multicides*.

Military deaths derive from four basic sources: killed as a result of hostile action (KIA), accidental deaths, deaths from disease, and homicides/suicides. Until recently, disease was a prime cause of death in war. In the U.S. Civil War, for example, more than 600,000 soldiers, Union and Confederate, died, an estimated 415,000 of them from dysentery, cholera, and malaria. Likewise, getting shot or having an arm or leg blown off was almost certain death, given the sketchy state of knowledge about how to effectively treat massive, life-threatening traumas. Thus, the ratio of those wounded in action (WIA) to those KIA gives some indication of advances in military medicine.

Table 12.2 summarizes these statistics for current generations of U.S. veterans as of 2015. Since the advent of WWII, just over 500,000 American military servicemembers have died in combat, about 80% of those in WWII. In 2015, 6,764 of those were in the 9/11 wars. (The calculation, if civilians and all others are included, is around 900,000 war-related deaths for Afghanistan, Iraq, Syria, and Yemen.)[38] With the exception of the Persian Gulf war, three-fourths or more of military deaths in theater are combat deaths, that is, KIAs. The percentage of those who died or were wounded are 10 to 14 for WWII, Korea, and Vietnam, but much lower for the wars in the Persian Gulf, Afghanistan, and Iraq. And, the ratio of WIA to KIA has been rising, indicating an increasing ability to keep those wounded in battle alive. The ratio of 7.69:1 for the post-9/11 wars means there are about 53,900 wounded U.S. veterans of those wars, many with traumatic amputations, severe burns, and other extensive bodily damage.

Not shown are the changing causes of death in combat. WWII and Korea were wars among near-peer militaries; hence artillery and bombs were the major killers. In Vietnam, South Vietnamese civilian insurgents and the North Vietnamese Army dictated the war's tempo, so the U.S. military and South Vietnamese Army mostly engaged in guerrilla

Table 12.2 War Casualties for Current Generations of U.S. Veterans, 2015[a]

War	Number in Theater of Combat	Total Number of Deaths in Theater	KIA (%)[b]	Ratio of WIA[c] to KIA	Deaths/ Wounded in Theater (%)
World War II	8,913,000	405,399	71.9	1.65-to-1	13.9
Korea	1,789,000	36,574	92.2	3.06-to-1	14.0
Vietnam	3,403,000	58,220	81.4	5.22-to-1	10.6
Persian Gulf	695,000	383	38.6	1.22-to-1	< 0.01
Post-9/11	1,900,000	6,764[d] 507,340	78.7	7.69-to-1	3.1

a Nese E. DeBruyne and Anne Leland, 2015, *American War and Military Operations Casualties: Lists and Statistics* (Congressional Research Service: January 2), downloaded from www.crs.gov. Table entries constructed from data in Tables 1, 6–9, and 12. Expanded version of this table presented in Wilbur J. Scott, 1992, "French Veterans of the War in Algeria: A Description and Comparison with America's Veterans of the Vietnam War," unpublished manuscript.

b KIA (Killed in Action), i.e., lethal injuries as a result of hostile action.

c WIA (Wounded in Action), i.e., nonlethal injuries as a result of hostile action.

d The number of U.S. servicemember deaths by 2021 was 7,052.

warfare. Here small-arms fire was the principal cause of combat deaths. In Afghanistan and Iraq, fighting against small, scattered paramilitary groups resulted in combat deaths produced primarily by IEDs, that is, homemade bombs or artillery shells detonated remotely or by pressure plates. Concussions among U.S. soldiers wounded in these encounters added a new medical diagnosis: Traumatic Brain Injury (mTBI).[39]

Also not noted in the table are the psychological consequences of surviving the killing zone. As noted earlier, the American Psychiatric Association in 1980 added a diagnosis whose etiology may lie in war trauma, PTSD. There remains a stigma attached to receiving the diagnosis. However, a highly respected military psychiatrist, Dr. Charles Hoge, has summarized a trove of research and his own clinical experience to illustrate the diagnosis' legitimacy.[40] His work thus addresses both theoretical and practical concerns about PTSD. We turn our attention now to another manifestation, *moral injury*.

Spotlight on Social Science Thinking and Research

Natalie Purcell, Kristine Burkman, Jessica Keyser, Phillip Fucella, and Shira Maguen, "Healing from Moral Injury"[41]

Natalie Purcell holds a Ph.D. in Sociology and is Assistant Professor of Behavioral Science in the School of Nursing at the University of California-San Francisco (UCSF). She also is the Integrative Health Program Director for the San Francisco Veterans Affairs Health Care System (SFVAHCS), where she has focused on evaluating programs addressing the impact of violence and combat trauma on military veterans. Kristine Burkman and Jessica Keyser are both Associate Professors in UCSF Medical School's Department of Psychiatry, and Staff Psychologists at SFVAHCS. Phillip Fucella is a counseling psychologist at SFVAHCS. And, Shira Maguen is Professor of Psychiatry in UCSF's Medical School and Director of

SFVAHCS's Post-911 Integrative Care Clinic. This team has developed SFVAHCS's innovative Impact of Killing (IOK) program, a psychological intervention for treating *moral injury* and trauma associated with killing in war.

Much research has shown that "going to war" itself incurs many stressors for soldiers, from completely reorienting their lives, being separated from familiar places and support systems, and moving into a world of dangerous unknowns over which they have little control. However, the single *most important stressor* associated with war is *combat and exposure to killing*. Since it is highly adaptive while in the killing zone to put aside thinking about what is taking place, veterans usually return home changed persons carrying inordinate amounts of unprocessed trauma.

Post-traumatic stress, the team notes, usually has its roots in *having felt intense fear or loss* because of brushes with death, being party to horrible incidents, or witnessing the gruesome deaths of comrades. Having *participated in killing* – one of the main marks of combat – may create its own specific set of unresolved issues.[42] All these events may be *repetitively relived* through obtrusive thoughts, startle reactions, interruptions of sleep, phobias, hair-trigger reactions, and the like. Coping may take the form of soldiers' all-time adaptive favorite, *self-medication*, along with other *maladaptive behaviors* such as self-blame, self-isolation, avoidance of certain places, sounds, and smells, or an abnormal desire to return to combat, this time to die.

Some post-traumatic stress, in a sense, is a fairly normal response to the experience of combat. However, such symptoms and behaviors merit a *PTSD diagnosis* when they interrupt a veteran's life to the point where he or she can no longer function, or the veteran commits some troubling or criminal act that attracts the attention of authorities. Purcell and associates cite a longitudinal study in which about one-fourth of veterans of the war in Iraq met the PTSD diagnostic criteria at least one time after their return home.[43]

In 1994, clinical psychiatrist Jonathan Shay introduced the term *moral injury* to describe the spiritual upheaval he noted in Vietnam veterans.[44] Unlike PTSD, *moral injury* does not have its roots in fear or loss, but in a sense of remorse or *guilt for having violated deeply held moral beliefs*, in this case the feeling that some killing in which they had participated was not legitimate or justifiable. Purcell notes that moral injury is not a formal diagnosis like PTSD, but rather some veterans' description of what they think lies at the heart of their war-related problems. From a clinical point of view, the symptoms and maladaptive behaviors observed in cases of "moral injury" may overlap substantially with those for PTSD. However, since the etiology is different, many clinicians have concluded that some treatment modalities other than those usually used for PTSD should be employed.

Currently, the "gold standard" treatments for PTSD are Prolonged Exposure Therapy (PE) and Cognitive Processing Therapy (CPT). Common to both is an *attempt to have the veteran-patient confront the original traumatic events* to process the emotions and pain invested in them. In PE, veterans work up to the point where they can record their account of the events which can be replayed repeatedly as a desensitizing mechanism. With CPT, veterans develop worksheets detailing traumatic events and surrounding issues. They then attempt resolution of them through writing their own care plans. The basic utility of both therapies, Purcell states, is supported by research evidence, but these therapies do not work for all veterans and do not contain elements addressing moral injury.[45]

Two of the authors, Shira Maguen and Kristine Burkman, thus developed the IOK program as a supplement to PE and CPT. IOK specifically focuses on moral injury by *acknowledging upfront that some of the killing* which took place in Iraq or other war zones may cross *personal moral boundaries*, resulting in guilt, shame, or self-condemnation. IOK thus provides a space within which veterans can address this issue head-on for purposes of reaching *self-acceptance* and *self-forgiveness*.

The study reported here by Purcell and her associates is an assessment of IOK from the standpoint of veterans in the program. They recruited participants through referrals from clinicians treating veterans at VA hospitals, clinics, and Vet Centers in the San Francisco area whom they thought exhibited moral injury. The veteran must already have had a diagnosis of PTSD. Excluded were those with recent episodes requiring hospitalization, suicidal or homicidal behavior, or an untreated substance abuse problem. Interested veterans were briefed and signed informed consent statements.

Twenty-eight were selected for the study. Two had served in Korea, 20 in Vietnam, one in the Persian Gulf war, and five in Iraq and/or Afghanistan. They ranged in age from 26 to 80, with a median age of 64. Seventeen were Non-Hispanic Whites, one was Hispanic White, five were Black, one was Pacific Islander, and four were mixed race. The ages of these veterans in the IOK program underscores that the simple passage of time does not heal all wounds.

The study details the program's activities which cover an intensive, highly structured six to eight week period.[46] Working with specially trained therapists, veterans identify their own sources of moral injury and develop their own specific self-care plans. The issue of self-forgiveness is tricky since, as Purcell and associates state, "when it comes to killing and violence in war, it is not clear who is authorized to forgive or whose forgiveness is needed and meaningful."[47] The goal is for the veterans to find some way both to honor their moral convictions and move beyond feelings of guilt and shame.

Evaluation of the study consists of qualitative analyses of the veterans' assessments of their own progress. Overall, most of the 28 veterans in the study felt the IOK program helped them focus on the source of their problem. For example, one Vietnam veteran revealed, "I never talked about it [killing] with anybody else ... not even other veterans," and another asked rhetorically, "Why do we keep reliving this fire-fight? [Because] we *need* to make sense of it" (emphasis in original).[48] Another Vietnam veteran spoke of his transformation this way: "You look at the mirror in the Funny House, and you look in that mirror and see how warped you can be ... After [IOK] I look in the mirror straight, and it's clear, you can see yourself for real."[49] Yet, many acknowledged there was much work to be done. A Korean war vet concluded[50]:

> When we got to the issue of self-forgiveness, I thought, "oh, this is going to be easy." [But] that has really thrown me. I have not been able to solve it and am still struggling with it. I try to work on it a lot. IOK has been a big help, but ... [voice trails off, end of sentence]

Questions for Discussion

1   French veterans of the war in Algeria and U.S. veterans of the Vietnam war both struggled for recognition and assistance. This may seem unusual these days when people, superficially at least, laud veterans' service. Does this surprise you? Explain. And, how might one account for the difference?

2   Timothy Ashplant and associates describe war memory and commemoration as a contest between official versions of what a war was all about and "lived" versions of what a war's participants experienced. Explain, in your view, how this has played out for veterans of Afghanistan – a war that ended rather disastrously.

3   The chapter presents two studies of the perceptions of veterans and veterans' issues, one in the U.K., the other in the U.S.A. Both employ clever research methods for getting at the question. Pick either one, describe the issue being studied, and comment upon the use of methods to do the investigation.

4   War trauma these days often is discussed in terms of *post-traumatic stress disorder* and/or *moral injury*. What is the difference between the two and what are the implications for devising treatment plans for each? What applications do you see for trauma originating in civilian life?

Notes

1   Wilbur J. Scott, 1993, *The Politics of Readjustment: Vietnam Veterans Since the War*, New York: Aldine de Gruyter, later reprinted as Scott, 2004, *Vietnam Veterans Since the War: The Politics of PTDS, Agent Orange, and the National Memorial*, Norman, Okla.: University of Oklahoma Press. Unless otherwise noted, citations in this chapter for this book are from the 2004 edition.

2   Bernard W. Sigg, 1989, *Le Silence et la Honte: Névroses de la Guerre d'Algérie*. Paris: Messidor/Editions Sociales.

3   See, Anndal Narayanon, 2014, "'Ready to Fight': Veterans of the Algerian War Take the Battle to France, 1958–1974," *Journal of the Western Society for French History*, 42: 137–149.

4   Benjamin Stora, 1991, *La Gangrène et L'Oubli: La Memoire de la Guerre d'Algérie*, Paris: Editions La Decouverte), pp. 266–267.

5   Jeffrey I. Olick and Joyce Robbins, 1998, "Social Memory Studies: From 'Collective Memory' to the Historical Sociology of Mnemonic Practices," *Annual Review of Sociology*, 24: 105–140.

6   Quoted in Olick and Robbins, pp. 109–110.

7   T.G. Ashplant, Graham Dawson, and Michael Roper (eds.), 2000, *The Politics of War Memory and Commemoration*, London and New York: Routledge.

8   Ibid., Section 1.2, "Politics of Mourning: Current Approaches to the Study of War Memory and Commemoration," pp. 7–15.

9   Eric Hobsbawm and Terence Ranger (eds.), 1983, *The Invention of Tradition*, Cambridge, UK: Cambridge University Press, cited and discussed in Ashplant et al., loc. cit.

10   Jay Winter, 1995, *Sites of Memory, Sites of Mourning: The Great War in European Cultural History*, Cambridge, UK: Cambridge University Press, cited and discussed in Ashplant et al., loc. cit.

11   D.W. Lloyd, 1998, *Battlefield Tourism: Pilgrimage and Commemoration of the Great War in Britain*, Oxford, UK: Oxford University Press, cited and discussed in Ashplant et al., loc. cit.

12   Alistair Thomson, 1994, *Anzac Memories: Living with the Legend*, Oxford, UK: University of Oxford Press, cited and discussed in Ashplant et al., loc. cit.

13   Ashplant et al., op. cit., p. 14.

14   Joel Brende and Edwin Randolph Parson, 1985, *Vietnam Veterans: The Road to Recovery*. New York: Signet Books.

15   Wilbur J. Scott, 1992, "French Veterans of the War in Algeria: A Description and Comparison with America's Veterans of the Vietnam War," unpublished manuscript. Gielt Algra and associates at the Dutch Veterans Institute also have written about "good war"/"bad war" scenarios in the case of the Netherlands: Gielt Algra, Martin Elands, and Jan René Shoeman, 2007,

"The Media and the Public Images of Dutch Veterans from World War II to Srebrenica," *Armed Forces & Society*, 33 (April): 396–413.

16 One of Jean-Marie Le Pen's daughters, Marine Le Pen, took over as president of the FN in 2011. In 2021, she renamed the party *Rassemblement national* (RN – the National Rally) to create more distance between the historical policies of the FN and her updated policies for another run for the President of France in 2022.

17 David Bonior, Stephen Champlin, and Timothy Kolly, 1984, *The Vietnam Veteran: A History of Neglect*, New York: Praeger, pp. 131–133; and Scott, op. cit., chapter 2.

18 Scott, 2004, op. cit., chapters 2 and 3.

19 Ibid., chapter 5.

20 Ibid., chapter 6.

21 Ibid., chapter 8. An earlier lawsuit filed by Vietnam-veteran-led Agent Orange Victims International had produced a settlement on 7 May 1984 of $180 million, to be paid by the manufacturers of Agent Orange to veterans and their children who suffered from diseases related to exposure to the herbicide. See also James B. Jacobs and Dennis McNamara, 1986, "Vietnam Veterans and the Agent Orange Controversy," *Armed Forces & Society*, 13 (Fall): 57–80, and Wilbur J. Scott, 1988, "Competing Paradigms in the Assessment of Latent Disorders: The Case of Agent Orange," *Social Problems*, 35 (April): 145–161.

22 Afghanistan/Iraq veterans have gravitated toward their own organizations such as the Iraq and Afghanistan Veterans of America (IAVA). Founded in 2004 by Paul Reichoff, who served as an infantry platoon leader in Iraq, IAVA's online website reports that 425,000 eligible veterans have logged in to accept its free membership. Cf., https://iava.org.

23 See https://www.va.gov/health.

24 National World War II Museum, "WWII Veterans Statistics, 2021," http://www.nationalww2museum.org/war/wwii-veterans-statistics.

25 Alair MacLean and Glen Elder, 2007, "Military Service in the Life Course," *Annual Review of Sociology*, 33: 175–196.

26 Irene Padavic and Anastasia Prokus, 2017, "Aiming High: Explaining the Earning Advantage for Female Veterans," *Armed Forces & Society*, 43 (April): 358–386, pp. 377, 381–382.

27 Rita Phillips, Vincent Connelly, and Michael Burgess, 2022, "Exploring the Victimization of British Veterans: Comparing British Beliefs about Veterans with Beliefs about Soldiers," *Armed Forces & Society*, 48 (April): 385–409. In the interest of full disclosure, author Wilbur Scott served as the External Reader for Phillips's doctoral dissertation examination at Oxford Brookes University.

28 Meredith Kleykamp, Crosby Hipes, and Alair MacLean, 2018, "Who Supports U.S. Veterans and Who Exaggerates Their Support?" *Armed Forces & Society*, 44 (January): 92–115.

29 Phillips et al., op. cit., pp. 386–389.

30 Ibid., pp. 390–392.

31 Ibid., pp. 398–402.

32 The brevity of the war should not detract from the gravity of the fact 383 Americans died during the 1990–1991 war.

33 See Elizabeth Samet, 2011, "On War, Guilt, and 'Thank You for Your Service,'" *Washington Monthly*, August 2, downloaded from http://washingtonmonthly.com/2011/08/02/war-guilt-and-thank-you-for-your-service/.

34 Kleykamp et al., op. cit., pp. 95–96.

35 Ibid., pp. 97–100.

36 Ibid., pp. 104–108.

37 Alair MacLean and Meredith Kleykamp, 2014, "Coming Home: Attitudes toward U.S. Veterans Returning from Iraq," *Social Problems*, 61: 131–154, cited in Kleykamp et al., op. cit., pp. 94–95.

38 Watson Institute of International & Public Affairs, 2021, "Costs of War," Providence, Rhode Island: Brown University, September. Downloaded from Human Costs of U.S. Post-9/11 Wars: Direct War Deaths in Major War Zones | Figures | Costs of War (brown.edu). The table lists the following estimates of deaths other than U.S. military for post-9/11 wars in Afghanistan, Iraq, Syria, Yemen, and Others: 21 U.S. DoD civilians, 8,189 U.S. contractors, 204,645 to 207,845 National Military and Police, 14,845 Other Allied troops, 363,939 to 387,072 Civilians, 296,858 to 301,933 Opposition fighters, 680 Journalists/and media workers, 892 Humanitarian/NGO workers, for a Total Death Count of 897,150 to 928,558.

39 For a fuller discussion, see Scott, 2016, op. cit.

40 Charles W. Hoge, *Once a Warrior, Always a Warrior: Navigating the Transition from Combat to Home – Including Combat Stress, PTSD, and mTBI*, Guilford, Conn.: Globe Pequot Press, 2010, Kindle edition. Relying on his own extensive clinical evidence and a great deal of reputable research evidence, he has carefully laid out the symptoms required for diagnosis, the physiological and psychological bases for the integrity of the diagnosis, and recommended paths for treatment and eventual post-traumatic growth.

41 Natalie Purcell, Kristine Burkman, Jessica Keyser, Phillip Fucella, and Shira Maguen, 2018, "Healing From Moral Injury: A Qualitative Evaluation of the *Impact of Killing* Treatment for Combat Veterans," *Journal of Aggression, Maltreatment, & Trauma*, DOI: 10.1080.10926771.2018.1463582, pp. 1–29, available for download at https://doi.org/10.1080/1 0926771/2018/1463582.

42 For an analysis of how soldiers (and police officers) may prepare, carry out, and make sense of killing, see Katherine T. Baggaley, Phillip C. Shon, and Olga Marques, 2019, "'I Was There' and 'It Happened to Me': An Exploratory Study of Killing as an Adventure Narrative," *Armed Forces & Society*, 45 (July): 511–531.

43 Purcell et al., op. cit., pp. 1–2.

44 Jonathan Shay, 1994, *Achilles in Vietnam: Combat Trauma and the Undoing of Character*, New York: Simon and Schuster. Cited in Purcell et al., loc. cit.

45 Purcell et al., op. cit., p. 3.

46 Ibid., pp. 5–6.

47 Ibid., p. 4.

48 Ibid., p. 13.

49 Ibid., p. 14.

50 Ibid., p. 16.

Recommendations for Additional Reading/Viewing

David Finkel, 2009, *The Good Soldiers*, New York: Farrar, Straus and Giroux; and, 2013, *Thank You for Your Service*, New York: Farrar, Straus and Girous.

(Embedded reporter follows the 2/16th Infantry, 1st Infantry Division on their deployment in 2007 to Iraq – *The Good Soldiers* – and then follows up on their readjustments following that deployment. Heart wrenching.)

Charles W. Hoge, M.D., Colonel, U.S. Army (ret.), 2010, *Once a Warrior Always a Warrior: Navigating the Transition from Combat to Home – Including Combat Stress, PTSD, and mTBI*. Guilford, Conn.: GPP Life.

(An essential resource by a seasoned military professional for identifying post-deployment problems and difficulties and for seeking appropriate assistance)

Wilbur J. Scott, 2004, *Vietnam Veterans Since the War: The Politics of PTSD, Agent Orange, and the National Memorial*, Norman, Okla.: University of Oklahoma Press.

(Narrative accounts of three central Vietnam veterans' issues based on interviews with the activists and those who opposed them, a study in the "politics of readjustment")

"Served Like a Girl" (film), 2017, directed by Lysa Heslov, available at www.servedlikeagirl.com.

(Touching, memorable documentary on women veterans who stage a beauty pageant to bring attention to their military service and the lack of support programs available for women who served)

Viet Thanh Nguyen, 2015, *The Sympathizer*, New York: Grove Press; and 2016, *Nothing Ever Dies: Vietnam and the Memory of War*, Cambridge, Mass. and London: Harvard University Press.

(A clever Pulitzer-Prize winning novel and follow-up non-fiction exploration of the South Vietnamese experience in the Vietnam war and its aftermath – a voice seldom heard in reminisces about that war)

Bao Ninh, [1993] 2018, *A Novel of North Vietnam*, translated from Vietnamese by Phan Thanh Hao, edited by Frank Palmos, New York: Anchor Books.

(Autobiographical novel about Kien, lone survivor of the North Vietnamese 27th Youth Brigade after ten years of combat and his return to Hanoi after the war – a striking look at a young man and a young country forever changed)

13 Parting Thoughts
What May Lie Ahead

We have given the reader a whiff of where military sociology as a field of study came from and where it has traveled since then. It is a remarkable journey that has enriched the social science disciplines from which it sprang, and the knowledge base policy makers could draw upon in making critical decisions about war and the military. We will now speculate where the field, and the objects of its study, might go in the near future. In many respects, the future already is here.

Dragons and Snakes

Military sociology developed after World War II in the fairly simple era of conventional warfare when the dominant threat was the bipolar Cold War. On one side was the U.S.A. and its immediate allies, the *First World*, and the Soviet Union and its allies on the other, the *Second World*. All others were the *Third World*. Things got more complex in the 1980s with the waning of the Cold War and the move toward multinational peacekeeping and similar *military operations other than war* (MOOTW, *moot-wah*). This was a marked change in the strategic environment, which the American military resisted, and European members of the North Atlantic Treaty Organization (NATO) largely followed its lead.

The 1991 Persian Gulf war, triggered by Saddam Hussein's invasion of Kuwait on 2 August 1990, lasted about 100 hours once underway. The U.S.A.'s post-Vietnam military mindset and awesome display of technological wizardry crushed the Iraqi military and its antiquated Russian doctrine and military equipment-. It thus was an extension of the Cold War. The message the U.S.A. took from this war was, this definitely is the way to go – *no more Vietnams*. However, adversaries – notably, Russia, China, Iran, North Korea – concluded that taking on the U.S.A. straight up meant certain, ignominious defeat. Thus, they have redirected their strategic plans since then toward what national security strategist David Kilcullen has called *liminal warfare*, that is, a suite of belligerent activities, "neither fully overt nor fully clandestine," which relies upon this ambiguity as a cover for "action without retaliation."[1]

Kilcullen is a former Australian infantry officer who holds a Ph.D. in Political Anthropology from the University of New South Wales in Canberra. For his doctoral dissertation on insurgency warfare, he did fieldwork in 1996 with the Dar'ul Islam insurgency movement in West Java.[2] In 2005, at the request of U.S. Secretary of State Condoleezza Rice, the Australian Army "seconded" him to the U.S. State Department as

DOI: 10.4324/9781003282549-13

a special consultant on counterinsurgency warfare. Reflecting in 2019 on his years of advising the U.S. military, he wrote:[3]

> If there is one takeaway …, it is that the military model pioneered by US forces in the 1991 Gulf War – the high-tech, high-precision, high-cost suite of networked systems that won the Gulf War so quickly and brought Western powers such unprecedented battlefield dominance … – is no longer working. Our enemies have figured out how to render it irrelevant.

So, who are these enemies? Kilcullen plucked the terms "dragons" and "snakes" from a national security summary provided by James Woolsey in his 1993 Senate confirmation hearings for appointment to the head of the Central Intelligence Agency. Woolsey pictured the two years since the demise of the Soviet Union this way, "Yes, we have slain a large dragon, but we live now in a jungle filled with a bewildering array of poisonous snakes. And in many ways the dragon was easier to keep track of."[4] These *snakes* included non-state paramilitaries brandishing military-grade weapons, others forming transnational terrorist cells, organized crime syndicates and narcotics traffickers, and groups conflicted by ethno-sectarian strife. From a defense policy point of view, all this percolated in an unfocused stew until 9/11.

The attacks on the World Trade Center and the Pentagon in 2001 changed everything for the U.S.A. With President George W. Bush's declaration of a *War on Terror*, the U.S. security picture narrowed. The focus shifted to just one snake (*terrorism*) and to just one subset of terrorism (*Salafi jihadism*), the variant of Islamic extremism associated with Osama bin Laden and fellow travelers. State-on-state conflicts became peripheral concerns, giving the *dragons* – Russia et al. – free rein to maneuver *liminally*. Further, the terrorist threat after 9/11 proved to be much more varied and multifaceted than Salafi jihadism. Still, the U.S.A. was slow to widen its lens.[5]

The snakes, too, knew better than to attack straight up, or if not, quickly perished, victims of *insurgent Darwinism*. Adept ones, too, quickly took advantage of "the explosion of new, mostly Western-designed technologies" and social media. Kilcullen observed how quickly and seamlessly some unsavory insurgents, including ones railing against Western modernism, integrated the use of iPhones and Android smartphones to recruit, derive intelligence, pass on information, conduct cyberattacks, coordinate missions, and post lessons learned.[6]

Threat Picture, I

In 2017, Paul Cornish and Kingsley Donaldson published, *2020: World of War*, in which they laid out likely sources of global war three years hence.[7] At that writing, Cornish, a former officer in the Royal Tank Regiment, was Professorial Fellow at Australian National University's National Security College; he currently is Professor of History at Cambridge University's Centre for Science and Policy. Kingsley Donaldson is a retired British Army officer who has served as a national security analyst for the U.K.'s Ministry of Defence. They devote the first chapter to "the Russian strategic challenge," prescient, as that has turned out to be the most pressing current threat.

Cornish and Donaldson begin with a review of General Sir John Hackett's *The Third World War, August 1985*.[8] Written in 1978, the novel presages a fictional World War III. In it, the Politburo determines the Soviet Union is slipping while the West and the U.S.A. are surging ahead, so it orders an invasion of Western Europe to recalibrate the balance.

The campaign moves quickly and pushes all the way into West Germany before it bogs down. Complications ensue, the invasion deteriorates, and the Soviets exercise the nuclear option, striking and leveling Birmingham, England. The U.K. and the U.S.A. respond with a nuclear blasting of Minsk, Belarus. More things happen, but this nuclear exchange leads to the war's end. Hackett's warning, Cornish and Donaldson state, is that deterrence deters only if predicated on "unflinching political will."

As in the novel, Soviet leaders in real life were concerned about the U.S.S.R.'s decline in power, relative to that of the U.S.A., well before it folded in 1991. Mikhail Gorbachev, for example, General Secretary of the Soviet Communist Party from 1985 to 1991 and President of the Soviet Union in 1990 to 1991, sought to address the situation *from within* by instigating changes in the Soviet economy and polity – hence *perestroika* and *glasnost*. It proved to be too little, too late. The result was catastrophe and Vladimir Putin's eventual rise to power in 2000. Putin's approach has been to re-exert strict control within, but to focus on Russia's *external environment* – meaning to him the meddling of Western powers in Russian affairs. To this end, Putin has orchestrated a strategic, well-formulated doctrine of *hybrid warfare* – turn everything into a weapon.[9]

In terms we advanced in chapters 8 and 10, *hybrid warfare* might be thought of as a combination of *Big Wars* and *New Wars*. It calls for the routine use of *non-kinetic and semi-kinetic actions*, from cyber-disruptions, disinformation campaigns, organized crime forays, interference with infrastructure systems, electronic jamming, sabotage and espionage, use of private and proxy military forces, and the like. This does not rule out the use of conventional military actions or threats of nuclear war, but overt, *kinetic actions* are part of the backup plan.[10]

Cornish and Donaldson in 2017 identified Ukraine as crucial in Putin's plan.[11] For starters, its long, shared border with Russia makes it a key piece in re-establishing Russia's control in countries that served as buffers against western Europe in Soviet days. Russia has long been obsessed with equating security with command over this buffer zone. As analyst Keir Giles (whom we introduced in a *New Wars* Spotlight) has observed, "Russians are only half joking when they say that the only secure Russian border is one with a Russian soldier standing on *both* sides of it" (emphasis in original).[12]

Still, Russia's invasion of Ukraine on 24 February 2022 came as a bit of a shock. A "military operation" that was to last only days and result in the installation of a Russia-friendly regime, this was not an impulsive action. Russia had spent months amassing more than 100,000 troops and war materiel at several points along the shared border with Ukraine. NATO's European member states, the U.S.A., and Canada disagreed on how they might respond, in part because of doubt Russia would carry through with its threat and also because Ukraine is not a NATO member. In retrospect, the invasion should not have come as such a surprise, given Russia's smaller military forays since 1991 into Chechnya, Georgia, Crimea, and eastern Ukraine. At the time of writing, the current war in Ukraine has ground on for more than 100 days, with enormous destruction of infrastructure, loss of life, and displacement of millions of Ukrainians, but with little end in sight.

An important implication is that the declaration of the Cold War's end, sounded almost unanimously in 1991, was a bit premature. Stephen Kotkin, Senior Fellow at Stanford University's Hoover Institution, admits the events of 1989–1991 were a mirage, "consequential, just not as consequential as most observers – myself included – took them to be."[13] The demise of the Soviet Union, Kotkin continues, has lulled many American policy makers into an unrealistic vision of a single, unified, and U.S.-led liberal international order. Putin (and the other dragons and snakes) have no intention of being a player in any such *one-worldism* defined by the West.

Still, it is surprising how badly Putin miscalculated this invasion. He clearly expected an outcome measured in hours rather than in months and only fragmented support by NATO for Ukraine. The result instead has been the staggering loss of 20,000 to 30,000 Russian troops to this point, and what can only be described as an embarrassing performance by the Russian military. Ominously, Putin himself has notified the world that the nuclear option is on the table.

Threat Picture, II

Militaries have more than a tendency to view security issues solely in military terms. But there always is a bigger picture, to include fundamental, trend-setting dynamics. We already have seen many instances of this in the previous chapters. Turning to the near future, what are some of these *megatrends*? Kilcullen has pondered this question in a context beyond the one already described above. The impending conflict ecosystem, he states, will be a "nonlinear, many-sided, wild, and messy world" marked by four trends: *rapid population growth, accelerating urbanization, littoralization* (concentration in coastal areas), and *networked interconnectivity*.[14] These four, he argues, define that ecosystem, and its characteristics point to a proliferation of the *snakes* and an increase in threat-complexity for those seeking to alleviate the problem.[15]

We already have used the phrase, *demography is destiny* (see Chapter 5). At the beginning of industrialization in the West around 1750, there were only about 0.75 billion people on all of Earth. It took 150 years to double in size by 1900 (1.5 billion), but then took only 60 years to reach 3 billion – that despite some 75 million people dying in World Wars I and II. Only 39 years later, in 1999, there were 6 billion.[16] The world's population stands at 7.9 billion at the time of writing. Demographers project that this growth will level off at around 9 billion, with famine, pestilence, and war, the so-called "positive" or natural checks on population growth kicking in substantially.

Further, this population growth is crammed into extremely large cities struggling to absorb their numbers. In 1800 only 3% of the world lived in cities of more than 1 million inhabitants; in 2020, almost 50% did. In 1950, there were only 83 cities of that size in the world; in 2017, there were 483. The most intense areas of urban growth are in the poorer, developing parts of the world. Richer, fully industrialized areas had their population explosion and attendant misery a century and a half ago – think Charles Dickens's London. This situation, Kilcullen states, increases the potential for many social ills, terrorism only one among them:[17]

> [O]verwhelmed by rapidly growing population, unplanned slum development, political instability, violent crime, conflict, disease, increased vulnerability to natural disaster, and shortages of energy, food, and water, ... the fact there also are extremists out there ... will be far from the main threat. ... The main cluster of threats ... will come from the environment itself, not from any one group within it.

All of this is compounded by *littoralization*, an odd word meaning that most of these huge cities are in coastal areas. As Kilcullen points out, three-fourths of the world's largest cities in 2012 were so located, and, because population growth concentrates in urban areas, 80% of people on Earth now live within 60 miles of a sea or its main waterways. These also are the areas most vulnerable to rises of up to 8 feet in sea level by 2100 anticipated by current levels of *global warming*.[18] The word *littoral*, however, implies more than "near a coast." It also connotes in military terms a land area that can be contested by sea-based weapons, and the surrounding sea area controlled by land-based armaments. With today's weapons' systems, a *littoral zone* itself may extend 100 miles or more inland and 200 miles out to sea, and its cyberspace can be hit from anywhere.[19]

Finally, the millions of people in littoral spaces are *networked* these days with each other and many parts of the outside world through the internet and social media. Jessica Mathews, a highly respected American expert on national security, conflict, and governance, years ago argued that this *interconnectivity* posed the greatest threat to nation-states' autonomy over what takes place within their own borders.[20] Interconnectivity undermines nation-states, particularly wobbly ones, she writes, because it loosens their control over "the collection and management of information," and links people across borders, "amplifying political and social fragmentation by enabling more and more identities and interests scattered around the globe to coalesce and thrive."

Where national and local governments already are challenged, computers, the internet, and now ever-present smartphones allow people who otherwise would be isolated from the world outside their neighborhood to communicate, stay informed, and conduct business. It also *empowers illicit and radical activities* that crop up when existing arrangements and governments cannot meet everyday wants and needs legitimately, and extends their reach across national borders.

It is readily apparent from this discussion that problematic conflicts in *littoral, urban, inter-connected sprawl* do not have purely military solutions. Still, there is a role to play by armed law enforcement, gendarmes, and savvy military forces, for administrative and ameliorative responses do not have a chance to work without stabilizing security. Paraphrasing a legendary military advisor from the Vietnam war, John Paul Vann, Kilcullen states, "security may be only 10 percent of the problem, or it may be 90 percent, but whatever it is, it is the *first* 10 percent or the *first* 90 percent [emphasis in original]."[21]

Implications for Military Organization

The most obvious starting point is that any hopes of a simplified threat picture have been dashed. Modern militaries truly must be poised to counter a range of military missions calling for varied, seemingly at odds, capabilities. Everything potentially is on the table – big wars, small wars, new wars, cyber- and information-wars, chemical and biological warfare, tactical nuclear warfare, and perhaps even the "unthinkable," the ultimate "big one." Each of these is not equally likely, so this messy, multi-linear threat picture portends areas of organization and re-organization.

The first line of strategic thinking ought to address *conflict prevention*, that is, how to address the environments that give rise to the cluster of threats described above. An important step at this stage is *pre-conflict sensing* – a smart, specific inventory of the context, make-up, and issues for each slice of the threat picture.[22] Thought thus must be given to the kind of mission capabilities needed to make grounded assessments that address the most pressing military threats but extend beyond strictly military considerations. A classified version of the U.S. National Defense Strategy 2022 has been presented to Congress, but the unclassified report is not yet available at the time of writing. However, a pre-release blurb emphasizes a defense posture that incorporates the relevant capacities of agencies outside the military with those of the Department of Defense.[23]

Not to be overlooked in such efforts is the path-breaking work of Valerie Hudson, Professor at Texas A&M University's George H.W. Bush School of Governance and Public Service (see our Spotlight in Chapter 2). The most recent work by Hudson's team confirms their earlier research findings, perhaps startling to some: how women are treated in societies across the globe has direct implications for bellicosity on the world stage.[24] They show that the more systematically women are subjugated in a society, the worse that society's governance and the less peaceful its affairs with other countries; conversely, steps to alleviate gender inequities improve these shortcomings. So, *more secure women, more secure world*. Addressing this concern is largely beyond the military's sphere of control, but for policy makers, that is one place, if not to start, not to neglect.

Militaries, however, as Kilcullen notes, are like firefighters. While fire prevention is always a good thing, firefighters ultimately are paid to fight fires. So too with the military.[25] A crucial component of *pre-conflict sensing* thus addresses the composition and directives for the *advance party* and *follow-on forces*. Much of the U.S. military's infantry and armor divisions already have gone to a *brigade combat team* (BCT) format. BCTs are self-contained, brigade-size units able to deploy and stand on their own. Still, future missions may require more precise *modular structures*, that is, unit elements and skill sets assembled with a very *specific battlespace* in mind. Again, the term battlespace should not conjure up solely kinetic capabilities, but kinetic ones within a weave of skill sets. The very words *air, land, and sea* associated with littoral areas imply *joint composition and capabilities*. Such more specialized cross-service teams may be particularly advantageous.

Urban warfare usually has a disaggregating effect, so some of these unit elements must be prepared to operate as even smaller, largely self-directed *splinter teams*. Instead of a Big War's military adversaries fighting toe-to-toe along a line in the sand, the modus operandi here might be, to borrow a term from former Marine Commandant General Charles Krulak, *three-block warfare*. General Krulak indicated Marines in such settings should expect to *wage a firefight* on block one, *switch to a peacekeeping mode* in block two, and *render humanitarian assistance* on block three. He offered the concept back in the 1990s, but its relevance is even more apropos here.

In any case, culturally astute officers and enlisted personnel are a must. The U.S Air Force recently converted the 61-B occupational specialty for officers – behavioral scientist, who essentially handled tasks more in line with human factors engineering – to 14-F (Information Operations, IO). Entry into the IO career field requires at least a baccalaureate degree in psychology, sociology, or anthropology. The "14" label is important because career fields in the Air Force beginning with "1" denote operational rather than support status. As such, 14-Fs are expected to represent socio-cultural considerations integral in the planning and execution of Air Force missions.

Cultural competencies can be exploited in two additional ways. One, culturally astute military advisers can play an especially effective role in training and advising confederates in allied militaries. Conversely, such advisors are likely to be the very ones most receptive to being tutored themselves by foreign militaries. Two, facilitating multi-national units to accomplish special-niche missions adds to the settings in which mission capabilities may be applied. Here, too, culturally competent servicemembers increase the likelihood of effective use of such units.

Still, these cannot be a contemporary military's only focus. Putin, after all, has put old-fashioned, interstate wars back on the table, to include the nuclear option. This does not necessarily imply however a return to conventional warfare as done in the past. As the current war in Ukraine has shown, both sides have made extensive use of cyber-assaults on web-based control systems, electronic jamming of communication equipment, and aerial drones for reconnaissance and targeting. In addition, Ukraine in particular has tapped social media capabilities for publicizing advances, exposing war crimes, eliciting sympathies, and combating Russian propaganda. And both sides have worked diplomatic networks to identify parameters that, if exceeded, might trigger escalation of the war to the nuclear level.

Some traditional concerns in military sociology remain front and center. The U.S. Army, for example, assigns soldiers to BCTs for a minimum of three years to increase and maintain *unit cohesion*. As we have shown in multiple places, but especially Spotlights on the work of Edward Shils and Morris Janowitz in Chapter 1 and of Nora Stewart in Chapter 5, unit cohesion is the basic glue holding military units together in the face of adversity. *Time together* and *shared experience* are crucial. However, Eyal Ben-Ari's study of *instant units* in the Israeli Defense Forces – what we above called *modular* units – found that military members over time develop the capacity to cohere easily, not just to one particular unit, but to most any military grouping, an ability they call *swift trust* (see our Spotlight in Chapter 10).[26] This merits further research attention and monitoring.

Likewise, smart *leadership* is critical to unit effectiveness and morale. Modern militaries require highly trained service members, and old-style transactional leadership is limited in its ability to encourage voluntary compliance and initiative. One of the more serious of the Russian military's breakdowns in the Ukraine has been its cumbersome and stultifying top-down leadership style. Not only has this gotten a dozen or so of its generals killed (as they raced to a front to issue orders a clued-in and motivated junior officer or sergeant could be giving), but it has also undermined morale. As Phillip Payson O'Brien, Professor of Strategic Studies at the University of St. Andrews, Scotland, wrote about the woes of the Russian military in its first month in Ukraine, wrote:[27]

> To predict [how any army will perform in a war], you must analyze not only its equipment and doctrine but its ability to undertake complex operations, … and the commitment of its soldiers to fight and die in the specific war being waged. Most important, you have to think about how it will perform when a competent enemy fires back. As [heavyweight boxing champion] Mike Tyson so eloquently put it, "Everyone has a plan until they are punched in the mouth."

Bad leadership plus conscripts clearly is a losing combination, but off-the-mark leadership is problematic in any scenario. Continued study of the science and art of leadership is warranted to enhance the motivation and morale of highly trained soldiers who must carry on in the operationally complex environments we have described above.

Finally, some of the more striking developments in the future will stem from the rapidly expanding military investment in *robotics* and *artificial intelligence*.[28] As we showed in Chapter 10, drones (remotely piloted aircraft, RPAs) initially were employed for surveillance and reconnaissance missions, but soon were eased into deadly kinetic ones before resolving the implications of using them. Similarly, updated, highly sophisticated versions of today's unmanned ground vehicles (UGVs) and "killer robots" conceivably could be parachuted into tomorrow's hotspots to carry out combat roles now done by infantry and armored units. Like today's RPAs, these future UGVs and robots no doubt would be radio-controlled in early missions by operators located elsewhere. However, recent advances in artificial intelligence almost guarantee that *autonomously operating vehicles* and *autonomous weapons systems* will populate future battle spaces. Since such technological innovations tend to be implemented more rapidly than the social and cultural adaptations to their use, employing autonomous systems surely will create a whole new category of tricky legal and ethical dilemmas.

Threat Picture, III

The U.S.A.'s 2022 National Defense Strategy Factsheet lists China as its "most consequential strategic competitor."[29] Of the 7.9 billion people on Earth, about 37% of them live in only two countries: China and India, who in 2022 have 1.44 billion and 1.38 billion inhabitants, respectively. China, the giant of East Asia, and India, the giant of South Asia, are both industrializing and have substantial militaries (and nuclear capabilities), China more so than India. In *The Economist*'s rating of 167 countries on its Democracy Index, India is the world's largest *democracy* (but in the *flawed* category); China (and Russia) both fall in the *authoritarian* category, though Russia has a slightly higher score.[30] During the Cold War, relations between India and the U.S.A. were distant. Since the collapse of the Soviet Union, the security interests of the two countries have converged.

Cornish and Donaldson considered China's threat potential.[31] On the one hand, China is an amazing success story. After the successful Communist revolution led by Mao Zedong in 1949 – and the retreat of Chairman Chiang Kai-shek and his followers to the island of Taiwan – the country stagnated disastrously for thirty-plus years. Mao died in 1976 and his eventual replacement, Deng Xiaoping, instituted "red-bloodedly capitalist" market reforms. In 2010, China became the world's second largest economy behind only the European Union. (The U.S.A. has the third largest.) China is the U.S.A.'s single largest trading partner, while India is the U.S.A.'s seventh largest. Conversely, China and India's single largest trading partner is the U.S.A. However, the U.S.A.'s economic relationship with China has been marred by China's use of economic espionage and copyright infringement.

Xi Jinping, China's president since 2013, has not presided over eye-popping economic growth. As a developing economy matures, its meteoric rise naturally levels off. Xi thus has replaced the simple goal of economic prosperity with fervent *Chinese nationalism* and *regional, eventually, world dominance*. In short, he seeks to restore China's place as a feared and respected world power.

Relative to our discussion, this policy has two dimensions. The first is the *extension of China's lawful maritime borders* in the East and South China Seas. To this end, China since 2015 has been "building" islands – putting permanent structures where before there were only partially submerged rocks – that may serve as small military bases. It routinely challenges naval vessels who sail these open waters, so the U.S. navy routinely cruises them to remind of their openness. The second of these is the *reclamation of Taiwan*. The island nation has an iron-clad defense treaty with the U.S.A. The enduring question for China is whether the U.S.A. really would come to the military defense of Taiwan, should China try to take it back by force.

Hence, China poses a serious threat. It is second only to the U.S.A. in defense spending and has formidable cyber and nuclear capabilities. Cornish and Donaldson do have a scenario of how this might play out. They suggest that an outright frontal attack will not be the Chinese style, given their economic entanglements with the U.S.A. They think instead that other ploys from the deep trick bag of *hybrid warfare* will come into play first, so that the final steps to any take-over of Taiwan are more in the form of *capitulation* than *surrender*. Of course, that is what many thought of Russia's approach to Ukraine before February of 2022.

A Concluding Tale

Elliot Ackerman and Admiral Jim Stavridis, USN (ret.) offer a doomsday scenario in their 2021 book of "speculative fiction," *2034: A Novel of the Next World War*. Ackerman is a former Marine who served five tours in Iraq and Afghanistan and a former White House Fellow. He also is an accomplished writer who knows how to tell a good story. Admiral Stavridis is a naval officer of more than 30 years and holds a Ph.D. from the Fletcher School of Law and Diplomacy at Tufts University. He is a former Supreme Allied Commander of NATO and has commanded an aircraft carrier battle group in combat. His attention to detail in describing how national defense networks interact and ships work at sea is very evident in the book.

The novel spins a tale of how a series of routine military encounters, combined with a few miscalculations, technological wizardry gone awry, and tunnel vision that delimits the choices, leads to catastrophic nuclear war between China and the U.S.A. It all begins innocently enough with a flotilla of three U.S. Navy *Arleigh Burke*-class destroyers passing through the waters off Mischief Reef in the South China Sea. They are on a freedom of navigation patrol, a standard, cheeky ploy to remind China these are international waters. The flotilla encounters a fishing trawler on fire, and the lead destroyer stops to render assistance. Sailors from the destroyer board the creaking vessel, the *Wén Ruì* (in English, *Mosquito*), and discover it has a suspicious vault of sophisticated radio communications equipment. Meanwhile, a navy pilot, call sign Wedge, in an *F-35 E Lightning* is flying a reconnaissance mission over Iran. Because of the plane's stealth technology, it is almost impossible to detect or track.

National security advisors at the White House are monitoring the two events. Suddenly, both situations turn ominous. The White House security shop loses all communication with the *Aleigh-Burke* destroyers, and they receive a report from their Pacific office indicating six Chinese warships have altered course toward Mischief Reef. On top of that, their monitoring devices indicate the autopilot in Wedge's aircraft is no longer operational. The plane seems to have been taken over by some outside agency and has begun to descend. It eventually lands by remote control in Bandar Abbas, an Iranian port facility. About this time, an admiral-rank Chinese defense attaché stationed in Washington, D.C. arrives unannounced at the White House gates and demands to speak with a national security advisor. A stupefied advisor comes to the gate and is told politely: "You have taken possession of one of our small ships, the *Wén Ruì*; we would like it back; do so, and we'll arrange a swap for your *F-35*; you will regret it if you don't."

The purpose of this, from the Chinese point of view, is to create a situation for asserting their primacy over the South China Sea and to push the envelope. On the U.S. side, their drawn-out response is calibrated to keep the South China Sea an open waterway and China in its place. Routine stuff. But, the exchange does not proceed smoothly, and in the confusion two U.S. destroyers are sunk by Chinese torpedoes. Alternating among decision-makers in China, the U.S.A., Iran, India, and Russia, events steadily escalate from there. Each encroachment that could have been defused is not. The words "military action" and "Taiwan" soon appear in connection with each other in the same chapter. The step-by-step escalation makes the eventual use of tactical nuclear weapons almost logical. And then, the "ultimate big ones." Confounding the picture at almost every step is the effective use, or misuse, of cyber-warfare to interrupt communication, obscure the locations and activities of ships and planes, and shut down web-based infrastructure.

In the last scenes, China and the U.S.A. are semi-prostrate with swaths of smoking ruins and radiation illnesses. Russia is still second-rate, so India is the only big power left standing.

Ackerman and Stavridis were interviewed about their intentions in writing the book.[32] Admiral Stavridis indicated that he had read General Hackett's novel (discussed above) early in his career and it often *gave him pause to think* about where crises he faced as a military man might go. He hoped this novel would do the same for others. General Hackett was at one point Principal of King's College in London and Admiral Stavridis served as Dean of the Fletcher School of Law and Diplomacy. They are, in military sociology parlance, *soldier-scholars* who have attempted to initiate a wider discussion about the threat of nuclear war.

The interviewer asks the authors, "With all these characters, as I read this book, I had a strong feeling, well, I kept asking: *why don't they just stop?* Just: don't hit the button." Stavridis responded,

> I wouldn't say this is a military thing. It's a ... human thing. Just look at the last hundred years or so, ... all the marvelous things of the last 100 years. Yet we stumbled into two massive world wars. ... We have to worry about this sense you get of the US and China sleepwalking potentially into a world war.

The final takeaway thus is that decisions to do war are made by humans – and, though events sometimes take on a life of their own, humans still make the decisions. This is where the decision-maker about to go cyber-max or nuclear must be deterred by a perceived certainty: *mutually assured destruction*. With this and a bit of luck, neither hits the button.

Notes

1 David Kilcullen, 2020, *The Dragons and the Snakes: How the Rest Learned to Fight the West*, Oxford, U.K. and New York: Oxford University Press, pp. 119–120.
2 David Kilcullen, 2009, *The Accidental Guerrilla: Fighting Small Wars in the Midst of a Big One*, Oxford, U.K. and New York: Oxford University Press, "Prologue".
3 Kilcullen, 2020, op. cit., p. 6.
4 Ibid., p. 10.
5 Ibid., pp. 12–14.
6 Ibid., pp. 62–63.
7 Paul Cornish and Kingsley Donaldson, 2017, *2020: World of War*, London: Hodder & Stoughton.
8 General Sir John Hackett & Other Top-Ranking NATO Generals & Advisors, 1978, *The Third World War, August 1985: A Future History*, London: Hutchinson & Company. The 1979 version of the book is available through Macmillan, London.
9 Cornish and Donaldson, op. cit., pp. 16–19, 30–31.
10 Of course, Russia, like any other nation, has the right to pursue and protect its interests within the contours of international law. Many of these actions, while shady, are not patently illegal. Russian leaders usually deny doing these kinds of things as a matter of policy but contend these are the very operations Western powers direct at them.
11 Cornish and Donaldson, op. cit., pp. 28–31.
12 Keir Giles, 2019, *Moscow Rules*, Chatham House Insight Series, New York: Brookings Institution Press, Kindle edition, location 36. Quoted in David Kilcullen, 2020, op. cit,, p. 134.
13 Stephen Kotkin, 2002, "The Cold War Never Ended," *Foreign Affairs*, May/June, https://www.foreignaffairs.com/reviews/review-essay/2022-04-06/cold-war-never-ended-russia-ukraine-war.
14 David Kilcullen, 2013, *Out of the Mountains: The Coming of Age of the Urban Guerrilla*, Oxford, U.K. and New York: Oxford University Press, p. 17.
15 Cornish and Donaldson, op. cit., provide a similar list: *climate change*, *demography* (population growth, migration, regional population change, and urbanization), *resource security* (water, food, energy), *health security*, and *financial security*. See chapter 2, "Strategy in Breadth."
16 Ibid., p. 28.
17 Ibid., p. 30.

18 Earth Science Communication Team, National Aeronautics and Space Administration (NASA), California Institute of Technology, 2022, "Facts," *Global Climate 2021 Change: Vital Signs of the Planet*, https://climate.nasa.gov/effects.
19 Kilcullen, 2013, op. cit., pp. 31–32.
20 Jessica Mathews, 1997, "Power Shift," *Foreign Affairs*, January/February, https://www.foreignaffairs.com/articles/1997-01-01/power-shift?utm_medium=newsletters&utm_source=-fa1008&utm_content=20220611&utm_campaign=FA%20100_061122_Power%20Shift&utm_term=fa-100.
21 Kilcullen, 2013, *op, cit.*, Appendix, pp. 263–264.
22 Ibid., pp. 274–275.
23 U.S. Department of Defense, 2022, *National Defense Strategy, Spotlight*, April 11, https://www.defense.gov/Spotlights/National-Defense-Strategy/.
24 Valerie M. Hudson, Donna Lee Bowen, and Perpetua Lynne Nielson, 2020, *The First Political Order: How Sex Shapes Governance and National Security Worldwide*, New York: Columbia University Press.
25 Kilcullen, 2013, op. cit., pp. 265, 283–289.
26 Eyal Ben-Ari, Zev Lerer, Uzi Ben-Shalom, and Ariel Vainer, 2010, *Rethinking Contemporary Warfare: A Sociological View of the Al-Aqsa Intifada*, Albany, New York: State University of New York Press, pp. 72, 79–84.
27 Phillip Payson O'Brien, 2022, "How the West Got Russia's Military So, So Wrong," *The Atlantic*, March 31, https://www.theatlantic.com/ideas/archive/2022/03/russia-ukraine-invasion-military-predictions-629418/.
28 Christa Gelb, 2011, "Salient But Unappreciated National and International Security and Defense Policy for the Next Decade," *Journal of Terrorism and Security Analysis*, 6 (Spring): 1–11.
29 U.S. Department of Defense, 2022, *Fact Sheet: 2022 National Defense Strategy*, March 28, https://media.defense.gov/2022/Mar/28/2002964702/-1/-1/1/NDS-Fact-Sheet.PDF.
30 The Democratic Index is based on 60 criteria related to political and civil life, and countries are categorized as *full democracies*, *flawed democracies* (have fair elections but significant faults in other areas), *hybrid regimes* (regularly have election frauds, restrictions on liberties), or *authoritarian regimes* (monarchies, dictatorships, or have some democratic-appearing institutions of little significance), https://www.economistgroup.com/group-news/economist-intellilgence/democracy-index-2021-less-than-half-the-world-lives-in-a-democracy.
31 Cornish and Donaldson, op. cit., chapter 3, "Unraveling Imperiums: China and the US in the Southeast Asian Region."
32 Maria Streshinsky, 2021, "What Did I Just Read? A Conversation with the Authors of *2034*," *Wired Magazine*, 2 May, https://www.wired.com/story/2034-novel-authors-conversation/.

Index

Pages in *italics* refer figures, **bold** refer tables and pages followed by n refer notes.